SPECTRA OF PARTIAL DIFFERENTIAL OPERATORS

BY

MARTIN SCHECHTER

Belfer Graduate School of Science,

Yeshiva University

1971

NORTH-HOLLAND PUBLISHING COMPANY – AMSTERDAM · LONDON
AMERICAN ELSEVIER PUBLISHING COMPANY, INC. – NEW YORK

PUBLISHERS:

NORTH-HOLLAND PUBLISHING CO. – AMSTERDAM

NORTH-HOLLAND PUBLISHING COMPANY LTD. – LONDON

SOLE DISTRIBUTORS FOR U.S.A. AND CANADA:

AMERICAN ELSEVIER PUBLISHING COMPANY, INC.

52 VANDERBILT AVENUE

NEW YORK, N.Y. 10017

Library of Congress Catalog Card Number: 76-157010

North-Holland ISBN: 0 7204 2366 X

Elsevier ISBN: 0 444 10109 8

PRINTED IN THE NETHERLANDS

NORTH-HOLLAND SERIES IN

APPLIED MATHEMATICS AND MECHANICS

VOLUME 14

NORTH-HOLLAND PUBLISHING COMPANY – AMSTERDAM · LONDON
AMERICAN ELSEVIER PUBLISHING COMPANY, INC. – NEW YORK

SPECTRA OF PARTIAL DIFFERENTIAL OPERATORS

BE''H

To the memory of my father, Joshua, HK''M,
my first and foremost teacher.

To the memory of my mother, Rose, Z''L,
who sacrificed so much for my education.

To my wife, Deborah, who has been my
constant source of inspiration.

To my children, who represent my hope
for the future.

To my people, the People of the Book, who
have cherished learning throughout the ages.

To all of them I humbly dedicate this book.

PREFACE

Roughly speaking, this book is addressed to the problem of describing the spectrum of a linear partial differential operator in $L^p(E^n)$, where $1 \leqslant p \leqslant \infty$ and E^n is Euclidean n-dimensional space. Simple as the problem sounds, the complete solution is far from known. We have tried to assemble a significant amount of material on the topic, but no attempt at completeness can be made at this time.

Much of what was known before 1962 is contained in the book by Glazman [1965]. However, there has been very rapid development since then, and the present volume is devoted primarily to some of these advances.

The underlying impetus for the study of the problem is not new; it stems from the study of the N-particle Hamiltonian

$$H = \sum_{k=1}^{N} \left\{ \frac{h^2}{8\pi^2 m_k} \sum_{j=1}^{3} \left[\frac{\partial}{i\partial x_{3k+j-3}} + b_j(x^{(k)}) \right]^2 + q_k(x^{(k)}) \right. + V(x)$$

on the Hilbert space $L^2(E^{3N})$ (see ch. 10). We believe that the recent surge in activity is due mainly to two factors. First, new mathematical methods have been developed which have become very useful in treating the classical problems. Secondly, new applications have arisen with complicated potentials $V(x)$. An additional factor is the theoretical interest coming from the study of general operators.

We have searched for a specific area which would

a) include many recent advances,
b) illustrate some of the useful mathematical tools,
c) give indication as to what one can expect in general,
d) be within the grasp of a non-specialist or a graduate student.

The theme we have chosen is the following: Given a partial differential operator $P(\mathrm{D})$ with constant coefficients on E^n and a partial differential operator $Q(x, \mathrm{D})$ with variable coefficients, we consider $P(\mathrm{D})+Q(x, \mathrm{D})$ as an operator on $L^p(E^n)$ (the precise definition of this operator requires care). Our aim is to describe the effect on the spectrum of $P(\mathrm{D})$ which is produced by the addition of the "perturbation" $Q(x, \mathrm{D})$. Our main concern will be to find conditions on $Q(x, \mathrm{D})$ which will insure that the spectrum of $P(\mathrm{D})$ is not "appreciably" disturbed (of course this requires a precise definition as well).

Since many of the topics treated here are of interest to chemists, engineers, mathematicians and physicists, we have attempted to make the book readable to all. For this purpose we have devoted the first three chapters to background material. Chapter 1 is the largest of the three and contains those topics from functional analysis which are used in the book. Most proofs are omitted in this chapter since the material is readily available in books on functional analysis (see ch. 11). One versed in functional analysis will find little new here. The only sections deserving slight special attention are §§3, 4, 6.

The second chapter deals with those classes of function spaces which are used in the book. Here we merely state the theorems; complete proofs would take us too far afield. References and some proofs are given in ch. 11. Chapter 3 collects the relatively few facts from the theory of partial differential operators which are needed.

It may be a bit surprising that the bulk of the basic information required is not from the theory of partial differential operators, but rather from functional analysis. One reason for this is the fact that we have chosen to avoid boundary value problems. This decision is based on several factors. The theory of boundary value problems is developed to a fairly complete degree only for second order operators or for elliptic operators. Since we have tried to treat a large class of non-elliptic operators of arbitrary order, we would have been unable to carry out the program with respect to boundary value problems. Even more important is the fact that boundary value problems involve much more in the way of "hard analysis" devoted to the derivation of "a priori estimates". Detailed analysis of such problems might obscure the essential aspects of the problems. Moreover, those regular boundary value problems which have been investigated do not give results materially different from the corresponding problems in all space. On the other hand, problems in which degeneracy develops at the boundary have not been studied to the degree that can be used here.

The first three chapters are not to be read as such, but merely used as a reference when needed in the main body of the book. The reader can either assume the theorems stated in them or obtain proofs from the references given in ch. 11. Thus in reality the book begins with chapter 4. The reader should begin there and refer to chs. 1–3 only when the need arises.

The main material is contained in chs. 4–10. The general theory for constant coefficient operators is given in ch. 4 while relative compactness is studied in ch. 5. Elliptic operators are considered in ch. 6 and the L^2 theory for operators bounded from below is given in ch. 7. Chapter 8 treats self-adjoint operators while ch. 9 gives a comprehensive theory for second order operators. In ch. 10 we apply this theory to quantum mechanical systems of particles. The last chapter, ch. 11, gives references, background material and discussions of related work.

Each chapter is divided into sections. In reference, ch. 6, §4 means section 4 of chapter 6. Theorems and lemmas are numbered consecutively in the section where they are found without distinguishing between them. Lemma 2.4 of ch. 7 is the fourth lemma or theorem given in §2 of ch. 7. The chapter number is omitted when reference is made to theorems, lemmas or sections of the same chapter.

I would like to thank my students S. Kohn and H. Koller for helpful suggestions and D. Wilamowsky for correcting parts of the manuscript. I am also indebted to Drs. T. Kato, P. Rejto and I. Segal for interesting conversations and F. Brownell for very helpful correspondence. Many thanks are due to Magdalene McNamara, Barbara Morris, Barbara Roberts and Vinnie Sommerich for typing the manuscript so beautifully. Also deserving of thanks are Bobbie Friedman and Florence Schreibstein for their cheerful help.

The research for this book was sponsored in part by the Air Force Office of Scientific Research, Air Force Systems Command, United States Air Force, under AFOSR grant 69-1786. In particular I want to thank Lt. Col. W. Trott and Dr. R. Pohrer for their encouragement.

Finally I come to my wife, Deborah, whose contributions to this effort cannot be fully appreciated.

In addition, I wish to thank Isaac Bulka for helping to correct the proofs.

New York M.S.
November, 1970

CONTENTS

CHAPTER 1

FUNCTIONAL ANALYSIS

§1. Banach and Hilbert spaces

In this chapter we give a brief review of those concepts and results of functional analysis which will be used in the book. Proofs will be supplied only when they are not readily available in a text on the subject or when they are very simple. Specific references will be given in ch. 11. For the convenience of the reader we start from the basic definitions. However, it will be difficult to follow this outline without some prior knowledge of the subject.

A *complex Banach space* B is a set of elements $a, b, c, \ldots$ (sometimes called vectors) for which there is defined an operation of addition and multiplication by complex numbers $\alpha, \beta, \gamma, \ldots$ (sometimes called scalars) such that the following statements hold for all vectors and scalars.

1. $a+b \in B$.
2. $\alpha a \in B$.
3. $a+b=b+a$.
4. $a+(b+c)=(a+b)+c$.
5. $\alpha(a+b)=\alpha a+\alpha b$.
6. $(\alpha+\beta)a=\alpha a+\beta a$.
7. $\alpha(\beta a)=(\alpha\beta)a$.
8. There is an element $0 \in B$ such that $a+0=a$ for all $a \in B$.
9. For each $a \in B$ there is an element $-a \in B$ such that $a+(-a)=0$.
10. To each $a \in B$ there corresponds a real number $\|a\|$ such that
    (a) $\|\alpha a\|=|\alpha|\,\|a\|$,
    (b) $\|a\|=0$ if and only $a=0$,
    (c) $\|a+b\| \leqslant \|a\|+\|b\|$,
    (d) if $\{a_n\}$ is a sequence of elements of B such that $\|a_n-a_m\| \to 0$ as $m, n \to \infty$, then there is an element $a \in B$ such that $\|a_n-a\| \to 0$ as $n \to \infty$.

We usually use $a-b$ as an abbreviation of $a+(-b)$, and $a_n \to a$ as an abbreviation of $||a_n - a|| \to 0$. In the latter case we say that the sequence $\{a_n\}$ *converges to* a.

The number $||a||$ is called the *norm* of a. It is non-negative since

$$0 = ||0|| = ||a-a|| \leqslant ||a|| + ||-a|| = 2||a|| .$$

Statement 10(d) is usually referred to as the completeness property of B. A sequence $\{a_n\}$ satisfying $||a_n - a_m|| \to 0$ as $n, m \to \infty$ is called a *Cauchy sequence*.

A subset S of B is called *closed* if $a \in S$ whenever there is a sequence $\{a_n\}$ of elements of S such that $a_n \to a$. A subset V of B is called a *subspace* if $\alpha a + \beta b$ is in V for all $a, b \in V$ and all scalars α, β. An element a is called a *linear combination* of the elements $a_1, \ldots, a_n$ if there are scalars $\alpha_1, \ldots, \alpha_n$ such that

$$a = \alpha_1 a_1 + \ldots + a_n a_n . \tag{1.1}$$

The set of elements $a_1, \ldots, a_n$ is called *linearly dependent* if there are scalars $\alpha_1, \ldots, \alpha_n$ not all zero such that

$$\alpha_1 a_1 + \ldots + \alpha_n a_n = 0 . \tag{1.2}$$

Otherwise it is called *linearly independent*. For n a non-negative integer, a subspace V is said to be of dimension $< n$ if every set of n vectors is linearly dependent. It is said to be of dimension n if it is of dimension $< n+1$ but not of dimension $< n$. If V is not of dimension n for any finite n, we say that it has infinite dimension. The dimension of a subspace V is sometimes denoted by dim V.

A set of elements $a_1, \ldots, a_n$ contained in a subspace V of B is called a *basis* of V if it is linearly independent and every element $a \in V$ can be expressed in the form (1.1). The following is an obvious consequence of the definition.

Lemma 1.1. *If the dimension of V is n, then every linearly independent set of n elements is a basis of V. Thus V has a basis.*

A subset S of B is called *bounded* if there is a (finite) number M such that $||a|| \leqslant M$ for all $a \in S$. It is called *compact* if every sequence $\{a_k\}$ of elements of S has a subsequence converging to an element of S.

Lemma 1.2. *A subspace V of B has finite dimension if and only if all of its bounded closed subsets are compact.*

Lemma 1.3. *For any n elements $a_1, \ldots, a_n$ of B, the set of their linear combinations forms a subspace of B of dimension $\leqslant n$.*

A subset S of B is called *dense* if for each $a \in B$ there is a sequence $\{a_n\}$ of elements of S converging to a. If a subspace is not dense, there is a closed subspace containing it which is not the whole of B.

If M, N are subspaces of B such that $M \cap N = \{0\}$, we let $M \oplus N$ denote the set of all vectors $a \in B$ which can be written in the form $b + c$ with $b \in M$ and $c \in N$. Clearly $M \oplus N$ is a subspace of B.

Lemma 1.4. *Suppose M, N, S are subspaces of B such that $M \cap N = M \cap S = \{0\}$. If $S \oplus M \subset N \oplus M$, then* $\dim S \leqslant \dim N$.

Proof. We shall show that for each n, $\dim N < n$ implies $\dim S < n$. Suppose that $\dim N < n$, and let $a_1, \ldots, a_n$ be any n elements of S. We can write

$$a_j = a_{j1} + a_{j2}\,, \qquad 1 \leqslant j \leqslant n\,,$$

where $a_{j1} \in N$ and $a_{j2} \in M$. Since $\dim N < n$, there are scalars $\alpha_1, \ldots, \alpha_n$ not all zero such that

$$\sum_1^n \alpha_j a_{j1} = 0\,.$$

Hence

$$\sum_1^n \alpha_j a_j = \sum_1^n \alpha_j a_{j2} \in M\,.$$

Since $S \cap M = \{0\}$, we have

$$\sum_1^n \alpha_j a_j = 0\,,$$

and the proof is complete.

A collection of elements which satisfy statements 1–10(c) is called a (complex) *normed vector space* (or *normed linear space*). The following theorem is important in many applications.

Theorem 1.5. *If X is a normed vector space, then there exists a Banach space B containing X such that the norm of B coincides on X with the norm of X, and X is a dense subspace of B.*

The Banach space B is called the *completion* of X.

If X and Y are normed vector spaces, we can form their *cartesian product* $X \times Y$ as follows. We consider all ordered pairs $\langle x, y \rangle$ of elements $x \in X$, $y \in Y$ and define

$$a\alpha_1 \langle x_1, y_1 \rangle + \alpha_2 \langle x_2, y_2 \rangle = \langle \alpha_1 x_1 + \alpha_2 x_2, \alpha_1 y_1 + \alpha_2 y_2 \rangle$$
$$\| \langle x, y \rangle \| = (\|x\|^2 + \|y\|^2)^{\frac{1}{2}} .$$

One checks easily that under these definitions $X \times Y$ becomes a normed vector space. Moreover, if X and Y are Banach spaces, the same is true of $X \times Y$.

A *scalar product* on a (complex) normed vector space X is an assignment which assigns a complex number (x, y) to each element $\langle x, y \rangle$ of $X \times X$ in such a way that

(i) $(\alpha x, y) = \alpha(x, y)$,
(ii) $(x, y) = \overline{(y, x)}$ (the complex conjugate),
(iii) $(x+y, z) = (x, y) + (y, z)$,
(iv) $(x, x) = \|x\|^2$.

A Banach space which has a scalar product is called a *Hilbert* space.

In a Hilbert space X we shall say that elements x, y are *orthogonal* and write $x \perp y$ if $(x, y) = 0$. If S is a subset of X we let $S^\perp$ denote the set of all elements of X which are orthogonal to all elements of S. If V is a closed subspace of X, we have $X = V \oplus V^\perp$.

A sequence $\{x_k\}$ of elements of a Hilbert space X is said to *converge weakly* to an element $x \in X$ if $(x_n - x, y) \to 0$ as $n \to \infty$ for each $y \in X$. We write $x_n \rightharpoonup x$ in this case. We shall need

Theorem 1.6. *If $\{x_k\}$ is a sequence of elements of a Hilbert space X such that*

$$\|x_k\| \leqslant C, \qquad k = 1, 2, \ldots,$$

then there is a subsequence $\{y_n\}$ of $\{x_k\}$ and an element $z \in X$ such that $y_n \rightharpoonup z$.

A sequence $\{x_k\}$ of elements of a Hilbert space is called *orthonormal* if

$$(x_j, x_k) = \delta_{jk}, \qquad j, k = 1, 2, \ldots,$$

where δ_{jk} is the Kronecker delta ($\delta_{jk} = 0$ for $j \neq k$, $\delta_{kk} = 1$).

Lemma 1.7. *Every infinite dimensional subspace of a Hilbert space contains an orthonormal sequence.*

Lemma 1.8. *Every orthonormal sequence converges weakly to zero.*

Let M be a closed subspace of a Banach space X. For any $x \in X$ let $[x]$ denote the set of all $y \in X$ such that $x - y \in M$. $[x]$ is called a *coset*. Let X/M denote the collection of all cosets and define

$$[x]+[y]=[x+y], \qquad \alpha[x]=[\alpha x]$$

$$\|[x]\| = \inf_{y \in [x]} \|y\| .$$

Theorem 1.9. *Under the above definitions X/M is a Banach space.*

Lemma 1.10. *A normed vector space is complete if and only if for each sequence $\{x_j\} \subset X$ such that*

$$\sum_1^\infty \|x_j\| < \infty ,$$

the sum $\sum_1^N x_j$ converges in X.

§2. Linear operators

Let X, Y be Banach spaces. A *linear operator* A from X to Y is an assignment which takes each element x in a subspace V of X into an element Ax of Y in such a way that

$$A(\alpha_1 x_1 + \alpha_2 x_2) = \alpha_1 A x_1 + \alpha_2 A x_2 . \tag{2.1}$$

The subspace V is called the *domain* of A and is denoted by $D(A)$. The set of all $y \in Y$ which satisfy $Ax = y$ for some $x \in D(A)$ is called the *range* of A and is denoted by $R(A)$. If $R(A) = Y$, we say that A is *onto*. The set of all $x \in D(A)$ such that $Ax = 0$ is called the null space of A and is denoted by $N(A)$. The operator A is called *bounded* or *continuous* if there is a constant C such that

$$\|Ax\| \leqslant C\|x\| , \qquad x \in D(A) .$$

This is equivalent to saying that $x_n \to x$ implies $Ax_n \to Ax$. A is called *closed* if whenever there is a sequence $\{x_n\}$ of elements of $D(A)$ such that $x_n \to x$ in X and $Ax_n \to y$ in Y, one has $x \in D(A)$ and $Ax = y$. It is called *closable* or *preclosed* if $y = 0$ whenever there is a sequence $\{x_n\}$ of elements of $D(A)$ such that $x_n \to 0$ in X and $Ax_n \to y$ in Y. A linear operator B from X to Y is called an *extension* of A if $D(A) \subset D(B)$ and $Bx = Ax$ for $x \in D(A)$. In this case we write $A \subset B$.

Theorem 2.1. *Every closable linear operator A from X to Y has a closed extension. Moreover, there is a "smallest" closed extension $\bar{A}$ (called the closure of A) in the sense that every closed extension of A is an extension of $\bar{A}$. An element $x \in X$ is in $D(\bar{A})$ if and only if there is a sequence $\{x_k\} \subset D(A)$ such that $x_k \to x$ in X and Ax_k converges in Y.*

Theorem 2.2 (the closed graph theorem). *If A is a closed operator from X to Y and $D(A)$ is a closed subspace of X, then A is bounded.*

A linear operator is said to be *one–to–one* if the only solution of $Ax = 0$ is $x = 0$. In this case it is said to have an *inverse* A^{-1} from Y to X defined as follows. $D(A^{-1}) = R(A)$ and if $y = Ax$, we set $A^{-1}y = x$. One checks easily that A^{-1} is a closed operator if and only if A is closed. As a corollary of Theorem 2.2 we have

Theorem 2.3 (the bounded inverse theorem). *If A is a one–to–one closed linear operator from X to Y, then $R(A)$ is closed in Y if and only if A^{-1} is a bounded linear operator from Y to X.*

Let $B(X, Y)$ denote the set of bounded linear operators A from X to Y with $D(A) = X$. Put

$$\|A\| = \sup_{x \neq 0} \frac{\|Ax\|}{\|x\|}. \tag{2.3}$$

Then $\|A\|$ is a finite real number for each $A \in B(X, Y)$. Under obvious definitions of $A + B$ and αA one checks easily that statements 1–10(c) of §1 are satisfied. In fact $B(X, Y)$ is complete as well, and we have

Theorem 2.4. *$B(X, Y)$ is a Banach space.*

A very important special case of $B(X, Y)$ is when Y is the Banach space of complex numbers (see §1). In this case the operators are called *functionals* and $B(X, Y)$ is denoted by X'. If x' is a functional in X', the complex number into which it maps an element $x \in X$ is sometimes denoted by $x'(x)$. A very useful theorem is

Theorem 2.5. *If V is a closed subspace of X and x_0 is an element of X not in V, then there is a functional $x'_0 \in X'$ such that $x'_0(x_0)=1$ and $x'_0(x)=0$ for all $x \in V$.*

A sort of dual to this is

Theorem 2.6. *If W is a finite dimensional subspace of X' and x'_0 is a functional in X' which is not in W, then there is an element $x_0 \in X$ such that $x'_0(x_0)=1$ while $x'(x_0)=0$ for all $x' \in W$.*

A linear operator A in $B(X, Y)$ is said to be *compact* (or *completely continuous*) if for each bounded sequence $\{x_k\}$ in X, $\{Ax_k\}$ has a subsequence convergent in Y. We denote the set of compact operators from X to Y by $K(X, Y)$. An operator $A \in B(X, Y)$ with $R(A)$ finite dimensional is called an operator of *finite rank*. From Lemma 1.2 one obtains immediately

Lemma 2.7. *Operators of finite rank are compact.*

Let A be a linear operator from X to Y with $D(A)$ dense in X. We define an operator A' from Y' to X' as follows. y' is in $D(A')$ if there is an $x' \in X'$ such that

$$y'(Ax) = x'(x), \qquad x \in D(A). \tag{2.4}$$

We define $A'y'$ to be x'. In order for this definition to make sense one must make sure that x' is unique. This follows from the density of $D(A)$. For if there were another $x'_1 \in X'$ satisfying (2.4), then we would have $(x'_1 - x')(x)=0$ for all $x \in D(A)$. Since $D(A)$ is dense in X and $x'_1 - x'$ is a bounded linear functional, this holds for all $x \in X$. Consequently $x'_1 = x'$. If $A \in B(X, Y)$, then $D(A') = Y'$ and $A' \in B(Y', X')$. The operator A' is called the *conjugate* of A.

If A and B are linear operators from X to Y we define $A+B$ as follows. $D(A+B)=D(A) \cap D(B)$ and $(A+B)x = Ax + Bx$ for $x \in D(A+B)$. If A is a linear operator from X to Y and E is a linear operator from Y to Z the operator EA is defined as a linear operator from X to Z as follows. $D(EA)$ consists of those $x \in D(A)$ such that $Ax \in D(E)$. We set $(EA)x = E(Ax)$.

Lemma 2.8. *Let A, B be closed operators from X to Y such that $D(A) \subset D(B)$. Then there is a constant C such that*

$$\|Bx\| \leqslant C(\|Ax\| + \|x\|), \qquad x \in D(A). \tag{2.5}$$

Proof. For $x \in D(A)$ introduce the norm

$$\|x\|_A = \|Ax\| + \|x\| . \tag{2.6}$$

It follows from the fact that A is a closed operator that $D(A)$ is complete with respect to this norm. Thus $D(A)$ can be considered as a Banach space Z with norm (2.6). Moreover, the restriction $\hat{B}$ of B to $D(A)$ is a linear operator from Z to Y with $D(\hat{B}) = Z$. The fact that B is a closed operator from X to Y implies that $\hat{B}$ is a closed operator from Z to Y. Hence by Theorem 2.2

$$\|\hat{B}x\| \leqslant C\|x\|_A , \qquad x \in Z ,$$

which is (2.5).

An operator B such that $D(A) \subset D(B)$ and (2.5) holds will be called *A-bounded.*

Theorem 2.9. *Let A and B be operators from X to Y such that A is closed, $D(A) \subset D(B)$ and*

$$\|Bx\| \leqslant a\|Ax\| + b\|x\| , \qquad x \in D(A) , \tag{2.7}$$

for some constants a, b with $a < 1$. Then $A+B$ is a closed operator. If A is the closure of an operator A_0, then $A+B$ is the closure of A_0+B.

Proof. First we note that (2.7) implies

$$(1-a)\|Ax\| \leqslant \|(A+B)x\| + b\|x\| , \qquad x \in D(A) . \tag{2.8}$$

Thus if $\{x_k\}$ is a sequence in $D(A)$ such that $x_k \to x$ and $(A+B)x_k \to y$, then $\{Ax_k\}$ is a Cauchy sequence by (2.8). Thus $Ax_k \to z$. Since A is closed, $x \in D(A)$ and $Ax = z$. Moreover $Bx_k \to y - z$. But by (2.7)

$$\|B(x_k - x)\| \leqslant a\|A(x_k - x)\| + b\|x_k - x\| \to 0 ,$$

showing that $Bx = y - z$. Hence $(A+B)x = y$, and the first assertion is proved. To prove the second, let x be any element in $D(A)$. Then there is a sequence $\{x_k\} \subset D(A_0)$ such that $x_k \to x$ and $A_0 x_k \to Ax$ (Theorem 2.1). By (2.7), Bx_k converges to some element h in Y. Thus $(A_0+B)x_k \to Ax + h$. This shows that x is in the domain of the closure of A_0+B (Theorem 2.1). Since $A+B$ is a closed extension of A_0+B, it must coincide with the closure of A_0+B. This completes the proof.

A *conjugate linear* functional F on X is a mapping which assigns to each $x \in X$ a complex number Fx in such a way that

$$F(\alpha_1 x_1 + \alpha_2 x_2) = \bar{\alpha}_1 F x_1 + \bar{\alpha}_2 F x_2 .$$

If we define $\alpha_1 F_1 + \alpha_2 F_2$ by

$$(\alpha_1 F_1 + \alpha_2 F_2)(x) = \alpha_1 F_1 x + \alpha_2 F_2 x ,$$

and $\|F\|$ by

$$\|F\| = \sup_{x \neq 0} \frac{|Fx|}{\|x\|} ,$$

then the set of conjugate linear functionals satisfies the statements 1–10(c) of §1. If we let $\hat{X}'$ denote this set, we have

Theorem 2.10. *$\hat{X}'$ is a Banach space.*

Let A be a linear operator from X to Y with $D(A)$ dense in X. We define an operator A^* from $\hat{Y}'$ to $\hat{X}'$ as follows. A functional $\hat{y}' \in \hat{Y}'$ is in $D(A^*)$ if there is an $\hat{x}' \in \hat{X}'$ such that

$$\hat{y}'(Ax) = \hat{x}'(x) , \qquad x \in D(A) . \tag{2.9}$$

Note that there is at most one $\hat{x}' \in \hat{X}'$ satisfying (2.9) (this is where the density of $D(A)$ in X is used). We define $A^* \hat{y}'$ to be $\hat{x}'$.

Lemma 2.11. *A^* is a closed linear operator from $\hat{Y}'$ to $\hat{X}'$.*

The operator A^* is called the *adjoint* of A. Note the distinction between the conjugate A' defined earlier and A^*. In particular note that

$$(\alpha A)' = \alpha A' , \qquad (\alpha A)^* = \bar{\alpha} A^* . \tag{2.10}$$

Let A be a linear operator from X to Y. A linear operator B from X to Y is called *A-compact* if $D(A) \subset D(B)$ and whenever a sequence $\{x_k\}$ of elements of $D(A)$ satisfies

$$\|x_k\| + \|Ax_k\| \leqslant C , \qquad k = 1, 2, \dots , \tag{2.11}$$

then $\{Bx_k\}$ has a subsequence convergent in Y.

Theorem 2.12. *If A is a closed linear operator from X to Y and B is A-compact, then*

(a) $\|Bx\| \leqslant C(\|Ax\| + \|x\|)$, $\qquad x \in D(A)$,
(b) $\|Ax\| \leqslant C(\|(A+B)x\| + \|x\|)$, $\qquad x \in D(A)$,
(c) *$A + B$ is a closed operator,*
(*d*) *B is $(A+B)$-compact.*

Proof. If (a) were false, there would be a sequence $\{x_k\}$ of elements in $D(A)$ such that

$$\|Ax_k\| + \|x_k\| = 1\,, \qquad \|Bx_k\| \to \infty\,. \tag{2.12}$$

But this contradicts the fact that B is A-compact. For (2.11) holds while $\{Bx_k\}$ has no convergent subsequence.

If (b) were not true, there would be a sequence $\{x_k\}$ of elements of $D(A)$ such that

$$\|(A+B)x_k\| + \|x_k\| \to 0\,, \qquad \|Ax_k\| = 1\,. \tag{2.13}$$

For this sequence (2.11) holds. Hence there is a subsequence $\{u_n\}$ of $\{x_k\}$ such that $\{Bu_n\}$ converges to some element $y \in Y$. By (2.13) $Au_n \to -y$. But $u_n \to 0$. Thus the fact that A is closed implies that $y=0$. But this is impossible since $\|Au_n\|=1$ for each n. Hence (b) holds.

To prove (c) let $\{x_k\}$ be a sequence of elements of $D(A)$ such that $x_k \to x$ in X while $(A+B)x_k \to y$ in Y. By (b), Ax_k converges to some element $w \in Y$. Thus $Bx_k \to y-w$. Since A is closed, we see that $x \in D(A)$ and $Ax=w$. Now by (a)

$$\|B(x_k-x)\| \leqslant C(\|A(x_k-x)\| + \|x_k-x\|) \to 0\,.$$

Thus $Bx_k \to Bx$. This shows that $Bx=y-w$. Hence $(A+B)x=y$, and the proof of (c) is complete.

The proof of (d) is simple. In fact if

$$\|x_k\| + \|(A+B)x_k\| \leqslant C_1\,,$$

then (2.11) holds by (b), and consequently $\{Bx_k\}$ has a convergent subsequence. This completes the proof of Theorem 2.12.

A useful application of Theorem 2.12 is

Theorem 2.13. *Let A be a closed operator, and let B be a closable operator which is A-compact. Then for each $\varepsilon>0$ there is a constant K_ε such that*

$$\|Bx\| \leqslant \varepsilon\|Ax\| + K_\varepsilon\|x\|\,, \qquad x \in D(A)\,. \tag{2.14}$$

Proof. If the theorem were not true, there would exist an $\varepsilon>0$ and a sequence $\{x_k\}$ of elements of $D(A)$ such that $\|Ax_k\|=1$ and

$$\|Bx_k\| > \varepsilon + k\|x_k\|\,, \qquad k=1, 2, \ldots\,. \tag{2.15}$$

By (a) of Theorem 2.12 we have

$$k\|x_k\| \leqslant C(1+\|x_k\|),$$

and this implies that $x_k \to 0$. Since $\|Ax_k\|+\|x_k\|$ is bounded, there is a subsequence $\{z_j\}$ of $\{x_k\}$ such that Bz_j converges. Since B is closable, Bz_j must converge to 0. But this contradicts (2.15). The proof is complete.

If X is a Hilbert space, then there is a natural mapping T of X into $\hat{X}'$ as follows. For $x \in X$, (x, y) is an element of $\hat{X}'$. Set

$$Tx(y) = (x, y), \qquad y \in X. \tag{2.16}$$

Clearly T is linear and one–to–one. Moreover one checks easily that

$$\|Tx\| = \|x\|, \qquad x \in X. \tag{2.17}$$

We also have

Theorem 2.14. $R(T) = \hat{X}'$.

Thus we can identify $\hat{X}'$ with X when X is a Hilbert space. If A is densely defined linear operator from X to Y, we have defined the adjoint A^* of A as an operator from $\hat{Y}'$ to $\hat{X}'$. If X and Y are Hilbert spaces, we can identify $\hat{X}'$ with X and $\hat{Y}'$ with Y and consider A^* as an operator from Y to X. This is tantamount to replacing A^* by the operator $T^{-1}A^*T$. We shall use this convention throughout the book.

Lemma 2.15. *If F is a linear operator of finite rank and A is a linear operator, then AF is of finite rank. If B is bounded, then FB is of finite rank.*

Theorem 2.16. *If $T \in B(X, Y)$ and $\{K_n\}$ is a sequence of operators in $K(X, Y)$ such that $\|K_n - T\| \to 0$, then $T \in K(X, Y)$.*

Theorem 2.17. *If $T \in K(X, Y)$, then $T' \in K(Y', X')$ and $T^* \in K(\hat{Y}', \hat{X}')$.*

Theorem 2.18. *If $T \in B(X, Y)$, then $\|T'\| = \|T^*\| = \|T\|$.*

Theorem 2.19. *For any $x \in X$*

$$\|x\| = \sup_{\substack{x' \in X' \\ x' \neq 0}} \frac{|x'(x)|}{\|x'\|}.$$

§3. Fredholm operators

Let X, Y be Banach spaces. A linear operator A from X to Y is called *Fredholm* if

1. $D(A)$ is dense in X,
2. A is closed,
3. $N(A)$ is finite dimensional,
4. $R(A)$ is closed in Y,
5. $N(A')$ is finite dimensional.

The set of Fredholm operators from X to Y is denoted by $\Phi(X, Y)$. We also set $\alpha(A)=\dim N(A)$, $\beta(A)=\dim N(A')$ and $i(A)=\alpha(A)-\beta(A)$. $i(A)$ is called the *index* of A.

The classical theory of F. Riesz shows that for $K \in K(X)$ the operator $I-K$ is in $\Phi(X)$ and $i(I-K)=0$.

Important theorems on Fredholm operators are the following.

Theorem 3.1. *If $A \in \Phi(X, Y)$ and $K \in K(X, Y)$, then $A+K \in \Phi(X, Y)$ and*

$$i(A+K)=i(A). \tag{3.1}$$

Theorem 3.2. *Let X, Y, Z be Banach spaces. If $A \in \Phi(X, Y)$ and $B \in \Phi(Y, Z)$, then $BA \in \Phi(X, Z)$ and*

$$i(BA)=i(B)+i(A). \tag{3.2}$$

Theorem 3.3. *If $A \in \Phi(X, Y)$ and B is A-compact, then $A+B \in \Phi(X, Y)$ and*

$$i(A+B)=i(A). \tag{3.3}$$

Theorem 3.4. *Suppose A, B are operators in $\Phi(X, Y)$ such that A is an extension of B and $i(A)=i(B)$. Then $A=B$.*

Proof. Clearly $\alpha(A) \geqslant \alpha(B)$ and $\beta(B) \geqslant \beta(A)$. Since $i(A)=i(B)$, we have

$$\alpha(A)+\beta(B)=\alpha(B)+\beta(A). \tag{3.4}$$

The only way this can happen is if $\alpha(A)=\alpha(B)$ and $\beta(A)=\beta(B)$. Since $R(A) \supset R(B)$ and they are both closed, we must have $R(A)=R(B)$. Since $N(A) \supset N(B)$, we also have $N(A)=N(B)$. Since A agrees with B on $D(B)$, these facts imply that $A=B$.

A partial converse of Theorem 3.2 is given by

Theorem 3.5. *Let X, Y, Z be Banach spaces, and let A be an operator in $\Phi(X, Y)$. Suppose B is a closed, densely defined linear operator from Y to Z such that $BA \in \Phi(X, Z)$. Then $B \in \Phi(Y, Z)$ and* (3.2) *holds.*

A Banach space X is called *reflexive* if . or each bounded linear function x'' on X' there is an $x \in X$ such that

$$x''(x') = x'(x), \qquad x' \in X'. \tag{3.5}$$

Theorem 3.6. *If X and Y are reflexive, then any of the following statements implies the others:*

(a) $A \in \Phi(X, Y)$,
(b) $A' \in \Phi(Y', X')$,
(c) $A^* \in \Phi(Y', X')$.

In this case $i(A') = i(A^*) = -i(A)$.

Theorem 3.7. *A densely defined closed linear operator A from X to Y is in $\Phi(X, Y)$ if and only if there is an operator $A_0 \in B(Y, X)$ and operators $K_1 \in K(X)$, $K_2 \in K(Y)$ such that*

$$A_0 A = I - K_1 \text{ on } D(A), \qquad AA_0 = I - K_2 \text{ on } Y. \tag{3.6}$$

Moreover we can always choose K_1, K_2 to be of finite rank.

Theorem 3.8. *Let X, Y, Z be Banach spaces and suppose $B \in B(Y, Z) \cap \Phi(Y, Z)$. Assume that A is a closed, densely defined linear operator from X to Y such that $BA \in \Phi(X, Z)$. Then $A \in \Phi(X, Y)$.*

§4. The spectrum

Let X be a Banach space, and let A be a linear operator from X to itself. Let I be the "identity operator" on X, i.e., the operator which takes every element of X into itself. For α a scalar, we shall write α for αI. A scalar λ is said to be in the *resolvent set* $\rho(A)$ of A if $R(A-\lambda)$ is dense in X and there is a constant C such that

$$\|x\| \leqslant C\|(A-\lambda)x\|, \qquad x \in D(A). \tag{4.1}$$

The *spectrum* $\sigma(A)$ of A consists of those scalars not in $\rho(A)$.

Lemma 4.1. *$\rho(A)$ is an open set; $\sigma(A)$ is closed.*

Lemma 4.2. *If A is a closed operator, then $\lambda \in \rho(A)$ if and only if $A-\lambda$ is one–to–one and $R(A-\lambda)=X$.*

Proof. If $A-\lambda$ is one–to–one and $R(A-\lambda)=X$, then $(A-\lambda)^{-1}$ exists and is defined on the whole of X. Moreover, $(A-\lambda)^{-1}$ is a closed operator. Thus by Theorem 2.3 it is bounded on X. This gives (4.1), and consequently $\lambda \in \rho(A)$. Conversely, if $\lambda \in \rho(A)$, then $R(A-\lambda)$ is dense in X and (4.1) holds. In particular $A-\lambda$ is one–to–one. Let z be any element of X. Since $R(A-\lambda)$ is dense in X, there is a sequence $\{x_n\}$ of elements of $D(A)$ such that $(A-\lambda)\cdot x_n \to z$ in X. By (4.1) we have

$$\|x_n - x_m\| \leqslant C\|(A-\lambda)(x_n - x_m)\| \to 0 \text{ as } m, n \to \infty .$$

Thus by the completeness of X there is an element of X such that $x_n \to x$. This means that $Ax_n \to z+\lambda x$, and since A is closed, we know that $x \in D(A)$ and $Ax = z+\lambda x$. This shows that $z \in R(A-\lambda)$. Since z was arbitrary, we see that $R(A-\lambda)=X$, and the proof is complete.

We shall say that a scalar λ belongs to the *essential spectrum* $\sigma_e(A)$ of A if $\lambda \in \sigma(A+K)$ for every operator K compact on X. Thus

$$\sigma_e(A) = \bigcap_{K \in K(X)} \sigma(A+K) . \tag{4.2}$$

An obvious consequence of the definition is that $\sigma_e(A)$ is a closed set. Another immediate consequence is

Lemma 4.3. *If K is compact, then*

$$\sigma_e(A+K) = \sigma_e(A) .$$

We also have

Theorem 4.4. *If there is a sequence $\{x_n\}$ of elements of $D(A)$ such that*

a) $\|x_n\| = 1$,

b) $(A-\lambda)x_n \to 0$,

c) $\{x_n\}$ *has no convergent subsequence,*

then $\lambda \in \sigma_e(A)$.

Proof. Suppose $\lambda \notin \sigma_e(A)$. Then there is a compact operator K such that $\lambda \in \rho(A+K)$. In particular we have

$$\|x\| \leqslant C\|(A+K-\lambda)x\| , \qquad x \in D(A) . \tag{4.3}$$

There is a subsequence $\{z_k\}$ of $\{x_n\}$ such that $\{Kz_k\}$ converges. Thus

$$\begin{aligned}\|z_j - z_k\| &\leqslant C\,\|(A+K-\lambda)(z_j - z_k)\| \\ &\leqslant C\,\|(A-\lambda)z_j\| + C\,\|(A-\lambda)z_k\| + C\,\|K(z_j - z_k)\| \\ &\to 0 \text{ as } j, k \to \infty .\end{aligned}$$

Thus the sequence $\{z_k\}$ converges. But this contradicts the fact that $\{x_n\}$ has no convergent subsequence. This completes the proof.

Remark. A sequence $\{x_n\}$ having properties a)–c) is called a *singular sequence* for $A-\lambda$.

For any linear operator A on X we let Φ_A denote the set of those scalars λ such that $A-\lambda$ is in $\Phi(X) = \Phi(X, X)$.

Theorem 4.5. *If A is closed, then $\lambda \notin \sigma_e(A)$ if and only if $\lambda \in \Phi_A$ and $i(A-\lambda) = 0$.*

Proof. If $\lambda \notin \sigma_e(A)$, then there is a compact operator K on X such that $\lambda \in \rho(A+K)$. In particular $A+K-\lambda \in \Phi(X)$ and $i(A+K-\lambda)=0$. By Theorem 3.1 we see that $A-\lambda \in \Phi(X)$ and $i(A-\lambda)=0$. This proves the theorem in one direction. Now suppose $\lambda \in \Phi_A$ and $i(A-\lambda)=0$. Without loss of generality we may take $\lambda = 0$. Let $x_1, \ldots, x_n$ be a basis for $N(A)$ and let $y_1', \ldots, y_n'$ be a basis for $N(A')$ (see Lemma 1.1). By Theorems 2.5 and 2.6 there are functionals $x_1', \ldots, x_n'$ in X' and elements $y_1, \ldots, y_n$ such that

$$x_j'(x_k) = \delta_{jk}, \quad y_j'(y_k) = \delta_{jk}, \qquad 1 \leqslant j,\ k \leqslant n . \tag{4.4}$$

Set

$$Kx = \sum_1^n x_k'(x)\, y_k . \tag{4.5}$$

Clearly K is a linear operator defined everywhere on X. It is bounded, since

$$\|Kx\| \leqslant \|x\| \sum_1^n \|x_k'\|\, \|y_k\| .$$

Moreover, the range of K is contained in a finite dimensional subspace of X (Lemma 1.3). By Lemma 2.7, K is a compact operator in X. Consequently $A-K \in \Phi(X)$ and $i(A-K)=0$. Now suppose $x \in N(A-K)$. Since

$$y_k'(Az) = A' y_k'(z) = 0, \qquad z \in D(A),$$

by the definition of A', we see by (4.4) that no non-zero linear combination of the y_k is in $R(A)$. Consequently the only way we can have $Ax = Kx$ is if $Ax = Kx = 0$. Thus $x \in N(A)$ and

$$x_k'(x) = 0, \qquad 1 \leqslant k \leqslant n, \tag{4.7}$$

since the y_k are linearly independent by (4.4). Since $x_1, \ldots, x_n$ forms a basis for $N(A)$, x must be a linear combination of them:

$$x = \sum_1^n \alpha_k x_k . \tag{4.8}$$

But by (4.4) we have $x_j'(x) = \alpha_j$ and by (4.7) we have $x_j'(x) = 0$. Hence $x = 0$. We see therefore that $\alpha(A-K) = 0$. Since $i(A-K) = i(A) = 0$, we also have $\beta(A-K) = 0$. Thus $A-K$ is one–to–one and $R(A-K) = X$. We now see by Lemma 4.2 that $0 \in \rho(A-K)$. Hence $0 \notin \sigma_e(A)$ and the proof is complete.

Theorem 4.6. *If A is closed and B is A-compact, then*

$$\sigma_e(A+B) = \sigma_e(A) . \tag{4.9}$$

Proof. If $\lambda \notin \sigma_e(A)$, then $\lambda \in \Phi_A$ and $i(A-\lambda) = 0$ by Theorem 4.5. Thus $\lambda \in \Phi_{A+B}$ and $i(A+B-\lambda) = 0$ by Theorem 3.3. Hence $\lambda \notin \sigma_e(A+B)$. Conversely, if $\lambda \notin \sigma_e(A+B)$, then $\lambda \in \Phi_{A+B}$ and $i(A+B-\lambda) = 0$. By Theorem 3.3, B is $(A+B)$-compact. Hence $\lambda \in \Phi_A$ and $i(A-\lambda) = 0$ (Theorem 3.4). This completes the proof.

If there is an $x \neq 0$ in $D(A)$ such that $(A-\lambda)x = 0$, then λ is called an *eigenvalue* and x is the corresponding *eigenelement* (or *eigenvector*). Clearly an eigenvalue is in $\sigma(A)$. By multiplying x by a suitable scalar we can always arrange that $\|x\| = 1$. In this case the eigenelement is called *normalized.*

Lemma 4.6. *If $0 \in \rho(A)$, then $\lambda \neq 0$ is in Φ_A if and only if $1/\lambda \in \Phi_{A^{-1}}$ and*

$$i(A-\lambda) = i(A^{-1} - \lambda^{-1}) . \tag{4.10}$$

Proof. We note that

$$A - \lambda = -\lambda(A^{-1} - \lambda^{-1})A . \tag{4.11}$$

Now suppose $1/\lambda \in \Phi_{A^{-1}}$. Then both operators on the right-hand side of (4.11) are in $\Phi(X)$ (note that $0 \in \rho(A)$ implies $A \in \Phi(X)$ and $i(A) = 0$). Thus by Theorem 3.2 the product is in $\Phi(X)$ and

$$i(A-\lambda) = i(A^{-1} - \lambda^{-1}) + i(A) .$$

Since $i(A) = 0$, we obtain (4.10). Conversely, assume $\lambda \neq 0$ is in Φ_A. Then the product on the right-hand side of (4.11) is in $\Phi(X)$. Since $A \in \Phi(X)$ and $A^{-1} - \lambda^{-1}$ is in $B(X) = B(X, X)$, we have by Theorem 3.5 that $A^{-1} - \lambda^{-1}$ is in $\Phi(X)$. Consequently $1/\lambda$ is in $\Phi_{A^{-1}}$ and (4.10) holds. This completes the proof.

A consequence of Lemma 4.6 is

Theorem 4.7. *Let A, B be closed densely defined operators on X. If $\lambda\in\rho(A)\cap\rho(B)$ and $(A-\lambda)^{-1}-(B-\lambda)^{-1}$ is a compact operator on X, then*

$$\sigma_e(A)=\sigma_e(B). \tag{4.12}$$

Proof. We may assume $\lambda=0$. From the fact that $A^{-1}-B^{-1}$ is compact, it follows that $\Phi_{A^{-1}}=\Phi_{B^{-1}}$ and that $i(A^{-1}-\eta)=i(B^{-1}-\eta)$ for $\eta\in\Phi_{A^{-1}}$ (Theorem 3.1). If we now apply Lemma 4.6 to both A and B we see that $\Phi_A=\Phi_B$ and $i(A-\lambda)=i(B-\lambda)$ for $\lambda\in\Phi_A$. This implies (4.12) and the proof is complete.

Theorem 4.8. *Let A, B be operators in $\Phi(X)$ such that $i(A)=i(B)=0$. Suppose A_0, B_0 are operators in $B(X)$ such that*

$$AA_0=I-K_1 \tag{4.13}$$

$$BB_0=I-K_2 \tag{4.14}$$

where K_1 and K_2 are in $K(X)$. If $A_0-B_0\in K(X)$, then $\sigma_e(A)=\sigma_e(B)$.

Proof. By (4.13) and (4.14) we have for any scalar λ

$$(A-\lambda)A_0-(B-\lambda)B_0=K_2-K_1-\lambda(A_0-B_0). \tag{4.15}$$

Now assume $\lambda\in\Phi_B$ with $i(B-\lambda)=0$. Since B is a closed operator, we can make $D(B)$ into a Banach space by introducing the norm

$$\|x\|_B=\|Bx\|+\|x\|.$$

Let Z denote this Banach space. Clearly B is a bounded operator from Z to X. By Theorem 3.8, (4.14) implies that $B_0\in\Phi(X,Z)$. Since $B-\lambda\in\Phi(X)$, it is easily checked that $B-\lambda\in\Phi(Z,X)$. Thus $(B-\lambda)B_0\in\Phi(X)$ by Theorem 3.2. Since the right-hand side of (4.15) is in $K(X)$, we see by Theorem 3.1 that $(A-\lambda)A_0\in\Phi(X)$ with

$$i[(A-\lambda)A_0]=i[(B-\lambda)B_0]. \tag{4.16}$$

As before, (4.13) implies that $A_0\in\Phi(X,Y)$, where Y is $D(A)$ under the norm (2.6). Consequently $(A-\lambda)\in\Phi(Y,X)$ by Theorem 3.5. It is easily checked that this implies $A-\lambda\in\Phi(X)$. Now by (4.13) and (4.14) we have in view of Theorems 3.1, 3.2

$$i(A)+i(A_0)=i(I-K_1)=i(I)=0$$

$$i(B)+i(B_0)=i(I-K_2)=i(I)=0.$$

But by hypothesis $i(A)=i(B)=0$. Hence $i(A_0)=i(B_0)=0$. In view of this, (4.16) implies

$$i(A-\lambda)=i(B-\lambda).$$

Since we assumed $i(B-\lambda)=0$, we must have $i(A-\lambda)=0$. Thus we have proved $\sigma_e(A)\subset\sigma_e(B)$. The opposite inclusion follows from symmetry.

§5. Interpolation

In this section we shall give a brief presentation of an important method in functional analysis. We shall present only a small segment of the theory, namely that portion which will give us the results we need for the book.

Let X be a Banach space and let Ω be a domain (open connected set) in the complex plane. We can consider mappings from $\bar{\Omega}$ to X. We call such a mapping $f(\zeta)$ a vector valued function of a complex variable. We say that $f(\zeta)$ is analytic in Ω if for each $x'\in X'$ the complex valued function $x'[f(\zeta)]$ is analytic in Ω. $f(\zeta)$ is called continuous in $\bar{\Omega}$ if for each $\zeta_0\in\bar{\Omega}$, $\zeta_n\to\zeta_0$ implies $f(\zeta_n)\to f(\zeta_0)$ in X.

For any domain Ω in the complex plane and any Banach space X we let $W(X,\Omega)$ denote the set of functions from $\bar{\Omega}$ to X such that $f(\zeta)$ is continuous in $\bar{\Omega}$ and analytic in Ω with

$$\|f\|_W=\sup_{\zeta\in\bar{\Omega}}\|f(\zeta)\|_X<\infty. \tag{5.1}$$

It is easily checked that (5.1) is a norm and that $W(X,\Omega)$ is a normed vector space. We also have

Lemma 5.1. *$W(X,\Omega)$ is a Banach space.*

Proof. Suppose $\{f_n\}$ is a sequence of elements of $W(X,\Omega)$ such that $\|f_n-f_m\|_W\to 0$. By (5.1) for each $\zeta\in\bar{\Omega}$

$$\|f_n(\zeta)-f_m(\zeta)\|_X\to 0, \tag{5.2}$$

and this convergence is uniform in $\bar{\Omega}$. Consequently for each $\zeta\in\bar{\Omega}$ the sequence $\{f_n(\zeta)\}$ converges in X to an element $f(\zeta)\in X$, and $f(\zeta)$ is continuous as a function from $\bar{\Omega}$ to X. Furthermore for each $x'\in X'$ the function $x'[f_n(\zeta)]$ is analytic in Ω and converges uniformly in Ω to $x'[f(\zeta)]$. Consequently $x'[f(\zeta)]$ is analytic in Ω. Thus $f(\zeta)\in W(X,\Omega)$ and $\|f_n-f\|_W\to 0$.

Lemma 5.2. *If Ω is bounded and $f \in W(X, \Omega)$, then*

$$\|f\|_W = \max_{\zeta \in \partial\Omega} \|f(\zeta)\|_X , \tag{5.3}$$

where $\partial\Omega$ denotes the boundary of Ω.

Proof. Let x' be any functional in X'. Then $x'[f(\zeta)]$ is analytic in Ω and continuous on $\bar{\Omega}$. By the maximum modulus theorem

$$|x'[f(z)]| \leqslant \max_{\zeta \in \partial\Omega} |x'[f(\zeta)]| \leqslant \|x'\| \max_{\zeta \in \partial\Omega} \|f(\zeta)\|_X$$

for all $z \in \bar{\Omega}$. By Theorem 2.19 this gives

$$\|f(z)\|_X \leqslant \max_{\zeta \in \partial\Omega} \|f(\zeta)\|_X, \qquad z \in \bar{\Omega} .$$

This is equivalent to (5.3).

Lemma 5.3. *If Ω is the strip $0 < x < 1$ and $f \in W(X, \Omega)$, then*

$$\|f\|_W = \max_{k=0,1} \sup_{y} \|f(k+\mathrm{i}y)\| . \tag{5.4}$$

Proof. Let $\varepsilon > 0$ be given and let M denote the right-hand side of (5.4). Let N be any positive number such that

$$\|f\|_W \leqslant \mathrm{e}^{\varepsilon N^2} M . \tag{5.5}$$

Set

$$g(z) = \mathrm{e}^{\varepsilon z^2} f(z) . \tag{5.6}$$

Then $g \in W(X, \Omega)$ and

$$\|g(x+\mathrm{i}y)\| = \|f(x+\mathrm{i}y)\| \exp\{\varepsilon(x^2 - y^2)\} . \tag{5.7}$$

Consequently

$$\|g(k+\mathrm{i}y)\| \leqslant \mathrm{e}^{\varepsilon} M , \qquad k = 0, 1 , \tag{5.8}$$

and by (5.5)

$$\|g(x \pm \mathrm{i}N)\| \leqslant \mathrm{e}^{\varepsilon} M , \qquad 0 \leqslant x \leqslant 1 . \tag{5.9}$$

These inequalities show that $\|g(z)\| \leqslant \mathrm{e}^{\varepsilon} M$ for z on the boundary of the rectangle $0 \leqslant x \leqslant 1$, $|y| \leqslant N$. By Lemma 5.2 this inequality must hold inside as well. Letting $N \to \infty$, we have

$$\|g(z)\| \leqslant \mathrm{e}^{\varepsilon} M , \qquad z \in \bar{\Omega}$$

or

$$\|f(z)\| \leqslant M \exp\{\varepsilon(y^2 - x^2 + 1)\} , \qquad z \in \bar{\Omega} .$$

Since this is true for every $\varepsilon > 0$, we get

$$\|f(z)\| \leq M, \qquad z \in \Omega.$$

Thus $\|f\|_W \leq M$, and the proof is complete.

Next let X_0 be a Banach space with norm $\| \ \|_0$, and suppose that there is a dense subspace $X_1 \subset X_0$ and a norm $\| \ \|_1$ defined in X_1 such that X_1 is complete with respect to the norm $\| \ \|_1$ and

$$\|x\|_0 \leq C\|x\|_1, \qquad x \in X_1. \tag{5.10}$$

Let Ω be the strip $0 < x < 1$ in the complex plane and let $H(X_0, X_1)$ denote the set of those $f(\zeta) \in W(X_0, \Omega)$ such that $f(1+iy) \in X_1$ for each real y, and

$$\|f\|_H = \max_{k=0,1} \sup_y \|f(k+iy)\|_k < \infty. \tag{5.11}$$

By Lemma 5.3 and (5.10) we have

$$\|f\|_W \leq C\|f\|_H, \qquad f \in H(X_0, X_1). \tag{5.12}$$

This implies

Lemma 5.4. *$H(X_0, X_1)$ is a Banach space under the norm* (5.11).

Proof. Let $\{f_n\}$ be a Cauchy sequence in $H(X_0, X_1)$. Then by (5.12) it is a Cauchy sequence in $W(X_0, \Omega)$. By Lemma 5.1, f_n converges in $W(X_0, \Omega)$ to a function $f(\zeta)$. Furthermore

$$\|f(1+iy)\|_1 \leq \|f\|_H, \qquad f \in H. \tag{5.13}$$

Consequently $\{f_n(1+iy)\}$ is a Cauchy sequence in X_1 which converges uniformly in y. Its limit in X_1 must coincide with its limit $f(1+iy)$ in X_0. Thus $f(1+iy)$ is in X_1 and

$$\|f_n(1+iy) - f(1+iy)\|_1 \to 0,$$

uniformly in y. Thus $f \in H(X_0, X_1)$ and $\|f_n - f\|_H \to 0$. This completes the proof.

Next let θ be a real number satisfying $0 \leq \theta \leq 1$. We let $X_\theta = [X_0, X_1]_\theta$ denote the set of those $x \in X_0$ such that $x = f(\theta)$ for some $f \in H(X_0, X_1)$. Put

$$\|x\|_\theta = \inf_{\substack{f \in H(X_0, X_1) \\ f(\theta) = x}} \|f\|_H. \tag{5.14}$$

Lemma 5.5. *$[X_0, X_1]_\theta$ is a Banach space under the norm* (5.14).

Proof. Let N_θ denote the subspace of $H(X_0, X_1)$ consisting of those $f(\zeta)$

such that $f(\theta)=0$. By (5.12), N_θ is a closed subspace. We merely note that the norm of $[X_0, X_1]_\theta$ is that of $H(X_0, X_1)/N_\theta$ and apply Theorem 1.9.

Theorem 5.6. *Let Y_0, Y_1 be Banach spaces which satisfy the hypotheses of X_0, X_1. Suppose T is a bounded linear operator from X_0 to Y_0 such that $Tx \in Y_k$ for $x \in X_k$ with*

$$\|Tx\|_{Y_k} \leqslant M_k \|x\|_k, \qquad k=0, 1. \tag{5.15}$$

Then T maps X_θ into Y_θ with

$$\|Tx\|_{Y_\theta} \leqslant M_0^{1-\theta} M_1^\theta \|x\|_{X_\theta}, \qquad x \in X_\theta. \tag{5.16}$$

Proof. Let $x \in X_\theta$ and $\varepsilon > 0$ be given. Then there is an $f \in H(X_0, X_1)$ such that $x = f(\theta)$ and

$$\|f\|_H \leqslant \|x\|_\theta + \varepsilon.$$

Set

$$g(z) = M_0^{z-1} M_1^{-z} T[f(z)].$$

Since T is a bounded operator from X_0 to Y_0, g is clearly in $W(Y_0, \Omega)$. By (5.15) we see also that it is in $H(Y_0, Y_1)$ with

$$M_0^{\theta-1} M_1^{-\theta} \|Tx\|_{Y_\theta} = \|g(\theta)\|_{Y_\theta} \leqslant \|g\|_{H(Y_0, Y_1)} \leqslant \|x\|_\theta + \varepsilon.$$

Since ε was arbitrary, (5.16) holds and the proof is complete.

§6. Intermediate extensions

Let W, Z be Banach spaces. A *bilinear form* $a(u, v)$ defined on $W \times Z$ is an assignment of a complex number to each pair w, z of elements $w \in W$, $z \in Z$, in such a way that the assignment is linear in w and conjugate linear in z. By this we mean that

$$a(\alpha_1 w_1 + \alpha_2 w_2, \beta_1 z_1 + \beta_2 z_2) = \alpha_1 \bar{\beta}_1 a(w_1, z_1) + \alpha_1 \bar{\beta}_2 a(w_1, z_2) + \\ + \alpha_2 \bar{\beta}_1 a(w_2, z_1) + \alpha_2 \bar{\beta}_2 a(w_2, z_2).$$

The bilinear form $a(w, z)$ is called bounded if

$$|a(w, z)| \leqslant C \|w\|_W \|z\|_Z, \qquad w \in W, z \in Z. \tag{6.1}$$

We shall say that a Banach space W is *continuously embedded* in a Banach space X if there is an operator $E \in B(W, X)$ which is one–to–one. We call

E an embedding operator. We shall say that W is densely embedded if $R(E)$ is dense in X.

Suppose X, Y, W, Z are Banach spaces such that W is continuously and densely embedded in X with embedding operator E and Y is continuously and densely embedded in $\hat{Z}'$ with embedding operator F (here $\hat{Z}'$ denotes the set of bounded conjugate linear functionals on Z; §2). Suppose $a(w, z)$ is a bounded bilinear form on $W \times Z$. We define two linear operators connected with $a(w, z)$.

The first, which we shall denote by A, is an operator from X to Y. We say that $x \in D(A)$ and $Ax = y$ if $x \in R(E)$, $y \in Y$ and

$$a(E^{-1},x, z) = Fy(z), \qquad z \in Z. \tag{6.2}$$

Since $R(F)$ is dense in $\hat{Z}'$, the operator A is uniquely defined. We call A the operator *associated* with the bilinear form $a(u, v)$.

The second operator, which we denote by $\hat{A}$, is from W to $\hat{Z}'$. We define it as follows. For fixed $w \in W$, $a(w, z)$ is a conjugate linear functional on Z. It is bounded by (6.1). Hence there is a $\hat{z}' \in \hat{Z}'$ such that

$$a(w, z) = \hat{z}'(z), \qquad z \in Z. \tag{6.3}$$

We define $\hat{A}w$ to be $\hat{z}'$. Clearly $\hat{A}$ is uniquely defined and $D(\hat{A}) = W$. We call $\hat{A}$ the *extended operator* associated with the bilinear form $a(u, v)$.

An important relationship between the operators A and $\hat{A}$ is given by

Proposition 6.1. $A = F^{-1}\hat{A}E^{-1}$.

Proof. Let B be the operator $F^{-1}\hat{A}E^{-1}$. Then $x \in D(B)$ and $Bx = y$ if $x \in R(E)$ and $\hat{A}E^{-1}x = Fy$. By (6.3) this means that

$$a(E^{-1}x, z) = Fy(z), \qquad z \in Z,$$

which is precisely (6.2). Thus $x \in D(A)$ and $Ax = y$. Conversely suppose $x \in D(A)$ and $Ax = y$. Then (6.2) holds. This means that $E^{-1}x \in D(\hat{A})$ and $\hat{A}E^{-1}x = Fy$. Applying F^{-1} to both sides we see that $Bx = y$, and the proof is complete.

Theorem 6.2. *If* $\hat{A} \in \Phi(W, \hat{Z}')$, *then* $A \in \Phi(X, Y)$ *and*

$$i(A) = i(\hat{A}). \tag{6.4}$$

Proof. Note that the operators E^{-1} and F^{-1} are densely defined, one–to–one and onto. Thus

$$E^{-1} \in \Phi(X, W), \qquad F^{-1} \in \Phi(\hat{Z}', Y) \tag{6.5}$$

and

$$i(E^{-1}) = i(F^{-1}) = 0\,. \tag{6.6}$$

We now apply Theorem 3.2 to the product $F^{-1}\hat{A}E^{-1}$. We conclude via (6.5) and (6.6) that this product is in $\Phi(X, Y)$ and has the same index as $\hat{A}$. The theorem now follows from Proposition 6.1.

The following is a simple consequence of Theorems 3.1 and 6.2.

Theorem 6.3. *Let $a(w, z)$ and $c(w, z)$ be bounded bilinear forms on $W \times Z$. Let $\hat{A}$ and $\hat{C}$ denote the corresponding extended operators associated with them. Suppose that $\hat{A} \in \Phi(W, \hat{Z}')$ and that $\hat{C} \in K(W, \hat{Z}')$. Let $\hat{B}$ denote the extended operator associated with the bilinear form*

$$b(w, z) = a(w, z) + c(w, z)\,. \tag{6.7}$$

Then $\hat{B} = \hat{A} + \hat{C} \in \Phi(W, \hat{Z}')$. Thus the operator B associated with $b(w, z)$ is in $\Phi(X, Y)$ and $i(B) = i(\hat{A})$.

Suppose that A_0 is a linear operator from X to Y such that $D(A_0) \subset R(E)$ and there is a bounded bilinear form $a(w, z)$ on $W \times Z$ satisfying

$$a(E^{-1}x, z) = F[A_0 x](z)\,, \qquad x \in D(A_0)\,,\ z \in Z\,. \tag{6.8}$$

If we let A denote the operator associated with the bilinear form $a(w, z)$, then A is obviously an extension of A_0. We call A an *intermediate extension* of A_0 *relative to W and Z* or briefly a W–Z extension.

Theorem 6.4. *Let U, V be dense subspaces of W and Z, respectively. Suppose that $a_0(u, v)$ is a bilinear form on $U \times V$ such that*

$$|a_0(u, v)| \leqslant C\|u\|_W \|v\|_Z\,, \qquad u \in U\,,\ v \in V\,. \tag{6.9}$$

If A_0 is a linear operator from X to Y such that $D(A) \subset E(U)$ and

$$a_0(E^{-1}x, v) = F[A_0 x](v)\,, \qquad x \in D(A),\ v \in V\,, \tag{6.10}$$

then A_0 has an intermediate extension relative to W and Z.

Proof. We merely extend $a_0(u, v)$ to $W \times Z$ by continuity using (6.9). If $a(w, z)$ denotes the extension, then (6.10) implies (6.8) by continuity.

We now consider a very important special case of an intermediate extension. Let X be a Hilbert space. An operator A on X is called *regularly accretive* if there is a Hilbert space W continuously and densely embedded in X with embedding operator E, a bounded bilinear form $a(u, v)$ on $W \times W$ and a constant $N \geqslant 0$ such that

$$\|u\|_W^2 \leqslant C[\operatorname{Re} a(u, u) + N\|Eu\|^2], \qquad u \in W, \tag{6.11}$$

and A is the operator associated with $a(u, v)$. In order that this definition make sense we need X to be continuously and densely embedded in $\hat{W}'$. This is given in

Lemma 6.5. *If X is a Hilbert space and W is a Banach space continuously and densely embedded in X with embedding operator E, then X can be continuously and densely embedded in $\hat{W}'$ with embedding operator F satisfying*

$$(x, Ew) = Fx(w), \qquad x \in X, \ w \in W. \tag{6.12}$$

Proof. For each x the expression (x, Ew) is a conjugate linear functional on W and

$$|(x, Ew)| \leqslant \|x\| \, \|E\| \, \|w\|_W.$$

Hence there is an element $\hat{w}' \in \hat{W}'$ such that $\|\hat{w}'\| \leqslant \|E\| \, \|x\|$ and

$$(x, Ew) = \hat{w}'(w), \qquad w \in W. \tag{6.13}$$

Define the operator F from X to $\hat{W}'$ by $Fx = \hat{w}'$. Clearly F is linear and bounded. It is also one–to–one since $R(E)$ is dense in X. Finally, suppose $\hat{w}'(w) = 0$ for all $\hat{w}' \in R(F)$. Then $(x, Ew) = 0$ for all $x \in X$. Thus $Ew = 0$ and consequently $w = 0$. This shows that $R(F)$ is dense, and the proof is complete.

From Lemma 6.5 we see that a regularly accretive operator is the operator associated with a bilinear form satisfying the additional condition (6.11). We shall find it convenient to write $a(u)$ in place of $a(u, u)$.

Theorem 6.6. *Let A_0 be an operator on a Hilbert space X. Let W be a Hilbert space continuously and densely embedded in X with embedding operator E. Let U be a dense subspace of W, and assume that $D(A_0) \subset E(U)$. Let $a_0(u, v)$ be a bilinear form defined on $U \times U$ which is bounded on $W \times W$ and satisfies*

$$\|u\|_W^2 \leqslant C[\operatorname{Re} a_0(u) + N\|Eu\|^2], \qquad u \in U, \tag{6.14}$$

for some constant N. Assume that

$$a_0(E^{-1}x, v) = (A_0 x, Ev), \qquad x \in D(A_0), \quad v \in V. \tag{6.15}$$

Then A_0 has a regularly accretive extension with corresponding bilinear form $a(u, v)$ satisfying (6.11).

Proof. By continuity we can extend $a_0(u, v)$ to $W \times W$, and the extension $a(u, v)$ will satisfy (6.11). Let A be the operator associated with the bilinear form $a(u, v)$. By definition A is regularly accretive and

$$a(E^{-1}x, v) = F[Ax](v), \qquad x \in D(A),\ v \in W, \tag{6.16}$$

where F is the embedding given by Lemma 6.5. By that lemma we have

$$a(E^{-1}x, v) = (Ax, Ev), \qquad x \in D(A),\ v \in W. \tag{6.17}$$

Clearly $D(A_0) \subset D(A)$ and if $x \in D(A_0)$, we have by (6.15) and (6.17)

$$(Ax - A_0 x, Ev) = 0, \qquad v \in U. \tag{6.18}$$

I claim that $E(U)$ is dense in X. To see this let $\varepsilon > 0$ and $x \in X$ be given. By hypothesis there is a $w \in W$ such that $\|x - Ew\| < \varepsilon$. Since U is dense in W, there is a $u \in U$ such that $\|u - w\|_W < \varepsilon$. Hence

$$\|x - Eu\| \leqslant \|x - Ew\| + \|Ew - Eu\| \leqslant \varepsilon + \|E\|\ \|w - u\|_W \leqslant (1 + \|E\|)\varepsilon.$$

Hence $E(U)$ is dense in X. Thus by (6.18), $Ax = A_0 x$. Hence A is an extension of A_0, and the proof is complete.

An important property of regularly accretive operators is given by

THEOREM 6.7. *If A is regularly accretive and its corresponding bilinear form $a(u, v)$ satisfies (6.11), then $\sigma(A)$ is contained in the half-plane* $\operatorname{Re}\lambda + N > 0$.

To prove Theorem 6.7 we make use of

LEMMA 6.8 (Lax-Milgram). *Let H be a Hilbert space and suppose that $a(u, v)$ is a bilinear form on $H \times H$ such that for some $M \geqslant m > 0$*

$$m\|u\|^2 \leqslant |a(u)| \leqslant M\|u\|^2, \qquad u \in H. \tag{6.19}$$

Then for each bounded linear functional F on H there are elements $w, u \in H$ such that

$$Fv = a(v, w) = \overline{a(u, v)}, \qquad v \in H. \tag{6.20}$$

Proof of Theorem 6.7. For $\operatorname{Re}\lambda + N \leqslant 0$ and $x \in D(A)$ we have by (6.11)

$$\begin{aligned} \operatorname{Re}([A - \lambda]x, x) &= \operatorname{Re} a(E^{-1}x) - \operatorname{Re}\lambda\|x\|^2 \\ &\geqslant \operatorname{Re} a(E^{-1}x) + N\|x\|^2 \geqslant \|E^{-1}x\|_W^2/C. \end{aligned}$$

Hence $A - \lambda$ is one–to–one. To show that it is onto, let f be any element of X. Then for $v \in W$

$$|(Ev, f)| \leqslant \|E\| \, \|v\|_W \|f\| \, .$$

Hence (Ev, f) is a bounded linear function on W. Moreover, the bilinear form $a(u, v) - \lambda(Eu, Ev)$ satisfies the hypothesis of the Lax-Milgram lemma (Lemma 6.8). Hence there is a $u \in W$ such that

$$a(u, v) - \lambda(Eu, Ev) = (f, Ev) \, , \qquad v \in W$$

(we have removed the conjugation bars). From this we see that $Eu \in D(A)$ and $A\,Eu = f + \lambda\,Eu$. Hence $x = Eu$ is a solution of $(A - \lambda)x = f$. Since f was arbitrary, we see that $R(A - \lambda) = X$. Thus $\lambda \in \rho(A)$. Since the half-plane $\operatorname{Re} \lambda + N \leqslant 0$ is contained is $\rho(A)$, we see that $\sigma(A)$ is contained in the half-plane $\operatorname{Re} \lambda + N > 0$. The proof is complete.

Lemma 6.9. *If A is regularly accretive and W is the Hilbert space for which (6.11) holds, then $E^{-1}[D(A)]$ is dense in W.*

Proof. Suppose $u \in W$ is such that

$$(u, E^{-1}x)_W = 0 \, , \qquad x \in D(A) \, ,$$

where $(\; , \;)_W$ denotes the scalar product of W. By (6.11) and the Lax-Milgram lemma (Lemma 6.8) there is a $w \in W$ such that

$$(u, h)_W = a(w, h) + N(w, h) \, , \qquad h \in W \, . \tag{6.21}$$

Thus

$$a(w, E^{-1}x) + N(w, E^{-1}x) = 0 \, , \qquad x \in D(A) \, .$$

By the definition of A this implies

$$(w, [A + N]x) = 0 \, , \qquad x \in D(A) \, .$$

By Theorem 6.7, $R(A + N)$ is the whole space. Consequently $w = 0$. We now have by (6.21) that $(u, h)_W = 0$ for all $h \in W$. Thus $u = 0$. This shows that $D(A)$ is dense in W.

The following lemma is easily proved.

Lemma 6.10. *If $a(u, v)$, $b(u, v)$ are bilinear forms and $|b(u)| \leqslant a(u)$ for all u, then*

$$|b(u, v)|^2 \leqslant 4a(u)\,a(v) \, , \qquad \text{all } u, v \, . \tag{6.22}$$

Theorem 6.11. *Let A be a regularly accretive operator on X with corresponding bilinear form $a(u, v)$ and Hilbert space W. Let $c(u, v)$ and $e(u, v)$ be bilinear forms bounded in $W \times W$ such that*

a) $|e(u)| \leqslant c(u)$, $\quad u \in W$;

b) $\|u_k\|_W \leqslant C$ *implies that there is a subsequence* $\{v_n\}$ *of* $\{u_k\}$ *such that* $c(v_m - v_n) \to 0$ *as* $m, n \to \infty$;

c) *if* $Eu_k \to 0$ *and* $c(u_j - u_k) \to 0$ *as* $j, k \to \infty$, *then* $c(u_k) \to 0$.

Set $b(u, v) = a(u, v) + e(u, v)$. *Then the operator* B *associated with* $b(u, v)$ *is regularly accretive and*

$$\sigma_e(B) = \sigma_e(A). \tag{6.23}$$

Proof. First we show that for any $\varepsilon > 0$ there is a C_ε such that

$$c(u) \leqslant \varepsilon \|u\|_W^2 + C_\varepsilon \|Eu\|^2, \qquad u \in W. \tag{6.24}$$

If (6.24) did not hold, there would be a sequence $\{u_k\}$ of elements of W such that

$$\|u_k\|_W = 1, \qquad c(u_k) > \varepsilon + k\|Eu_k\|^2, \qquad k = 1, 2, \ldots. \tag{6.25}$$

By hypothesis b) there is a subsequence $\{v_n\}$ of $\{u_k\}$ such that $c(v_m - v_n) \to 0$. Since $c(u, v)$ is bounded on $W \times W$, we see by (6.25) that $Eu_k \to 0$. Thus by hypothesis c) we have $c(v_n) \to 0$. But $c(v_n) > \varepsilon$ for each n. This contradiction proves (6.24).

By (6.11) and (6.24) we have

$$\operatorname{Re} b(u) = \operatorname{Re} a(u) + \operatorname{Re} e(u) \geqslant C^{-1}\|u\|_W^2 - N\|Eu\|^2 - c(u)$$

$$\geqslant \frac{1}{2C}\|u\|_W^2 - N'\|Eu\|^2,$$

where we have taken $\varepsilon = \frac{1}{2}C$. This inequality shows that the operator associated with $b(u, v)$ is regularly accretive.

To prove (6.23), let λ be any real number $\geqslant N' \geqslant N$. Set

$$a_\lambda(u, v) = a(u, v) + \lambda(Eu, Ev), \qquad b_\lambda(u, v) = b(u, v) + \lambda(Eu, Ev).$$

Thus

$$\|u\|_W^2 \leqslant C \operatorname{Re} a_\lambda(u), \qquad \|u\|_W^2 \leqslant C \operatorname{Re} b_\lambda(u), \qquad u \in W. \tag{6.26}$$

By the Lax-Milgram lemma (Lemma 6.8) for each $u \in W$ there is an element $Gu \in W$ such that

$$e(u, v) = a_\lambda(Gu, v), \qquad v \in W. \tag{6.27}$$

Clearly G is a linear operator in W. I claim that it is compact. To show this first note that G is bounded. For by (6.26), (6.27) and Lemma 6.10

$$\|Gu\|_W^2 \leqslant C \operatorname{Re} a_\lambda(Gu) = C \operatorname{Re} e(u, Gu)$$
$$\leqslant 4C\, c(u)^{\frac{1}{2}} c(Gu)^{\frac{1}{2}} \leqslant C'\, c(u)^{\frac{1}{2}} \|Gu\|_W .$$

Whence

$$\|Gu\|_W \leqslant C'\, c(u)^{\frac{1}{2}}, \qquad u \in W. \tag{6.28}$$

Now suppose that $\{u_k\}$ is a sequence of elements of W such that $\|u_k\|_W \leqslant C$. Then by hypothesis b) there is a subsequence $\{v_n\}$ such that $c(v_m - v_n) \to 0$. By (6.28) we see that Gv_n converges in W. Hence G is a compact operator.

Now by (6.27)

$$b_\lambda(u, v) = a_\lambda(u, v) + e(u, v) = a_\lambda(u + Gu, v).$$

This shows that $Eu \in D(B)$ if and only if $E(I+G)u \in D(A)$ and

$$(B-\lambda)Eu = (A-\lambda)E(I+G)u, \qquad u \in W.$$

Hence

$$(A-\lambda)^{-1} = E(I+G)E^{-1}(B-\lambda)^{-1}$$

(recall that $\lambda \in \rho(A) \cap \rho(B)$). This gives

$$(A-\lambda)^{-1} - (B-\lambda)^{-1} = EGE^{-1}(B-\lambda)^{-1}. \tag{6.29}$$

By (6.26) we have

$$\begin{aligned} \|E^{-1}(B-\lambda)^{-1}f\|_W^2 / C &\leqslant \operatorname{Re} b_\lambda(E^{-1}(B-\lambda)^{-1}f) \\ &= \operatorname{Re}(f, (B-\lambda)^{-1}f) \leqslant \|f\|\, \|(B-\lambda)^{-1}f\| \\ &= \|f\|\, \|EE^{-1}(B-\lambda)^{-1}f\| \\ &\leqslant \|f\|\, \|E\|\, \|E^{-1}(B-\lambda)^{-1}f\|_W, \end{aligned}$$

whence

$$\|E^{-1}(B-\lambda)^{-1}f\|_W \leqslant C\|f\|\, \|E\|, \qquad f \in X.$$

Thus $E^{-1}(B-\lambda)^{-1}$ is a bounded operator from X to W. Since G is a compact operator on W and E is a bounded operator from W to X, we see that $EGE^{-1}(B-\lambda)^{-1}$ is a compact operator on X. If we now apply Theorem 4.7 to (6.29), we obtain (6.23). The proof is complete.

Theorem 6.12. *Let X, W, Z be Banach spaces such that W can be densely and continuously embedded in X with embedding operator E and X can be densely and continuously embedded in $\hat{Z}'$ with embedding operator F. Let $a(u, v)$, $b(u, v)$ be bilinear forms continuous in $W \times Z$ and let A, B be the* *operators and* $\hat{A}$, $\hat{B}$ *be the extended operators associated with* $a(u, v)$, $b(u, v)$, *respectively. Suppose* $\hat{A}$ *and* $\hat{B}$ *are in* $\Phi(W, \hat{Z}')$ *with* $i(\hat{A}) = i(\hat{B}) = 0$. *Thus there are operators* A_1, B_1 *in* $B(\hat{Z}', W)$ *such that*

$$\hat{A}A_1 = I - K_1, \quad \hat{B}B_1 = I - K_2, \tag{6.30}$$

where K_1 *and* K_2 *are finite rank operators in* W (*Theorem* 3.7). *If* $A_1 - B_1$ *is in* $K(\hat{Z}', W)$, *then* $\sigma_e(A) = \sigma_e(B)$.

Proof. Set $A_0 = EA_1F$, $B_0 = EB_1F$. Then A_0, B_0 are bounded operators on X. By Proposition 6.1, $A = F^{-1}\hat{A}E^{-1}$ and $B = F^{-1}\hat{B}E^{-1}$. Hence

$$AA_0 = I - F^{-1}K_1F, \quad BB_0 = I - F^{-1}K_2F.$$

Now by Lemma 2.15, $F^{-1}K_jF$ are finite rank operators for $j = 1, 2$. Since $A_0 - B_0 = E(A_1 - B_1)F$, the desired result follows from Theorem 4.8.

Theorem 6.13. *Let* X, W, Z *be Banach spaces satisfying the hypotheses of the preceding theorem. Let* $a(u, v)$, $b(u, v)$ *be bilinear forms continuous in* $W \times Z$ *and let* A, B *be the operators and let* $\hat{A}$, $\hat{B}$ *be the extended operators associated with* $a(u, v)$, $b(u, v)$, *respectively. Assume that* $\hat{A}$ *is in* $\Phi(W, \hat{Z}')$ *with* $i(\hat{A}) = 0$. *If* $\hat{A} - \hat{B}$ *is in* $K(W, \hat{Z}')$, *then* $\sigma_e(A) = \sigma_e(B)$.

Proof. By Theorem 3.1, $\hat{B} \in \Phi(W, \hat{Z}')$ and $i(\hat{B}) = 0$. Thus by Theorem 3.7 there are operators $A_1, B_1 \in B(\hat{Z}', W)$ and finite rank operators K_1, K_2, K_3, K_4 such that

$$\hat{A}A_1 = I - K_1, \quad \hat{B}B_1 = I - K_2 \text{ on } \hat{Z}' \tag{6.31}$$

$$A_1\hat{A} = I - K_3, \quad B_1\hat{B} = I - K_4 \text{ on } W. \tag{6.32}$$

Moreover

$$\begin{aligned} A_1 - B_1 &= A_1\hat{B}B_1 - A_1\hat{A}B_1 + A_1B_1 + A_1K_2 - K_3B_1 \\ &= A_1(\hat{B} - \hat{A})B_1 + A_1K_2 - K_3B_1. \end{aligned} \tag{6.33}$$

I claim that the right-hand side of (6.33) is in $K(\hat{Z}', W)$. In fact, B_1 is in $B(\hat{Z}', W)$ while $\hat{B} - \hat{A}$ is in $K(W, \hat{Z}')$ by hypothesis. Since $A_1 \in B(Z', W)$, we see that $A_1(\hat{B} - \hat{A})B_1$ is in $K(\hat{Z}', W)$. Similarly, since $K_2 \in K(\hat{Z}')$, the operator A_1K_2 is in $K(\hat{Z}', W)$. Since $K_3 \in K(W)$, $K_3B_1 \in K(\hat{Z}', W)$. Hence $A_1 - B_1$ is in $K(\hat{Z}', W)$. We can now apply Theorem 6.12 to obtain the desired conclusion.

Theorem 6.14. *A regularly accretive operator is closed.*

Proof. Suppose A is regularly accretive on X with corresponding $a(u, v)$, W and E. If $\{x_n\}$ is a sequence of elements of $D(A)$ such that $x_n \to x$ and $Ax_n \to y$ in X, then by (6.11)

$$\begin{aligned}\|E^{-1}(x_n-x_m)\|_W^2 &\leqslant C[\operatorname{Re} a(E^{-1}[x_n-x_m])+N\|x_n-x_m\|^2] \\ &= C[\operatorname{Re}(A[x_n-x_m],[x_n-x_m])+N\|x_n-x_m\|^2] \\ &\to 0 \text{ as } m,n\to\infty .\end{aligned}$$

Hence $E^{-1}x_n$ converges in W to some element u. Since

$$a(E^{-1}x_n, v) = (Ax_n, Ev), \qquad v\in W,$$

we have in the limit

$$a(u, v) = (y, Ev), \qquad v\in W. \tag{6.34}$$

Set $x=Eu$. Then by definition $x\in D(A)$ and $Ax=y$. This shows that A is closed.

Theorem 6.15. *If A is regularly accretive and $\lambda\in\rho(A)$, then $E^{-1}(A-\lambda)^{-1}$ is a bounded operator from X to W.*

Proof. One checks easily that $(A-\lambda)E$ is a closed operator from W to X. Since it is onto, its inverse must be bounded (Theorem 2.3).

Theorem 6.16. *If A is regularly accretive and* $\operatorname{Re}\lambda+N\leqslant 0$, *then $\hat{A}-\lambda FE$ is one–to–one and onto, where $\hat{A}$ is the extended associated operator.*

Proof. By (6.11)

$$\begin{aligned}\|u\|_W^2 &\leqslant C\operatorname{Re}[a(u)-\lambda\|Eu\|^2] = C\operatorname{Re}[(\hat{A}-\lambda FE)u](u) \\ &\leqslant C\|(\hat{A}-\lambda FE)u\|_{\hat{W}'}\|u\|_W\end{aligned}$$

(see Lemma 6.5). Thus

$$\|u\|_W \leqslant C\|(\hat{A}-\lambda FE)u\|_{\hat{W}'}, \qquad u\in W. \tag{6.35}$$

This shows that $\hat{A}-\lambda FE$ is one–to–one and its range is closed (Theorem 2.3). Now by Theorem 6.7, $\lambda\in\rho(A)$. Hence for each $f\in X$ there is an $x\in D(A)$ such that

$$a(E^{-1}x, v)-\lambda(x, Ev) = (f, Ev), \qquad v\in W. \tag{6.36}$$

Setting $u=E^{-1}x$, we see that $(\hat{A}-\lambda FE)u=Ff$ (see Lemma 6.5). Thus $R(\hat{A}-\lambda FE)\supset F(X)$. Since $F(X)$ is dense in $\hat{W}'$ (Lemma 6.5), we see that $R(\hat{A}-\lambda FE)=\hat{W}'$. This completes the proof.

§7. Self-adjoint operators

A densely defined operator A in a Hilbert space X is called *self-adjoint* if $A^*=A$. The present section will be devoted to those properties of self-adjoint operators which will be used in the book. Note that we always work in a complex Hilbert space.

LEMMA 7.1. *If A is self-adjoint and* $\operatorname{Im} \lambda \neq 0$, *then* $\lambda \in \rho(A)$ *and*

$$\|(A-\lambda)^{-1}\| \leqslant 1/|\operatorname{Im} \lambda| . \tag{7.1}$$

Proof. We have

$$\operatorname{Im}(u,(A-\lambda)u) = (\operatorname{Im} \lambda)\|u\|^2 .$$

This implies:

$$|\operatorname{Im} \lambda| \, \|u\| \leqslant \|(A-\lambda)u\| , \tag{7.2}$$

from which we can conclude that $\alpha(A-\lambda)=0$. Since A is self-adjoint, we have $\beta(A-\lambda)=0$ as well. Consequently $R(A-\lambda)$ is dense in X. Thus by definition $\lambda \in \rho(A)$. Inequality (7.1) follows from (7.2).

An operator A on a Hilbert space X is called *symmetric* if $D(A)$ is dense in X and

$$(u, Av) = (Au, v) , \qquad u, v \in D(A) . \tag{7.3}$$

Clearly every self-adjoint operator is symmetric. A partial converse is

LEMMA 7.2. *Suppose A is a symmetric operator having a number λ such that $R(A-\lambda)=R(A-\bar{\lambda})=X$. Then A is self-adjoint.*

Proof. Let v be any element in $D(A^*)$. Since $R(A-\lambda)=X$, there is a $w \in D(A)$ such that $(A-\lambda)w=(A^*-\lambda)v$. Thus for $u \in D(A)$ we have

$$(v, (A-\bar{\lambda})u) = ((A^*-\lambda)v, u) = ((A-\lambda)w, u) = (w, (A-\bar{\lambda})u) .$$

Hence

$$(v-w, (A-\bar{\lambda})u) = 0 , \qquad u \in D(A) .$$

Since $R(A-\bar{\lambda})=X$, we see that $w=v$. This shows that $v \in D(A)$ and $Av=A^*v$. Hence A is self-adjoint.

LEMMA 7.3. *If A is self-adjoint and B is a symmetric operator such that $D(A) \subset D(B)$ and*

$$\|Bu\| \leqslant a\|Au\|+b\|u\| , \qquad u \in D(A) , \tag{7.4}$$

holds with $a<1$, then $A+B$ is self-adjoint.

Proof. Clearly $A+B$ is symmetric. Thus by Lemma 7.2 it suffices to show that there is a number λ such that λ and $\bar{\lambda}$ are in $\rho(A+B)$. Now for $\lambda \in \rho(A)$ we have

$$A+B-\lambda = [I+B(A-\lambda)^{-1}](A-\lambda). \tag{7.5}$$

It follows from this that if $\lambda \in \rho(A)$ is such that $B(A-\lambda)^{-1}$ is bounded with

$$\|B(A-\lambda)^{-1}\| < 1, \tag{7.6}$$

then $\lambda \in \rho(A+B)$. For then the operator $I+B(A-\lambda)^{-1}$ has a bounded inverse. Thus it suffices to prove (7.6) for some λ and its conjugate. Take λ pure imaginary. If $(A-\lambda)u=f$, then

$$\|Au\|^2 = \|f\|^2 + 2 \operatorname{Re} \lambda(u, f) + |\lambda|^2 \|u\|^2 .$$

But

$$\lambda(u, f) = \lambda(u, Au) - |\lambda|^2 \|u\|^2 ,$$

and since λ is pure imaginary

$$\operatorname{Re} \lambda(u, f) = -|\lambda|^2 \|u\|^2 .$$

Consequently

$$\|Au\|^2 = \|f\|^2 - |\lambda|^2 \|u\|^2 \leqslant \|f\|^2 . \tag{7.7}$$

Substituting into (7.4) we get by Lemma 7.1

$$\|Bu\| \leqslant a\|f\| + b\|u\| \leqslant (a + b\, |\operatorname{Im} \lambda|^{-1}) \|f\| .$$

If we now take $|\lambda|$ sufficiently large, we obtain (7.6). This completes the proof.

Note that a symmetric operator is closable and that its closure is also symmetric. A symmetric operator A_0 is called *essentially self-adjoint* if its closure A is self-adjoint. In this case there is only one self-adjoint extension of A_0. For suppose B is a self-adjoint extension of A_0. Then it is an extension of A (Theorem 2.1). Thus $B = B^* \subset A^* = A \subset B$ and consequently $B = A$.

Lemma 7.4. *Suppose that A_0 is essentially self-adjoint and let A be its closure. If B is an operator such that $D(A) \subset D(B)$ and* (7.4) *holds with $a < 1$, then $A_0 + B$ is essentially self-adjoint.*

Proof. By Theorem 2.9, the closure of $A_0 + B$ is $A + B$. We now apply Lemma 7.3 to conclude that $A + B$ is self-adjoint.

Lemma 7.5. *Let A_0 be a symmetric operator such that $R(A_0)$ is dense and*

$$\|u\| \leqslant C\|A_0 u\| , \qquad u \in D(A_0) . \tag{7.8}$$

Then A_0 is essentially self-adjoint.

Proof. Let A be the closure of A_0. Then A is symmetric and

$$\|v\| \leqslant C\|Av\|, \qquad v \in D(A).$$

This shows that $R(A)$ is closed (see Theorem 2.3). Since $R(A) \supset R(A_0)$ and the latter is dense, we see that $R(A)$ must be the whole space. We now apply Lemma 7.2.

A bilinear form $a(u, v)$ is called *symmetric* if $a(u, v) = \overline{a(v, u)}$ for all u, v. This is equivalent to saying that $a(u)$ is real for all u. We have

Theorem 7.6. *A regularly accretive operator is self-adjoint if and only if the corresponding bilinear form is symmetric.*

Proof. Let X, W be the Hilbert spaces, A the operator, $a(u, v)$ the bilinear form and E the embedding operator. If A is self-adjoint, we have for u, $v \in E^{-1}[D(A)]$

$$a(u, v) = (A\,Eu, Ev) = (Eu, A\,Ev) = \overline{a(v, u)}.$$

Since $E^{-1}[D(A)]$ is dense in W (Lemma 6.9), it follows that $a(u, v)$ is symmetric. Conversely, suppose $a(u, v)$ is symmetric. Then for $x, y \in D(A)$ we have

$$(Ax, y) = a(E^{-1}x, E^{-1}y) = \overline{a(E^{-1}y, E^{-1}x)} = \overline{(Ay, x)} = (x, Ay).$$

Thus A is symmetric and consequently A^* is an extension of A. Since A is regularly accretive, $A + N$ is one–to–one and onto for some N (Theorem 6.7). Thus the same must be true of $(A+N)^* = A^* + N$. Since $A^* + N$ is an extension of $A + N$, we must have $A + N = A^* + N$ or $A = A^*$. This completes the proof.

An operator A in a Hilbert space is said to be *bounded from below* if there is a constant N such that

$$\operatorname{Re}(Ax + Nx, x) \geqslant 0, \qquad x \in D(A). \tag{7.9}$$

Theorem 7.7. *A self-adjoint operator is regularly accretive if and only if it is bounded from below.*

Proof. Let A be a self-adjoint operator. If A is regularly accretive, let $a(u, v)$, W, E, N have their usual meanings. We have by (6.11)

$$(Ax, x) = a(E^{-1}x) \geqslant \|E^{-1}x\|_W^2 - N\|x\|^2, \qquad x \in D(A).$$

Thus (7.9) holds. Conversely, suppose (7.9) holds. Set

$$[x, y] = (Ax, y) + (N+1)(x, y), \qquad x, y \in D(A).$$

It is easily checked that $[x, y]$ has all of the properties of a scalar product (see §1). Let $|x|$ denote the corresponding norm. Then we have

$$|x|^2 = [x, x] \geqslant \|x\|^2, \qquad x \in D(A).$$

We let W denote the completion of $D(A)$ with respect to the norm $|\ |$ (see §1). Clearly W can be considered a subspace of X. Set

$$a_0(x, y) = (Ax, y), \qquad x, y \in D(A).$$

Then

$$a_0(x) + (N+1)\|x\|^2 = |x|^2, \qquad x \in D(A).$$

In particular, since $N \geqslant 0$ we have $a_0(x) \leqslant |x|^2$. By Lemma 6.10 we see that $a_0(x, y)$ is bounded in $W \times W$ and consequently it can be extended to a bilinear form $a(x, y)$ defined on the whole of $W \times W$ and satisfying

$$a(x) + (N+1)\|x\|^2 = |x|^2, \qquad x \in W. \tag{7.10}$$

Now if $x, y \in D(A)$, then

$$a(x, y) = (Ax, y).$$

By continuity this holds for $x \in D(A)$ and $y \in W$. Moreover, if $x \in W$, $f \in X$ and

$$a(x, y) = (f, y), \qquad y \in W,$$

then

$$(x, Ay) = (f, y), \qquad y \in D(A).$$

Since A is self-adjoint, we have $x \in D(A)$ and $Ax = f$. Thus A is regularly accretive, and the proof is complete.

Lemma 7.8 *Let A be a self-adjoint operator. Then $\lambda \notin \sigma_e(A)$ if and only if $\alpha(A-\lambda) < \infty$ and $R(A-\lambda)$ is closed in X.*

Proof. The "only if" part follows immediately from Theorem 4.5. The "if" is also immediate. For if Im $\lambda \neq 0$, then $\lambda \in \rho(A)$ (Lemma 7.1). If λ is real, then $\beta(A-\lambda) = \alpha(A-\lambda)$ by the self-adjointness of $A-\lambda$. Thus if $\alpha(A-\lambda) < \infty$ and $R(A-\lambda)$ is closed, we see that $\lambda \in \Phi_A$ and $i(A-\lambda) = 0$.

Theorem 7.9. *If A is self-adjoint, then $\lambda \in \sigma_e(A)$ if and only if there is a sequence $\{u_k\}$ of elements in $D(A)$ such that $\|u_k\| = 1$, $u_k \rightharpoonup 0$ and $(A-\lambda)u_k \to 0$.*

Proof. Suppose such a sequence exists. It cannot have a convergent subsequence, for a convergent subsequence would have to converge to 0, since $u_k \rightharpoonup 0$, but would also have to converge to an element of norm one, since $\|u_k\|=1$ for each k. Consequently $\{u_k\}$ is a singular sequence for $A-\lambda$. By Theorem 4.1, $\lambda \in \sigma_e(A)$. Conversely, suppose $\lambda \in \sigma_e(A)$. Then either $\alpha(A-\lambda)=\infty$ or $R(A-\lambda)$ is not closed in X (Lemma 7.8). If $\alpha(A-\lambda)=\infty$, let $\{u_k\}$ be an orthonormal sequence in $N(A-\lambda)$ (see Lemma 1.7). Then $\|u_k\|=1, u_k \rightharpoonup 0$ and $(A-\lambda)u_k=0$. Then $\{u_k\}$ satisfies the requirements of the theorem. If $\alpha(A-\lambda)<\infty$, there is a sequence $\{u_k\}$ of elements in $D(A) \cap N(A-\lambda)^\perp$ such that $\|u_k\|=1$, $(A-\lambda)u_k \to 0$. By Theorem 1.6, there is a subsequence (again denoted by $\{u_k\}$) which converges weakly in X to some element u. Consequently $Au_k \rightharpoonup \lambda u$. Thus if $v \in D(A)$, then $(u, Av)=\lim(u_k, Av)=\lim(Au_k, v)=\lambda(u, v)$. This gives

$$(u, (A-\lambda)v)=0, \qquad v \in D(A).$$

Since A is self-adjoint, this means that $u \in N(A-\lambda)$. Thus $(u_k, u)=0$ for each k. Whence

$$\|u\|^2 = \lim(u_k, u) = 0,$$

and $u=0$. Thus the sequence $\{u_k\}$ has the required properties, and the proof is complete.

Lemma 7.10. *If A is self-adjoint, then $\lambda \in \sigma(A)$ if and only if there is a sequence $\{u_k\}$ of elements of $D(A)$ such that $\|u_k\|=1$ and $(A-\lambda)u_k \to 0$.*

Proof. If such a sequence exists, then the inequality

$$\|u\| \leqslant C\|(A-\lambda)u\|, \qquad u \in D(A) \tag{7.11}$$

does not hold. Consequently $\lambda \in \sigma(A)$. Conversely, suppose $\lambda \in \sigma(A)$. If $\lambda \in \sigma_e(A)$ then $A-\lambda$ has a singular sequence (Theorem 4.4) which more than suffices. If $\lambda \notin \sigma_e(A)$, then $\alpha(A-\lambda)<\infty$ and $R(A-\lambda)$ is closed (Lemma 7.8). The only way we can have $\lambda \in \sigma(A)$ is if λ is an eigenvalue. In this case we can take all of the u_k equal to a normalized eigenelement. This completes the proof.

Theorem 7.11. *Suppose that A is self-adjoint and there is a sequence $\{\lambda_k\}$ of numbers in $\sigma(A)$ such that $\lambda_0 \neq \lambda_k \to \lambda_0$. Then $\lambda_0 \in \sigma_e(A)$.*

Proof. Since $\lambda_k \in \sigma(A)$, there is an element $u_k \in D(A)$ such that $\|u_k\|=1$ and $\|(A-\lambda_k)u_k\| \leqslant |\lambda_0-\lambda_k|/k$. By Theorem 1.6 there is a subsequence (also denoted by $\{u_k\}$) such that $u_k \rightharpoonup u$ for some $u \in X$. Thus for $v \in D(A)$ we have

$$\lambda_0(u, v) = \lim \lambda_k(u_k, v) = \lim (Au_k, u) = \lim (u_k, Av) = (u, Av).$$

Hence $u \in N(A - \lambda_0)$. Moreover

$$(\lambda_0 - \lambda_k)(u_k, u) = (u_k, Au) - \lambda_k(u_k, u) = ((A - \lambda_k)u_k, u) .$$

Thus

$$|\lambda_0 - \lambda_k| \, |(u_k, u)| \leqslant |\lambda_0 - \lambda_k| \, ||u|| / k .$$

This shows that $(u_k, u) \to 0$ as $k \to \infty$.
But

$$(u_k, u) \to ||u||^2 .$$

Thus $u = 0$ and $\{u_k\}$ satisfies the hypothesis of Theorem 7.9. Hence $\lambda \in \sigma_e(A)$, and the proof is complete.

Corollary 7.12. *If A is self-adjoint and $\lambda \in \sigma(A) \backslash \sigma_e(A)$, then it is an isolated eigenvalue of finite multiplicity.*

Theorem 7.13. *Let A be a self-adjoint operator which is the closure of its restriction to a subspace $W \subset D(A)$. Then $\lambda \in \sigma(A)$ if and only if there is a sequence $\{w_k\}$ of elements in W such that $||w_k|| = 1$ and $(A - \lambda)w_k \to 0$.*

Proof. The "if" part follows from Lemma 7.10. Moreover, if $\lambda \in \sigma(A)$, then by that lemma there is a sequence $\{u_k\}$ of elements of $D(A)$ such that $||u_k|| = 1$ and $(A - \lambda)u_k \to 0$. Since A is the closure of its restriction to W, for each k there is a $w_k \in W$ such that

$$||w_k - u_k|| < 1/k , \qquad ||Aw_k - Au_k|| < 1/k . \tag{7.12}$$

Thus

$$||w_k|| \to 1 , \qquad (A - \lambda)w_k \to 0 \text{ as } k \to \infty .$$

Setting $v_k = w_k / ||w_k||$, we obtain

$$||v_k|| = 1 , \quad (A - \lambda)v_k \to 0 . \tag{7.13}$$

This completes the proof.

Theorem 7.14. *Under the same hypotheses $\lambda \in \sigma_e(A)$ if and only if there is a sequence $\{w_k\}$ of elements of W such that $||w_k|| = 1$, $w_k \rightharpoonup 0$ and $(A - \lambda)w_k \to 0$.*

Proof. In this case the "if" part follows from Theorem 7.9. Moreover by the same theorem if $\lambda \in \sigma_e(A)$ there is a sequence $\{u_k\}$ of elements of $D(A)$ such that $||u_k|| = 1$, $u_k \rightharpoonup 0$ and $(A - \lambda)u_k \to 0$. As in the previous proof for each k there is a $w_k \in W$ satisfying (7.12). Setting $v_k = w_k / ||w_k||$ we see that (7.13) holds and

$$|(v_k, z)| = |(w_k, z)|/\|w_k\| \leqslant |(w_k - u_k, z)|/\|w_k\| + |(u_k, z)|/\|w_k\|$$
$$\leqslant \|z\|/k\|w_k\| + |(u_k, z)|/\|w_k\| \to 0 \text{ as } k \to \infty .$$

Hence $v_k \rightharpoonup 0$, and the proof is complete.

Until this point we have not used the *spectral theorem* for self-adjoint operators. This theorem states that a self-adjoint operator *A admits a spectral decomposition*

$$A = \int_{-\infty}^{\infty} \lambda \mathrm{d}E_\lambda , \tag{7.14}$$

where $\{E_\lambda\}$ is a family of projection operators (called the *spectral family* associated with A) monotonically non-decreasing in λ and tending to the zero operator as $\lambda \to -\infty$ and to the identity operator as $\lambda \to \infty$.

Theorem 7.15. *Suppose that there is an infinite dimensional manifold* $U \subset D(A)$ *such that*

$$\|(A-\lambda_0)u\| \leqslant M\|u\| , \qquad u \in U . \tag{7.15}$$

If A is self-adjoint, then the interval $\Delta = [\lambda_0 - M, \lambda_0 + M]$ *contains a point of* $\sigma_e(A)$.

Proof. If Δ did not contain a point of $\sigma_e(A)$, then it could have at most a finite number of isolated eigenvalues of finite multiplicity (Theorem 7.11 and Corollary 7.12). Thus there is an $\varepsilon > 0$ such that the same is true of the interval $|\lambda - \lambda_0| < M + \varepsilon$. Let G be the finite dimensional subspace spanned by the eigenvectors of these eigenvalues, and set $Y = G^\perp$. Then for $u \in Y$

$$\|(A-\lambda_0)u\|^2 = \int_{|\lambda-\lambda_0| \geqslant M+\varepsilon} (\lambda-\lambda_0)^2 \mathrm{d}(E_\lambda u, u)$$
$$\geqslant (M+\varepsilon)^2 \int_{-\infty}^{\infty} \mathrm{d}(E_\lambda u, u) = (M+\varepsilon)^2 \|u\|^2 ,$$

where $\{E_\lambda\}$ is the spectral family for A. Hence

$$\|(A-\lambda_0)u\| \geqslant (M+\varepsilon)\|u\| , \qquad u \in Y \cap D(A) . \tag{7.16}$$

If U is infinite dimensional, then $Y \cap U$ must contain a non-zero element (Lemma 1.4). This element would have to satisfy (7.15) and (7.16) simultaneously, which is impossible. This proves the theorem.

Theorem 7.16. *Let A be a self-adjoint operator, and suppose that* $\{u_k\}$ *is an orthonormal sequence of elements of* $D(A)$ *such that*

$$(Au_j, Au_k) = 0\,, \qquad j \neq k\,, \tag{7.17}$$

and

$$\|Au_k\| \leqslant M\,, \qquad k = 1, 2, \dots\,.$$

Then the interval $[-M, M]$ *contains a point of* $\sigma_e(A)$.

Proof. Let U be the set of linear combinations of the u_k. Clearly U is infinite dimensional and is contained in $D(A)$. If

$$u = \Sigma\, \alpha_k u_k\,,$$

then

$$\|Au\|^2 = \Sigma\, \alpha_j \bar{\alpha}_k (Au_j, Au_k) = \Sigma\, |\alpha_k|^2 \|Au_k\|^2$$
$$\leqslant M^2 \Sigma\, |\alpha_k|^2 = M^2 \|u\|^2\,.$$

We now apply Theorem 7.15 to obtain our conclusion.

Lemma 7.17. *If A is self-adjoint and $\sigma(A) \subset [v, \infty]$, then*

$$(Au, u) \geqslant v\,\|u\|^2\,, \qquad u \in D(A)\,.$$

Proof. Let $\{E_\lambda\}$ be the spectral family associated with A. Then

$$A = \int_v^\infty \lambda \mathrm{d}E_\lambda\,,$$

and consequently

$$(Au, u) = \int_v^\infty \lambda \mathrm{d}\|E_\lambda u\|^2 \geqslant v \int_v^\infty \mathrm{d}\,\|E_\lambda u\|^2$$
$$\geqslant v \int_{-\infty}^\infty \mathrm{d}\|E_\lambda u\|^2 = v\,\|u\|^2\,.$$

This is the desired result.

Theorem 7.18. *Let H be a Hilbert space, and suppose there is an operator J on H such that*

$$J(x+y) = Jx + Jy, \quad J(\alpha x) = \bar{\alpha} Jx, \quad J^2 x = x\,,$$
$$(Jx, Jy) = (y, x)\,, \qquad x, y \in H\,.$$

Suppose A is a symmetric linear operator in H such that $Jx \in D(A)$ whenever $x \in D(A)$ and $AJx = JAx$. Then A has a self-adjoint extension.

An operator A in a Hilbert space is called *normal* if $AA^* = A^*A$.

CHAPTER 2

FUNCTION SPACES

§1. Functions on E^n

By *Euclidean n-dimensional space* E^n we mean the set of vectors of the form $x=(x_1, \ldots, x_n)$, where the x_k are real numbers. The *magnitude* of a vector is given by

$$|x| = (x_1^2 + \ldots + x_n^2)^{\frac{1}{2}} . \tag{1.1}$$

One checks easily that $| \ |$ has all of the properties of a norm (see §1 of ch. 1). The functions we shall consider are mappings of E^n into the complex numbers. We shall always assume that the functions are measurable. For a function $\varphi(x)$ the closure of the set of points for which $\varphi(x) \neq 0$ is called the *support* of φ and denoted by supp φ. For any subset Ω of E^n we let $C^\infty(\Omega)$ denote the set of functions which are infinitely differentiable on Ω. By $C_0^\infty(\Omega)$ we shall denote the set of functions $\varphi \in C^\infty(\Omega)$ such that supp φ is bounded and contained in Ω. We set $C^\infty = C^\infty(E^n)$ and $C_0^\infty = C_0^\infty(E^n)$.

Lemma 1.1. *A subset of E^n is compact if and only if it is bounded and closed.*

Theorem 1.2. *Let U, V be two disjoint closed subsets of E^n. If V is compact, then there is a $\varphi \in C_0^\infty$ such that*

$$\begin{aligned} \varphi(x) &= 1 \quad \textit{for} \quad x \in U \\ &= 0 \quad \textit{for} \quad x \in V \\ 0 \leqslant \varphi(x) &\leqslant 1 \quad \textit{for} \quad x \in E^n . \end{aligned} \tag{1.2}$$

For a subset Ω of E^n and p a real number satisfying $1 \leqslant p < \infty$ we let $L^p(\Omega)$ denote the set of those complex valued functions $u(x)$ which are measurable on Ω and such that $|u(x)|^p$ is Lebesque integrable on Ω. We identify functions which are equal almost everywhere.

Theorem 1.3. *Under the norm*

$$\|u\|_p^\Omega = \left(\int_\Omega |u(x)|^p \,\mathrm{d}x \right)^{1/p}, \tag{1.3}$$

$L^p(\Omega)$ is a Banach space.

We let $L^\infty(\Omega)$ denote the set of measurable functions $u(x)$ on Ω such that for each $u(x)$ there is a finite constant C satisfying

$$|u(x)| \leqslant C \text{ almost everywhere in } \Omega. \tag{1.4}$$

We define $\|u\|_\infty^\Omega$ as the greatest lower bound of all constants C satisfying (1.4).

Theorem 1.4. *$L^\infty(\Omega)$ is a Banach space.*

When $\Omega = E^n$ we write $\|u\|_p$ in place of $\|u\|_p^\Omega$. For $1 < p < \infty$ we let p' denote $p/(p-1)$, for $p=1$ we set $p'=\infty$ and for $p=\infty$ we set $p'=1$. Thus

$$1/p + 1/p' = 1 \tag{1.5}$$

if we interpret $1/\infty$ to be 0.

Theorem 1.5. *If $u \in L^p(\Omega)$ and $v \in L^{p'}(\Omega)$, then $uv \in L^1(\Omega)$ and*

$$\|uv\|_1^\Omega \leqslant \|u\|_p^\Omega \|v\|_{p'}^\Omega. \tag{1.6}$$

If we set

$$(u, v)_\Omega = \int_\Omega u(x)\overline{v(x)}\,\mathrm{d}x, \tag{1.7}$$

then $(u, v)_\Omega$ exists for $u \in L^p(\Omega)$ and $v \in L^{p'}(\Omega)$ and

$$|(u, v)_\Omega| \leqslant \|u\|_p^\Omega \|v\|_{p'}^\Omega. \tag{1.8}$$

Thus for fixed $u \in L^p(\Omega)$ the expression $(u, v)_\Omega$ is a bounded conjugate linear functional F on $L^{p'}(\Omega)$ with $\|F\| \leqslant \|u\|_p^\Omega$ (see §2 of ch. 1).

Theorem 1.6. *If $1 < p \leqslant \infty$, then for each bounded conjugate linear functional F on $L^{p'}(\Omega)$ there is a $u \in L^p(\Omega)$ such that*

$$Fv = (u, v), \qquad v \in L^{p'}(\Omega), \tag{1.9}$$

and

$$\|u\|_p^\Omega = \|F\|. \tag{1.10}$$

Theorem 1.7. *If* $1 \leqslant p < \infty$, *then* $C_0^\infty(\Omega)$ *is dense in* $L^p(\Omega)$.

Let $j(x)$ be defined by

$$j(x) = a \exp\{(|x|^2 - 1)^{-1}\}, \qquad |x| < 1 \qquad (1.11)$$
$$= 0 \qquad |x| \geqslant 1,$$

where the constant a is chosen so that

$$\int j(x)\,dx = 1. \qquad (1.12)$$

Set

$$j_k(x) = k^n j(kx), \qquad k = 1, 2, \ldots. \qquad (1.13)$$

Then we have

$$j_k(x) > 0, \qquad |x| < 1, \quad k = 1, 2, \ldots \qquad (1.14)$$
$$= 0, \qquad |x| \geqslant 1, \quad k = 1, 2, \ldots$$

$$\int j_k(x)\,dx = 1, \qquad k = 1, 2, \ldots. \qquad (1.15)$$

When $\Omega = E^n$ we write L^p in place of $L^p(\Omega)$. For $u \in L^p$ set

$$J_k u(x) = \int j_k(x-y)\,u(y)\,dy, \qquad k = 1, 2, \ldots. \qquad (1.16)$$

We have

Theorem 1.8. *For each* $u(x) \in L^p$

$$\|J_k u\|_p \leqslant \|u\|_p, \qquad k = 1, 2, \ldots. \qquad (1.17)$$

If $1 \leqslant p < \infty$, *then*

$$\|u - J_k u\|_p \to 0 \text{ as } k \to \infty. \qquad (1.18)$$

Theorem 1.9. *If* $u \in L^1$, $v \in L^p$ *and*

$$w(x) = \int u(x-y)\,v(y)\,dy, \qquad (1.19)$$

then $w \in L^p$ *and*

$$\|w\|_p \leqslant \|u\|_1 \|v\|_p. \qquad (1.20)$$

The function $w(x)$ defined by (1.19) is called the *convolution* of u and v. It is sometimes written as $w = u * v$. Note that (1.17) is a special case of (1.20).

Let $\mu=(\mu_1, \ldots, \mu_n)$ be a multi-index of non-negative integers. We set $|\mu|=\mu_1+\ldots+\mu_n$ and define

$$D^{\mu}=(-i)^{|\mu|}\partial^{|\mu|}/\partial x_1^{\mu_1}\ldots\partial x_n^{\mu_n}. \tag{1.21}$$

Thus D^{μ} is essentially the derivative of order $|\mu|$ corresponding to the multi-index μ. The factor $(-i)^{|\mu|}$ is introduced for convenience. The reason will be apparent in the next section and in ch. 3.

Let A be a linear operator from $L^p(\Omega)$ to $L^q(G)$, where Ω, G are subsets of E^n and $1 \leqslant p, q < \infty$. By definition A^* is a mapping from the conjugate linear bounded functionals on $L^q(G)$ to those on $L^p(\Omega)$ (see §2 of ch. 1). However, by Theorem 1.6 these spaces can be identified with $L^{q'}(G)$ and $L^{p'}(\Omega)$, respectively. Thus we may consider A^* as an operator from $L^{q'}(G)$ to $L^{p'}(\Omega)$. Consequently $u \in L^{q'}(G)$ is in $D(A^*)$ if there is an $f \in L^{p'}(\Omega)$ such that

$$(u, Av)_G = (f, v)_{\Omega}, \qquad v \in D(A). \tag{1.22}$$

We then have $A^*u=f$.

In the sequel we shall write (u, v) in place of $(u, v)_{\Omega}$ when $\Omega=E^n$ and $\|u\|$ in place of $\|u\|_p$ when the value of p is clear from the context.

Theorem 1.10. *If $\varphi \in C_0^{\infty}$, $u \in L^1$ and* supp u *is bounded, then $\varphi * u$ is in C_0^{∞} and*

$$D^{\mu}[\varphi * u] = D^{\mu}\varphi * u. \tag{1.23}$$

Theorem 1.11. *If $u_k \to u$ in L^p, then there is a subsequence of $\{u_k\}$ which converges to u almost everywhere.*

Theorem 1.12 (Lusin). *Let u be a (measurable) function defined on a closed bounded set $K \subset E^n$. Then for each $\varepsilon > 0$ there is an open set U such that the measure $m(U)$ of U is less than ε and u is continuous on $K \setminus U$.*

Theorem 1.13. *For $1 \leqslant p \leqslant \infty$*

$$\|u\|_p = \sup_{v \in L^{p'}} \frac{|(u, v)|}{\|v\|_{p'}}, \qquad u \in L^p.$$

A function u is said to be locally in L^p if $\varphi u \in L^p$ for every $\varphi \in C_0^{\infty}$.

Theorem 1.14. *For $1 < p < \infty$ the space L^p is reflexive.*

Theorem 1.15. *Let $f(x)$ be a continuous function from E^n to E^m. If G is an open subset of E^m, then the set Ω of points mapped into G by f is open in E^n.*

§2. Fourier transforms

Let S denote the set of those functions $u(x) \in C^\infty$ such that

$$|x|^k |D^\mu u(x)|$$

is bounded on E^n for each $k \geq 0$ and μ. Note that $C_0^\infty \subset S$ and any derivative of a function in S is also in S. Likewise a polynomial multiplied by a function in S is in S.

For a function $u(x) \in S$ the Fourier transform of u is given by

$$Fu(\xi) = (2\pi)^{-\frac{1}{2}n} \int e^{-i\xi x} u(x) dx\,, \qquad \xi \in E^n\,, \tag{2.1}$$

where $\xi x = \xi_1 x_1 + \ldots + \xi_n x_n$.

Theorem 2.1. *If $u(x) \in S$ and $w(\xi) = Fu$, then*

(a) $w(\xi) \in S$

(b) $D_\xi^\mu w(\xi) = (-1)^{|\mu|} F[x^\mu u(x)]$, *where D_ξ^μ is the derivative D^μ with respect to the ξ variables and*

$$x^\mu = x_1^{\mu_1} \ldots x_n^{\mu_n} \tag{2.2}$$

(c) $\xi^\mu w(\xi) = F[D_x^\mu u(x)]$

(d) $u(x) = \bar{F} w$, *where $\bar{F}$ is the inverse Fourier transform given by*

$$\bar{F} w(x) = (2\pi)^{-\frac{1}{2}n} \int e^{ix\xi} w(\xi) d\xi\,. \tag{2.3}$$

Theorem 2.2. *If $w(\xi) \in S$, then $u(x) = \bar{F} w$ is in S and $w = Fu$.*

Theorem 2.3. *If $u \in L^p$ for $1 \leq p \leq 2$, then*

$$\int_{|x|<R} e^{-i\xi x} u(x) dx$$

converges in $L^{p'}$ as $R \to \infty$. If we let Fu denote the limit, then

$$\|Fu\|_{p'} \leq \|u\|_p$$

and

$$(Fu, v) = (u, \bar{F} v)\,, \qquad v \in S\,.$$

Theorem 2.4.

$$(Fu, Fw) = (u, w), \qquad u, w \in L^2,$$

$$\|Fu\|_2 = \|u\|_2, \qquad u \in L^2.$$

Theorem 2.5. *If* $v \in S$ *and* $u \in L^p$, *then* $v * u \in C^\infty$ *and*

$$D^\mu[v * u] = D^\mu v * u.$$

Theorem 2.6. *If* $u \in L^1$ *and* $v \in S$, *then*

$$F[u * v] = Fu\, Fv.$$

Theorem 2.7. *If* $g(t)$ *is a function of* $t \geqslant 0$ *and* $g(|\xi|) \in L^1(E^n)$, *then*

$$\bar{F}[g(|\xi|)] = (2\pi)^{\frac{1}{2}n} |x|^{\frac{1}{2}(2-n)} \int_0^\infty g(\rho) \rho^{\frac{1}{2}n} J_{\frac{1}{2}(n-2)}(|x|\rho) \mathrm{d}\rho,$$

where $J_\lambda(\sigma)$ *denotes the Bessel function of the first kind of order* λ.

Theorem 2.8. *For* $\lambda \geqslant 0$

$$\mathrm{d}[t^\lambda J_\lambda(t)]/\mathrm{d}t = t^\lambda J_{\lambda-1}(t)$$

$$J_\lambda(t) = O(t^\lambda) \quad \text{as} \quad t \to 0$$

$$J_\lambda(t) = O(t^{-\frac{1}{2}}) \quad \text{as} \quad t \to \infty.$$

Lemma 2.9. *For* $a > 0$, $b > 0$

$$\int_0^\infty t^{b-1} \exp\{-at\} \mathrm{d}t = \Gamma(b)/a^b.$$

Theorem 2.10. *For* $a > 0$, $\lambda > 0$

$$\int_0^\infty t^\lambda J_{\lambda-1}(t) \exp(-at^2) \mathrm{d}t = \exp\left(-\frac{1}{4a}\right) \bigg/ (2a)^\lambda.$$

§3. L^p multipliers

Let $w(\xi)$ be a function in L^∞. Suppose u, f are functions in L^2 such that

$$(w Ff, v) = (u, \bar{F}v), \qquad v \in S. \tag{3.1}$$

Then by Theorems 2.3 and 2.4 we have

$$(w\,Ff - Fu, v) = 0\,, \qquad v \in S\,,$$

and consequently $Fu = w\,Ff$. This implies that

$$\|Fu\|_2 \leqslant \|w\|_\infty \|Ff\|_2\,,$$

which implies via Theorem 2.4

$$\|u\|_2 \leqslant C\|f\|_2\,, \tag{3.2}$$

with $C = \|w\|_\infty$. Thus (3.1) implies (3.2) with the constant C depending only on $w(\xi)$. A function $w(\xi)$ having this property is called a *L^2 multiplier*. We have shown

Theorem 3.1. *Every function in L^∞ is an L^2 multiplier.*

In general a function $w(\xi)$ is called an *L^p multiplier* (or a *multiplier* in L^p) if (3.1) holding for $f \in S$ and $u \in L^p$ implies

$$\|u\|_p \leqslant C\|f\|_p\,, \tag{3.4}$$

where the constant C depends only on $w(\xi)$ and p. If $p \neq 2$, it is no longer true that every bounded function is an L^p multiplier. However, every L^p multiplier must be bounded.

Theorem 3.2. *If $w(\xi)$ is an L^p multiplier for some p, then $w(\xi)$ is in L^∞.*

We must now give several sufficient conditions on $w(\xi)$ in order that it be an L^p multiplier. The following theorem is suitable for our needs.

Theorem 3.3. *Suppose $1 < p < \infty$ and there exist non-negative integers m_1, m_2 and a real number θ such that*

$$0 \leqslant \theta \leqslant 1\,, \qquad (1-\theta)m_1 + \theta m_2 > n\,|\tfrac{1}{2} - 1/p|\,. \tag{3.5}$$

Assume that $w(\xi)$ is a function on E^n having continuous derivatives up to order $\max(m_1, m_2)$ satisfying

$$\max_{R \leqslant |\xi| \leqslant 2R} |D^\mu w|^{1-\theta} \max_{R \leqslant |\xi| \leqslant 2R} |D^\nu w|^\theta \leqslant C/R^{(1-\theta)|\mu| + \theta|\nu|}\,,$$

$$|\mu| \leqslant m_1,\ |\nu| \leqslant m_2\,. \tag{3.6}$$

Then $w(\xi)$ is an L^p multiplier.

Theorem 3.4. *Assume* $1<p<\infty$, *and let* l *be the smallest integer greater than* $n\,|\frac{1}{2}-1/p|$. *Suppose* $w(\xi)$ *has continuous derivatives up to order* l *and there are constants* a, b *such that*

$$a \leqslant 1\,, \qquad b>(1-a)n\,|\tfrac{1}{2}-1/p| \tag{3.7}$$

and

$$|D^{\mu} w(\xi)| \leqslant C|\xi|^{-a|\mu|-b}\,, \qquad |\mu| \leqslant l\,. \tag{3.8}$$

Then $w(\xi)$ *is an* L^p *multiplier.*

Proof. If $b \geqslant (1-a)l$, we can take $m_1=m_2=l$, $\theta=0$ in Theorem 3.3. Then $a|\mu|+b \geqslant |\mu|$ for $|\mu| \leqslant l$. Thus

$$|D^{\mu} w(\xi)| \leqslant C|\xi|^{-|\mu|}\,, \qquad |\mu| \leqslant l\,,$$

which implies (3.6) for $m_1=m_2=l$, $\theta=0$. *If* $b<(1-a)l$, then we must have $a<1$. Thus

$$0<l-[b/(1-a)]<l-n|\tfrac{1}{2}-1/p| \leqslant 1\,.$$

Consequently, there is a θ satisfying $0<\theta<1$ such that

$$l-[b/(1-a)]<\theta<l-n|\tfrac{1}{2}-1/p|\,.$$

This means that

$$n|\tfrac{1}{2}-1/p|<l-\theta<b/(1-a)\,.$$

Hence

$$\max_{R \leqslant |\xi| \leqslant 2R} |D^{\mu} w|^{1-\theta} \max_{R \leqslant |\xi| \leqslant 2R} |D^{\nu} w|^{\theta} \leqslant C/R^{a[(1-\theta)|\mu|+\theta|\nu|]+b}\,, \; |\mu|, |\nu| \leqslant l\,.$$

Take $m_1=l$, $m_2=l-1$. Then $(1-\theta)m_1+\theta m_2=l-\theta$ and

$$(1-\theta)|\mu|+\theta|\nu| \leqslant l-\theta<b/(1-a)\,, \qquad |\mu| \leqslant l\,, \quad |\nu| \leqslant l-1\,.$$

This implies that

$$a[(1-\theta)|\mu|+\theta|\nu|]+b \geqslant (1-\theta)|\mu|+\theta|\nu|\,.$$

Thus (3.6) holds and the result follows from Theorem 3.3.

§4. The spaces $H^{s,p}$

In this section we shall define classes of functions which are important in the theory of partial differential equations. For $v \in S$, s real and $1 \leqslant p \leqslant \infty$ we define

$$\|v\|_{s,p} = \|\bar{F}[(1+|\xi|^2)^{\frac{1}{2}s} Fv]\|_p . \tag{4.1}$$

One checks easily that this is a norm on S and that $\|v\|_{0,p}$ is merely the L^p norm of v. Thus for each s and p, S is a normed vector space under the norm (4.1). We let $H^{s,p}$ denote the completion of S with respect to the norm (4.1). (See Theorem 1.5 of ch. 1.)

For $u, v \in S$ we have by Theorem 2.4

$$(u, v) = (Fu, Fv) = (\bar{F}[(1+|\xi|^2)^{\frac{1}{2}s} Fu], \bar{F}[(1+|\xi|^2)^{-\frac{1}{2}s} Fv]) .$$

Consequently

$$|(u, v)| \leqslant \|u\|_{s,p} \|v\|_{-s,p'} . \tag{4.2}$$

Thus we can define (u, v) for $u \in H^{s,p}$ and $v \in H^{-s,p'}$. Thus for fixed $v \in H^{-s,p'}$ the expression (u, v) is a bounded linear functional on $H^{s,p}$. We also have

Theorem 4.1. *If $1 \leqslant p < \infty$ and F is a bounded linear functional on $H^{s,p}$, then there is a $v \in H^{-s,p'}$ such that*

$$Fu = (u, v) , \qquad u \in H^{s,p} , \tag{4.3}$$

$$\|v\|_{-s,p'} = \|F\| . \tag{4.4}$$

Corollary 4.2. *For $1 < p < \infty$, $H^{s,p}$ is reflexive.*

Theorem 4.3. *If $t < s$, then $H^{s,p} \subset H^{t,p}$ with*

$$\|u\|_{t,p} \leqslant C \|u\|_{s,p} , \qquad u \in H^{s,p} , \tag{4.5}$$

where the constant C depends only on t, s, p.

Theorem 4.4. *Suppose $\{u_k\}$ is a bounded sequence of elements of $H^{s,p}$. If $\varphi \in C_0^\infty$ and $t < s$, then there is a subsequence $\{v_j\}$ of $\{u_k\}$ such that $\{\varphi v_j\}$ converges in $H^{t,p}$.*

For $v \in S$ set

$$\|v\|_{T^{s,p}} = \|\bar{F}[(1+|\xi|^2)^{\frac{1}{2}s} v(\xi)]\|_p . \tag{4.6}$$

Then we have by Theorem 2.1

$$\|Fv\|_{T^{s,p}} = \|\bar{F}v\|_{T^{s,p}} = \|v\|_{s,p} , \tag{4.7}$$

$$\|v\|_{T^{s,p}} = \|Fv\|_{s,p} = \|\bar{F}v\|_{s,p} . \tag{4.8}$$

For $1 \leqslant p \leqslant \infty$ and real s, (4.6) is a norm on S. We let $T^{s,p}$ denote the completion of S with respect to this norm. $T^{s,p}$ is a Banach space.

Theorem 4.5. *We can extend F and $\bar{F}$ as bounded linear operators from $H^{s,p}$ to $T^{s,p}$ so that* (4.7) *holds for all $v \in H^{s,p}$. Similarly F and $\bar{F}$ can be extended to bounded linear operators from $T^{s,p}$ to $H^{s,p}$ with* (4.8) *holding for all $v \in T^{s,p}$. Both F and $\bar{F}$ are onto and*

$$F\bar{F}u = \bar{F}Fu = u\,, \quad u \in H^{s,p}\,, \qquad F\bar{F}v = \bar{F}Fv = v\,, \quad v \in T^{s,p}\,. \tag{4.9}$$

We let C_B^∞ denote the set of functions in C^∞ having all derivatives bounded in E^n.

Theorem 4.6. *If $1 < p < \infty$, $u \in H^{s,p}$ and $v \in C_B^\infty$, then $uv \in H^{s,p}$ and*

$$\|uv\|_{s,p} \leqslant C\|u\|_{s,p}\,, \tag{4.10}$$

where the constant C does not depend on u.

Theorem 4.7. *If $1 < p < \infty$, $u \in H^{s,p}$ and $v \in L^1$, then $u * v \in H^{s,p}$ and*

$$\|u * v\|_{s,p} \leqslant C\|u\|_{s,p}\,, \tag{4.11}$$

where the constant C does not depend on u.

Lemma 4.8. *For $v \in S$ and each s, p*

$$\|D^\mu v\|_{s,p} \leqslant \|v\|_{s+|\mu|,\,p}\,. \tag{4.12}$$

Thus D^μ can be extended to a bounded linear operator from $H^{s,p}$ to $H^{s+|\mu|,\,p}$.

Theorem 4.9. *If $u \in H^{s,p}$, then*

$$F[D^\mu u] = \xi^\mu Fu, \qquad \bar{F}[D^\mu u] = (-\xi)^\mu \bar{F}u \tag{4.13}$$

$$F[x^\mu u] = (-D)^\mu Fu, \; \bar{F}[x^\mu u] = D^\mu \bar{F}u\,. \tag{4.14}$$

Theorem 4.10. *If $u \in H^{s,p}$ and $v \in S$, then*

$$F(u * v) = Fu\,Fv \tag{4.15}$$

$$F(uv) = Fu * Fv\,. \tag{4.16}$$

Theorem 4.11. *For each s and $1 < p < \infty$, C_0^∞ is dense in $H^{s,p}$.*

Theorem 4.12. *For real s and* $1<p<\infty$

$$\|u\|_{s,p} = \sup_{v\in C_0^\infty} \frac{|(u,v)|}{\|v\|_{-s,p'}}, \qquad u\in H^{s,p}. \tag{4.17}$$

Theorem 4.13. *If* $v\in S$, *then for each s, t, p with* $1<p<\infty$ *there is a constant C such that*

$$\|v*u\|_{s+t,p} \leqslant C\|u\|_{s,p}, \qquad u\in S. \tag{4.18}$$

Theorem 4.14. *For* $s_0<s_1$, $0\leqslant\theta\leqslant 1$ *and* $s=(1-\theta)s_0+\theta s_1$

$$[H^{s_0,2}, H^{s_1,2}]_\theta = H^{s,2} \tag{4.19}$$

with equal norms (see ch. 1, §5).

Proof. Denote the space on the left-hand side of (4.19) by H_θ. Let $u\in H_\theta$ and $\varepsilon>0$ be given. Then there is an $f\in H(H^{s_0,2}, H^{s_1,2})$ such that $f(\theta)=u$ and $\|f\|_H\leqslant\|u\|_\theta+\varepsilon$. Set

$$\lambda_j(\xi) = (1+|\xi|^2)^{s_j,2}, \qquad j=0,1,$$

$\lambda=\lambda_1/\lambda_0$ and $g(z)=\bar{F}[\lambda_0\lambda^z Ff(z)]$. Since $f(z)\in H^{s_0,2}$ for $0\leqslant x\leqslant 1$, we see that $g(z)\in H^{s_0-s_1,2}$ for each z. Let v be any function in S. Then

$$h(z) = (g(z), v)$$

is continuous and bounded in the strip $0\leqslant x\leqslant 1$ and analytic in $0<x<1$. Moreover

$$|h(\mathrm{i}y)|=|(g(\mathrm{i}y),v)|\leqslant\|f(\mathrm{i}y)\|_{s_0,2}\|v\|$$

and

$$|h(1+\mathrm{i}y)|=|g(1+\mathrm{i}y),v)|\leqslant\|f(1+\mathrm{i}y)\|_{s_1,2}\|v\|.$$

Consequently, by Lemma 5.3 of ch. 1 (scalar case)

$$|h(z)|\leqslant\|f\|_H\|v\|, \qquad v\in S, \qquad 0\leqslant x\leqslant 1.$$

In particular this holds for $z=\theta$. Thus

$$|(g(\theta),v)|\leqslant(\|u\|_\theta+\varepsilon)\|v\|, \qquad v\in S.$$

Since S is dense in L^2 (Theorem 1.7), we see that $g(\theta)\in L^2$ and

$$\|g(\theta)\|\leqslant\|u\|_\theta+\varepsilon.$$

Since ε was arbitrary,

$$\|g(\theta)\| \leqslant \|u\|_\theta .$$

But

$$g(\theta) = \bar{F}[\lambda_0 \lambda^\theta Fu] = \bar{F}[(1+|\xi|^2)^{\frac{1}{2}s} Fu] .$$

This shows that $u \in H^{s,2}$ and

$$\|u\|_{s,2} = \|g(\theta)\| \leqslant \|u\|_\theta .$$

Next suppose $u \in H^{s,2}$. Put

$$f(z) = \bar{F}[\lambda^{\theta-z} Fu] .$$

Then $f(\theta) = u$ and

$$\lambda_0 |Ff(x+iy)| = \lambda_0 \lambda^{\theta-x} |Fu| \leqslant \lambda_0 \lambda^\theta |Fu|$$
$$= (1+|\xi|^2)^{\frac{1}{2}s} |Fu| ,$$

we see that $f(z)$ is in $H^{s_0,2}$ for $0 \leqslant x \leqslant 1$ and

$$\|f(z)\|_{s_0,2} \leqslant \|u\|_{s,2} .$$

Thus $f \in W(H^{s_0,2}, \Omega)$, where Ω is the strip $0 < x < 1$. Furthermore

$$\lambda_1 |Ff(1+iy)| = \lambda_1 \lambda^{\theta-1} |Fu| = (1+|\xi|^2)^{\frac{1}{2}s} |Fu| .$$

This shows that $f \in H(H^{s_0,2}, H^{s_1,2})$ and

$$\|f\|_H \leqslant \|u\|_{s,2} .$$

Thus $u \in H_\theta$ and

$$\|u\|_\theta \leqslant \|u\|_{s,2} .$$

This completes the proof.

Corollary 4.15. *Suppose* $s_0 < s_1$, $t_0 < t_1$, $0 \leqslant \theta \leqslant 1$, $s = (1-\theta)s_0 + \theta s_1$, $t = (1-\theta)t_0 + \theta t_1$. *If* T *is a bounded operator from* $H^{s_0,2}$ *to* $H^{t_0,2}$ *and its restriction to* $H^{s_1,2}$ *is a bounded operator from* $H^{s_1,2}$ *to* $H^{t_1,2}$, *then its restriction to* $H^{s,2}$ *is a bounded operator from* $H^{s,2}$ *to* $H^{t,2}$.

Proof. Apply Theorem 4.14 to Theorem 5.6 of ch. 1.

CHAPTER 3

PARTIAL DIFFERENTIAL OPERATORS

§1. Constant coefficient operators

In this chapter we summarize the basic properties of partial differential operators which will be needed in the book. Most of them are simple and easily proved.

Let

$$P(\xi) = \sum_{|\mu| \leqslant m} a_\mu \xi^\mu \tag{1.1}$$

be a polynomial of degree m. (Recall the notation: $\mu = (\mu_1, \ldots, \mu_n)$ is a multi-index of non-negative integers, $|\mu| = \mu_1 + \ldots + \mu_n$, $\xi = (\xi_1, \ldots, \xi_n)$ and $\xi^\mu = \xi_1^{\mu_1} \ldots \xi_n^{\mu_n}$.)

The coefficients of $P(\xi)$ are allowed to be complex numbers. Corresponding to the polynomial $P(\xi)$ we can form a partial differential operator as follows. Set

$$\mathrm{D}_k = -\mathrm{i}\, \partial/\partial x_k, \qquad 1 \leqslant k \leqslant n, \tag{1.2}$$

$\mathrm{D} = (\mathrm{D}_1, \ldots, \mathrm{D}_n)$ and

$$\mathrm{D}^\mu = \mathrm{D}_1^{\mu_1} \ldots \mathrm{D}_n^{\mu_n}. \tag{1.3}$$

We then define the partial differential operator of order m

$$P(\mathrm{D}) = \sum_{|\mu| \leqslant m} a_\mu \mathrm{D}^\mu. \tag{1.4}$$

$P(\xi)$ is called the polynomial corresponding to the operator $P(\mathrm{D})$.

The classical formula of Leibnitz implies that

$$\mathrm{D}_k(uv) = u\mathrm{D}_k v + v\mathrm{D}_k u, \qquad u, v \in C^\infty(\Omega). \tag{1.5}$$

There is a very useful generalization which is also known as Leibnitz's formula.

Theorem 1.1. *If $u, v \in C^\infty$, then*

$$P(\mathrm{D})[uv] = \sum_{|\mu| \leqslant m} P^{(\mu)}(\mathrm{D})u\,\mathrm{D}^{\mu}v/\mu!\,, \tag{1.6}$$

where $P^{(\mu)}(\mathrm{D})$ *corresponds to the polynomial*

$$P^{(\mu)}(\xi) = \mathrm{i}^{|\mu|}\,\mathrm{D}_{\xi}^{\mu}P(\xi) = \partial^{|\mu|}P(\xi)/\partial\xi^{\mu} \tag{1.7}$$

and

$$\mu! = \mu_1!\ldots\mu_n! \tag{1.8}$$

Proof. By repeated applications of (1.5) we have

$$P(\mathrm{D})[uv] = \sum_{|\mu| \leqslant m} Q_{\mu}(\mathrm{D})u\,\mathrm{D}^{\mu}v\,, \qquad u, v \in C^{\infty}\,,$$

for some operators $Q_{\mu}(\mathrm{D})$. The object is to show that

$$Q_{\mu}(\xi) = P^{(\mu)}(\xi)/\mu!\,. \tag{1.9}$$

Let $u = \mathrm{e}^{\mathrm{i}\xi x}$, $v = \mathrm{e}^{\mathrm{i}\eta y}$, where ξ, η are vectors and $\xi x = \xi_1 x_1 + \ldots + \xi_n x_n$. Thus $u, v \in C^{\infty}$ and

$$\mathrm{D}^{\mu}u = \xi^{\mu}u, \qquad \mathrm{D}^{\mu}v = \eta^{\mu}v\,, \tag{1.10}$$

and consequently

$$P(\mathrm{D})u = P(\xi)u, \qquad P(\mathrm{D})v = P(\eta)v\,. \tag{1.11}$$

Thus

$$P(\mathrm{D})[uv] = \sum_{|\mu| \leqslant m} Q_{\mu}(\xi)u\,\eta^{\mu}v\,. \tag{1.12}$$

But $uv = \mathrm{e}^{\mathrm{i}(\xi+\eta)x}$, and consequently

$$P(\mathrm{D})[uv] = P(\xi+\eta)uv\,. \tag{1.13}$$

Comparing (1.12) and (1.13) we have

$$P(\xi+\eta) = \sum_{|\mu| \leqslant m} Q_{\mu}(\xi)\eta^{\mu}\,. \tag{1.14}$$

But we know by Taylor's formula

$$P(\xi+\eta) = \sum_{|\mu| \leqslant m} P^{(\mu)}(\xi)\eta^{\mu}/\mu!\,. \tag{1.15}$$

Since the right-hand sides of (1.14) and (1.15) are identically equal, we see that (1.9) holds and the proof is complete.

Another convenient formula follows from Theorems 2.1 and 4.9 of ch. 2.

Theorem 1.2. *If* $u \in H^{s,p}$ *then*

$$F[P(\mathrm{D})u] = P(\xi)Fu \tag{1.16}$$

$$P(\mathrm{D})Fu = F[P(-x)u] \tag{1.17}$$

$$\bar{F}[P(\mathrm{D})u] = P(-\xi)\bar{F}u \tag{1.18}$$

$$P(\mathrm{D})\bar{F}u = \bar{F}[P(x)u]\,. \tag{1.19}$$

The following theorem will be needed

Theorem 1.3. *Let* μ *be any multi-index and let* $Q(\xi)$ *be any polynomial. Then there are vectors* $\theta^{(1)}, \ldots, \theta^{(r)}$ *in* E^n *and complex constants* $\gamma_1, \ldots, \gamma_r$ *such that*

$$t^{|\mu|}Q^{(\mu)}(\xi) = \sum_1^r \gamma_j Q(\xi + t\theta^{(j)})\,, \qquad \xi \in E^n\,,\ t \text{ real}\,. \tag{1.20}$$

Proof. Let r denote the number of distinct multi-indices ν such that $|\nu| \leqslant |\mu|$. There exist vectors $\theta^{(1)}, \ldots, \theta^{(r)}$ in E^n such that $\det(\theta^{(i)\nu}) \neq 0$. This is easily proved by induction on $|\mu|$. Now by Taylor's formula

$$Q(\xi + t\theta^{(j)}) = \Sigma\, Q^{(\nu)}(\xi)t^{|\nu|}\theta^{(j)\nu}/\nu!\,, \qquad 1 \leqslant j \leqslant r\,.$$

Thus for any scalars $\gamma_1, \ldots, \gamma_r$

$$\sum_{j=1}^r \gamma_j Q(\xi + t\theta^{(j)}) = \sum_\nu \left(\sum_{j=1}^r \gamma_j \theta^{(j)\nu} \right) t^{|\nu|} Q^{(\nu)}(\xi)/\nu!\,. \tag{1.21}$$

Since the matrix $(\theta^{(j)\nu})$ is non-singular, we can solve the system

$$\sum_{j=1}^r \gamma_j \theta^{(j)\nu} = \delta_{\mu\nu}\mu!\,, \qquad |\mu| \leqslant |\nu|\,,$$

where $\delta_{\mu\nu} = 0$ for $\mu \neq \nu$, $\delta_{\mu\mu} = 1$. Note that the solution does not depend on Q, t or ξ. Substituting the solution into (1.21) we obtain (1.20). This completes the proof.

§2. Operators with variable coefficients

The most general linear partial differential operator of order m is of the form

$$P(x, \mathrm{D}) = \sum_{|\mu| \leqslant m} a_\mu(x)\mathrm{D}^\mu\,, \tag{2.1}$$

where the coefficients $a_\mu(x)$ are now functions of x. The *principal part* of

$P(x, \mathrm{D})$ consists of the terms of highest order, i.e., the terms of order m:

$$P_m(x, \mathrm{D}) = \sum_{|\mu|=m} a_\mu(x)\mathrm{D}^\mu . \tag{2.2}$$

A very useful tool is the formula for *integration by parts*. It is based upon

Lemma 2.1. *If* $v \in S$, *then*

$$\int \mathrm{D}_k v \, \mathrm{d}x = 0 , \qquad 1 \leqslant k \leqslant n . \tag{2.3}$$

For $u, v \in S$ we have

$$\begin{aligned} \mathrm{D}_k(u\bar{v}) &= (\mathrm{D}_k u)\bar{v} + u\mathrm{D}_k\bar{v} \\ &= (\mathrm{D}_k u)\bar{v} - u\overline{\mathrm{D}_k v} . \end{aligned}$$

This gives

Corollary 2.2. *For* $u, v \in S$

$$(\mathrm{D}_k u, v) = (u, \mathrm{D}_k v) , \qquad 1 \leqslant k \leqslant n . \tag{2.4}$$

Corollary 2.3. *If the* $a_\mu(x)$ *are in* C_{B}^∞, *then*

$$(P(x, \mathrm{D})u, v) = (u, P^*(x, \mathrm{D})v) , \qquad u, v \in S , \tag{2.5}$$

where

$$P^*(x, \mathrm{D})v = \sum_{|\mu|\leqslant m} \mathrm{D}^\mu[\overline{a_\mu(x)}v] , \qquad v \in S . \tag{2.6}$$

Note that if we carry out all of the differentiations in (2.6) we obtain

$$P^*(x, \mathrm{D})v = \sum_{|\mu|\leqslant m} b_\mu(x)\mathrm{D}^\mu v , \qquad v \in S . \tag{2.7}$$

Thus $P^*(x, \mathrm{D})$ is a linear partial differential operator of order m. It is called the formal adjoint of $P(x, \mathrm{D})$. Note that $b_\mu(x)$ is the complex conjugate of $a_\mu(x)$ for $|\mu| = m$. Thus the principal part of $P^*(x, \mathrm{D})$ is given by

$$P_m^*(x, \mathrm{D}) = \sum_{|\mu|=m} \overline{a_\mu(x)}\mathrm{D}^\mu . \tag{2.8}$$

If $P(\mathrm{D})$ is a constant coefficient operator of the form (1.4), then $P^*(\mathrm{D}) = \bar{P}(\mathrm{D})$, where

$$\bar{P}(\mathrm{D}) = \sum_{|\mu|\leqslant m} \bar{a}_\mu \mathrm{D}^\mu \tag{2.9}$$

is the operator whose coefficients are the complex coefficients of those of

$P(\mathrm{D})$. In this case (2.5) becomes

$$(P(\mathrm{D})u, v) = (u, \bar{P}(\mathrm{D})v), \qquad u, v \in S. \tag{2.10}$$

Theorem 2.4. *If $P(x, \mathrm{D})$ is given by (2.1) and the $a_\mu(x)$ are in C_B^∞, then for each s, p there is a constant C such that*

$$\|P(x, \mathrm{D})v\|_{s,p} \leqslant C\|v\|_{s+m,p}, \qquad v \in S, \tag{2.11}$$

$$|(P(x, \mathrm{D})v, w)| \leqslant C\|v\|_{s,p}\|w\|_{m-s,p'}, \qquad v, w \in S. \tag{2.12}$$

§3. Elliptic operators

A very important class of partial differential operators is the set of *elliptic* operators. The operator $P(\mathrm{D})$ given by (1.4) is called elliptic if the only real solution of

$$P_m(\xi) = \sum_{|\mu|=m} a_\mu \xi^\mu = 0 \tag{3.1}$$

is $\xi = 0$. The operator $P(x, \mathrm{D})$ given by (2.1) is called elliptic at a point x if the only real solution of $P_m(x, \xi) = 0$ is $\xi = 0$. The operator $P(x, \mathrm{D})$ is called elliptic in a region $\Omega \subset E^n$ if it is elliptic and of order m at each point $x \in \Omega$.

Theorem 3.1. *If $P(\mathrm{D})$ is an elliptic operator of order m, then for each real s and $1 < p < \infty$ there is a constant C such that*

$$\|v\|_{s+m,p} \leqslant C(\|P(\mathrm{D})v\|_{s,p} + \|v\|_{s,p}), \qquad v \in S. \tag{3.2}$$

If $P(\xi) \neq 0$ for $\xi \in E^n$, then there is a constant C such that

$$\|v\|_{s+m,p} \leqslant C\|P(\mathrm{D})v\|_{s,p}, \qquad v \in S. \tag{3.3}$$

Proof. Consider first (3.3). Note that

$$(1 + |\xi|^2)^{\frac{1}{2}(s+m)} Fv = \{(1 + |\xi|^2)^{\frac{1}{2}m}/P(\xi)\}(1 + |\xi|^2)^{\frac{1}{2}s} F[P(\mathrm{D})v].$$

If we set $w(\xi) = (1 + |\xi|^2)^{\frac{1}{2}m}/P(\xi)$, a simple calculation shows that

$$|\mathrm{D}^\mu w(\xi)| \leqslant C/|\xi|^{|\mu|}. \tag{3.4}$$

In view of Theorem 3.3 of ch. 2 we see that w is a multiplier in L^p for $1 < p < \infty$. This gives (3.3). To prove (3.2) note that

$$(1 + |\xi|^2)^{\frac{1}{2}s}|\xi|^m Fv = \frac{|\xi|^m}{P_m(\xi)}(1 + |\xi|^2)^{\frac{1}{2}s} F[P_m(\mathrm{D})v].$$

If we now set $w(\xi)=|\xi|^m/P_m(\xi)$, another simple calculation shows that (3.4) holds. Another application of Theorem 3.3 of ch. 2 gives

$$\begin{aligned}\|\bar{F}[|\xi|^m Fv]\|_{s,p} &\leqslant C_1\|P_m(\mathrm{D})v\|_{s,p}\\ &\leqslant C_1(\|P(\mathrm{D})v\|_{s,p}+\|[P(\mathrm{D})-P_m(\mathrm{D})]v\|_{s,p})\\ &\leqslant C_2(\|P(\mathrm{D})v\|_{s,p}+\|v\|_{s+m-1,p})\end{aligned} \tag{3.5}$$

by Theorem 2.4. From the fact that $(1+|\xi|^2)^{\frac{1}{2}m}/(1+|\xi|^m)$ is a multiplier in L^p we conclude that

$$\|v\|_{m+s,p}\leqslant C_3(\|P(\mathrm{D})v\|_{s,p}+\|v\|_{m+s-1,p})\,. \tag{3.6}$$

If we replace s by $s-1$ in (3.6) we obtain

$$\|v\|_{m+s-1,p}\leqslant C_4(\|P(\mathrm{D})v\|_{s-1,p}+\|v\|_{m+s-2,p})\,. \tag{3.7}$$

Combined with (3.6) this implies

$$\|v\|_{m+s,p}\leqslant C_5(\|P(\mathrm{D})v\|_{s,p}+\|v\|_{m+s-2,p})\,.$$

Continuing in this way we eventually reach (3.2). This completes the proof.

Theorem 3.2. *Let $P(\mathrm{D})$ be an elliptic operator of order m and let Ω be an open subset of E^n. Suppose $u\in L^2(\Omega)$ satisfies*

$$|(u,P(\mathrm{D})\varphi)|\leqslant C\,\|\varphi\|_{t,2}\,,\qquad \varphi\in C_0^\infty(\Omega)\,,$$

where $t<m$. Then for each $\psi\in C_0^\infty(\Omega)$ the function ψu is in $H^{m-t,2}$ and

$$\|\psi u\|_{m-t,2}\leqslant C(C_0+\|u\|_{0,2}^{\Omega})\,,$$

where the constant C depends only on $P(\mathrm{D})$ and ψ.

An operator $P(\mathrm{D})$ is called *hypoelliptic* if there is an $a>0$ such that

$$P^{(\mu)}(\xi)/P(\xi)=\mathrm{O}(|\xi|^{-a|\mu|})\ \text{as}\ |\xi|\to\infty,\qquad \xi\in E^n,\ \text{each}\ \mu.$$

CHAPTER 4

GENERAL L^p THEORY

In this chapter we discuss constant coefficient operators $P(\mathrm{D})$ on C_0^∞ and their extensions in L^p. We first show that they are closable, and we describe their minimal and maximal extensions. In particular it is shown that they coincide for $1 \leqslant p < \infty$. Then we consider the spectra of these extensions. For $p=2$ the spectrum is completely determined for all operators $P(\mathrm{D})$ in terms of the corresponding polynomial. We then give a large class of operators for which the spectrum is independent of p for $1 < p < \infty$. Some important examples are given in §5. Finally we give a preliminary discussion of operators of the form $P(\mathrm{D}) + q(x)$, where $q(x)$ is a function (potential). General properties are discussed which do not depend on the form of the operator $P(\mathrm{D})$. In particular, a criterion is given for the essential spectrum of any extension of $P(\mathrm{D}) + q$ to contain the spectrum of $P(\mathrm{D})$. Generalizations are given to operators of the form $P(\mathrm{D}) + \Sigma\, q_j(x)\, Q_j(\mathrm{D})$.

§1. The minimal operator

Let $P(\mathrm{D})$ be a constant coefficient operator. For $1 \leqslant p \leqslant \infty$ this operator can be considered as an operator on L^p with domain consisting of C_0^∞. It is not a closed operator, but we have

Lemma 1.1. *The operator $P(\mathrm{D})$ defined on C_0^∞ is closable in L^p.*

Proof. Let $\{\varphi_k\}$ be a sequence of functions in C_0^∞ converging to 0 in L^p and such that $P(\mathrm{D})\varphi_k$ converges to a function $f \in L^p$. Then by formula (2.10) of ch. 3 we have

$$(P(\mathrm{D})\varphi_k, \psi) = (\varphi_k, \bar{P}(\mathrm{D})\psi), \qquad k = 1, 2, \ldots,$$

for each $\psi \in C_0^\infty$. Consequently

$$(f, \psi) = 0, \qquad \psi \in C_0^\infty. \tag{1.1}$$

I claim that this implies $f = 0$. Let $j_k(x)$ be the function defined by (1.13) of ch. 2. For each y, the function $j_k(y-x)$ is in C_0^∞. Hence

$$\int f(x) j_k(y-x) \mathrm{d}x = 0 , \qquad k=1, 2, \dots .$$

This shows that $J_k f$ vanishes identically for each k. But for $1 \leqslant p < \infty$ the functions $J_k f$ converge to f in L^p (Theorem 1.8 of ch. 2). Thus f must be identically zero in this case. When $p = \infty$, we note that f is the uniform limit of functions in C_0^∞. Thus f is continuous. A continuous function satisfying (1.1) is easily shown to vanish. Thus $P(\mathrm{D})$ is closable in L^p.

By Theorem 2.1 of ch. 1, $P(\mathrm{D})$ acting on C_0^∞ has a closure in L^p. Denote this closure by P_{0p}. When there is no doubt about the value of p, we shall denote it merely by P_0. By definition a function $u \in L^p$ is in $D(P_0)$ and $P_0 u = f$ if there is a sequence $\{\varphi_k\}$ of functions in C_0^∞ such that $\varphi_k \to u$ and $P(\mathrm{D}) \varphi_k \to f$ in L^p. The operator P_0 is called the *minimal* or *strong* extension of $P(\mathrm{D})$ on C_0^∞. The term minimal signifies the fact that every closed extension of an operator is an extension of its closure (see Theorem 2.1 of ch. 1). The term strong is to contrast P_0 with a weak extension to be discussed in the next section.

We shall make use of the following important facts.

Lemma 1.2. *If $v \in C^\infty$ and $P^{(\mu)}(\mathrm{D}) v \in L^p$ for each μ, then $v \in D(P_0)$ and*

$$P_0 v = P(\mathrm{D}) v . \tag{1.2}$$

In particular, $S \subset D(P_0)$ and (1.2) holds for $v \in S$.

Proof. Let ψ be a function in C_0^∞ such that

$$0 \leqslant \psi(x) \leqslant 1 , \qquad x \in E^n \tag{1.3}$$

$$\psi(x) = 1 , \qquad |x| \leqslant 1 . \tag{1.4}$$

(Such a function exists by Theorem 1.2 of ch. 2.) Let v be a function satisfying the hypotheses of the lemma. Then for each $R > 0$ the function

$$v_R(x) = \psi(x/R) v(x)$$

is in C_0^∞. Moreover

$$\|v_R - v\| \leqslant \|v\|^R \to 0 \text{ as } R \to \infty ,$$

where $\|v\|^R$ is the norm of $L^p(|x| > R)$. In addition, by Leibnitz's formula (see Theorem 1.1 of ch. 3)

$$\begin{aligned} P(\mathrm{D}) v_R(x) &= \Sigma \; P^{(\mu)}(\mathrm{D}) v D^\mu \psi(x/R)/\mu ! \\ &= \Sigma \; P^{(\mu)}(\mathrm{D}) v \psi_\mu(x/R)/\mu ! R^{|\mu|} , \end{aligned}$$

where $\psi_\mu(x) = D^\mu \psi(x)$. Thus

$$\|P(D)(v_R - v)\| \leqslant \|P(D)v\|^R + \sum_{\mu \neq 0} \|P^{(\mu)}(D)v\| \, \|\psi_\mu\|_\infty / \mu! R^{|\mu|} \to 0 \text{ as } R \to \infty.$$

This shows that $v \in D(P_0)$ and that (1.2) holds. The proof is complete.

Lemma 1.3. *If there is a constant $c_0 > 0$ such that*

$$|P(\xi)| \geqslant c_0, \quad \xi \textit{ real}, \tag{1.5}$$

then $N(P_0) = \{0\}$, and for each $f \in S$ there is a $v \in S$ such that $P(D)v = f$. Thus $S \subset R(P_0)$.

Proof. Let f be any function in S. Then $Ff \in S$ (Theorem 2.1 of ch. 2). By (1.5) the same is true of Ff/P. Hence there is a function $u \in S$ such that

$$Fu = Ff/P \tag{1.6}$$

(Theorem 2.2 of ch. 2). By the Fourier inversion formula this implies

$$P(D)u = f \tag{1.7}$$

(Theorem 1.2 of ch. 3). By Lemma 1.2, $u \in D(P_0)$ and $P_0 u = f$. Thus $f \in R(P_0)$.

Next suppose $w \in N(P_0)$. Then there is a sequence $\{\varphi_k\}$ of functions in C_0^∞ such that $\varphi_k \to w$ and $P(D)\varphi_k \to 0$ in L^p. Thus for each $v \in S$ and each k

$$(\varphi_k, \bar{P}(D)v) = (P(D)\varphi_k, v). \tag{1.8}$$

Taking the limit as $k \to \infty$, we have

$$(w, \bar{P}(D)v) = 0, \qquad v \in S.$$

Now (1.5) implies

$$|\bar{P}(\xi)| \geqslant c_0, \qquad \xi \text{ real}. \tag{1.9}$$

Hence by what was just proved, for each function $g \in S$ there is a $v \in S$ satisfying

$$\bar{P}(D)v = g.$$

Therefore

$$(w, g) = 0, \qquad g \in S. \tag{1.10}$$

As was shown in the proof of Lemma 1.1, this implies that $w = 0$. The proof is complete.

Lemma 1.4. *If*

$$\|v\| \leqslant K_0 \|P(\mathrm{D})v\|, \qquad v \in S, \tag{1.11}$$

then

$$1 \leqslant K_0 |P(\xi)|, \qquad \xi \in E^n. \tag{1.12}$$

Proof. Let ξ be any vector in E^n and let ψ be any function in C_0^∞ such that $\|\psi\| = 1$. Set

$$\varphi_k(x) = \psi(x/k) \exp\{\mathrm{i}x\xi\}/k^{n/p}, \qquad k = 1, 2, \ldots,$$

where $x\xi = x_1\xi_1 + \ldots + x_n\xi_n$. Then $\varphi_k \in C_0^\infty$ for each k and $\|\varphi_k\| = 1$. Furthermore by Leibnitz's formula (Theorem 1.1 of ch. 3)

$$P(\mathrm{D})\varphi_k = \exp\{\mathrm{i}\xi x\} \sum_\mu P^{(\mu)}(\xi)\psi_\mu(x/k)/\mu!\, k^{|\mu| + (n/p)},$$

where $\psi_\mu(x) = \mathrm{D}^\mu \psi(x)$. Since

$$\|\psi_\mu(x/k)/k^{n/p}\| = \|\psi_\mu\|,$$

we see that

$$\|P(\mathrm{D})\varphi_k\| \to |P(\xi)| \text{ as } k \to \infty.$$

By (1.11)

$$1 = \|\varphi_k\| \leqslant K_0 \|P(\mathrm{D})\varphi_k\| \to K_0 |P(\xi)|$$

as $k \to \infty$. This gives (1.12), and the proof is complete.

§2. The maximal extension

We can define another closed extension P_p of $P(\mathrm{D})$ on C_0^∞ as follows. We say that $u \in D(P_p)$ and $P_p u = f$ if u and f are in L^p and

$$(u, \bar{P}(\mathrm{D})\varphi) = (f, \varphi), \qquad \varphi \in C_0^\infty. \tag{2.1}$$

It is clear from the definition that P_p is a closed extension of $P(\mathrm{D})$ on C_0^∞. It is called the *maximal* or *weak* extension. It is maximal in the sense that it is the largest closed extension having C_0^∞ contained in the domain of its adjoint. It is weak in the sense that functions in its domain need not be differentiable in any way.

Clearly P_p is an extension of P_0. For if $u \in D(P_0)$ and $P_0 u = f$, then there is a sequence $\{\varphi_k\}$ of functions in C_0^∞ such that $\varphi_k \to u$ and $P(\mathrm{D})\varphi_k \to f$ in L^p.

By integration by parts (see (2.10) of ch. 3)

$$(P(\mathrm{D})\varphi_k, \psi) = (\varphi_k, \bar{P}(\mathrm{D})\psi), \qquad \psi \in C_0^\infty, \; k=1, 2, \dots,$$

so that in the limit we have

$$(f, \psi) = (u, \bar{P}(\mathrm{D})\psi), \qquad \psi \in C_0^\infty,$$

which means that $u \in D(P_p)$ and $P_p u = f$.

Let $\bar{P}_{0p'}$ denote the minimal extension of $\bar{P}(\mathrm{D})$ on C_0^∞ in $L^{p'}$. Then (2.1) implies

$$(u, \bar{P}_{0p'}v) = (f, v), \qquad v \in D(\bar{P}_{0p'}). \tag{2.2}$$

In fact, for each $v \in D(\bar{P}_{0p'})$ there is sequence $\{\varphi_k\}$ of functions in C_0^∞ such that $\varphi_k \to v$ and $\bar{P}(\mathrm{D})\varphi_k \to \bar{P}_{0p'}v$ in $L^{p'}$. By (2.1)

$$(u, \bar{P}(\mathrm{D})\varphi_k) = (f, \varphi_k), \qquad k=1, 2, \dots .$$

Taking the limit we obtain (2.2).

As a consequence of (2.2) we have

$$P_p = (\bar{P}_{0p'})^*, \qquad 1 < p \leqslant \infty . \tag{2.3}$$

Moreover from (2.2) and Lemma 1.2 we see that $u \in D(P_p)$ and $P_p u = f$ if and only if

$$(u, \bar{P}(\mathrm{D})v) = (f, v), \qquad v \in S . \tag{2.4}$$

One might wonder whether or not there are occasions when the minimal and maximal extensions coincide. The following theorem shows that this happens for $1 \leqslant p < \infty$.

Theorem 2.1. *For any polynomial* $P(\xi)$ *we have*

$$P_p = P_{0p}, \qquad 1 \leqslant p < \infty . \tag{2.4}$$

Proof. Since P_p is an extension of P_{0p}, it suffices to show that $D(P_p) \subset D(P_{0p})$. Suppose $u \in D(P_p)$. Then there is a $f \in L^p$ such that (2.1) holds. Take $\varphi(x) = j_k(y-x)$, where $j_k(x)$ is given by (1.13) of ch. 2. Then (2.1) becomes

$$P(\mathrm{D}) J_k u = J_k f, \qquad k=1, 2, \dots . \tag{2.5}$$

Thus by Theorem 1.8 of ch. 2

$$J_k u \to u, \quad P(\mathrm{D}) J_k u \to f \text{ in } L^p \text{ as } k \to \infty . \tag{2.6}$$

For each $R > 0$ define

$$u_R(x) = u(x), \qquad |x| \leqslant R \tag{2.7}$$
$$\quad = 0, \qquad |x| > R.$$

Moreover, for each k and R, $J_k u_R$ is a function in C_0^∞ (see Theorem 1.10 of ch. 2). I am going to show that for each k

$$J_k u_R \to J_k u, \quad P(\mathrm{D}) J_k u_R \to P(\mathrm{D}) J_k u \text{ in } L^p \text{ as } R \to \infty. \tag{2.8}$$

It will then follow that $u \in D(P_0)$. For if $\varepsilon > 0$ is given, we can take k so large that

$$\|u - J_k u\| < \tfrac{1}{2}\varepsilon, \qquad \|f - P(\mathrm{D}) J_k u\| < \tfrac{1}{2}\varepsilon.$$

Once k is fixed, we take R so large that

$$\|J_k(u - u_R)\| < \tfrac{1}{2}\varepsilon, \qquad \|P(\mathrm{D}) J_k(u - u_R)\| < \tfrac{1}{2}\varepsilon.$$

This gives the desired result. It thus remains to prove (2.8). Now by Theorem 2.4 of ch. 2

$$J_k w = j_k * w, \qquad P(\mathrm{D}) J_k w = P(\mathrm{D}) j_k * w.$$

Thus by Young's inequality (see Theorem 1.8 of ch. 2)

$$\|J_k w\| \leqslant \|w\|$$

and

$$\|P(\mathrm{D}) J_k w\| \leqslant \|P(\mathrm{D}) j_k\|_1 \|w\|.$$

Now $\|P(\mathrm{D}) j_k\|_1$ is finite for each k. Thus

$$\|J_k(u - u_R)\| \leqslant \|u - u_R\|$$

$$\|P(\mathrm{D}) J_k(u - u_R)\| \leqslant \|P(\mathrm{D}) j_k\|_1 \|u - u_R\|.$$

Since $\|u - u_R\| \to 0$ as $R \to \infty$, the proof is complete.

An immediate consequence of (2.3) and Theorem 2.1 is

Corollary 2.2. *For* $1 < p < \infty$

$$(P_{0p})^* = \bar{P}_{0p'}. \tag{2.10}$$

Corollary 2.3. *P_{02} is a normal operator. If the coefficients of $P(\xi)$ are real, then P_{02} is self-adjoint.*

Proof. By Corollary 2.2 we have

$$(P_{02})^* = \bar{P}_2 = \bar{P}_{02}. \tag{2.11}$$

This proves the second statement. To prove the first we note that for $\varphi \in C_0^\infty$

$$\|P(\mathrm{D})\varphi\|^2 = (P(\mathrm{D})\varphi, P(\mathrm{D})\varphi) = (\bar{P}(\mathrm{D})P(\mathrm{D})\varphi, \varphi)$$
$$= (P(\mathrm{D})\bar{P}(\mathrm{D})\varphi, \varphi) = \|\bar{P}(\mathrm{D})\varphi\|^2 .$$

Thus $D(\bar{P}_{02}) = D(P_{02})$ and

$$\|\bar{P}_{02}u\| = \|P_{02}u\| , \qquad u \in D(P_{02}) . \tag{2.12}$$

Hence P_{02} is normal.

§3. The spectrum of the minimal operator

In this section we discuss the spectrum of P_0. We first prove

Theorem 3.1. *Let $P(\xi)$ be a polynomial and λ a complex number. If $P(\xi)$ is not bounded away from λ for real vectors ξ, then $\lambda \in \sigma_e(P_0)$.*

Proof. Without loss of generality we may assume that $\lambda = 0$. If $P(\xi)$ is bounded away from zero, there is a sequence $\{\xi^{(k)}\}$ of real vectors such that $P(\xi^{(k)}) \to 0$ as $k \to \infty$. Let ε_k be a sequence of positive numbers tending to zero and such that

$$\varepsilon_k^{|\mu|} P^{(\mu)}(\xi^{(k)}) \to 0 \text{ as } k \to \infty \tag{3.1}$$

holds for each μ. Such a sequence exists because there is only a finite number of derivatives $P^{(\mu)}(\xi)$, and by assumption $P(\xi^{(k)}) \to 0$ (recall that $P^{(\mu)}(\xi) = P(\xi)$ for $|\mu| = 0$). Let $\psi(x)$ be a function in C_0^∞ which vanishes near zero and such that $\|\psi\| = 1$ (see Theorem 1.2 of ch. 2). Set

$$\varphi_k(x) = \varepsilon_k^{n/p} \psi(\varepsilon_k x) \exp\{i\xi^{(k)} x\} , \qquad k = 1, 2, \ldots , \tag{3.2}$$

where $1/\infty$ is to be interpreted as 0. Thus

$$\|\varphi_k\| = 1, \qquad k = 1, 2, \ldots . \tag{3.3}$$

Now by Leibnitz's formula (Theorem 1.1 of ch. 3)

$$P(\mathrm{D})\varphi_k(x) = \varepsilon_k^{n/p} \sum_{\mu} \varepsilon_k^{|\mu|} P^{(\mu)}(\xi^{(k)}) \psi_\mu(\varepsilon_k x) \exp\{i\xi^{(k)} x\}/\mu! ,$$

where $\psi_\mu(x) = \mathrm{D}^\mu \psi(x)$. Since

$$\|\varepsilon_k^{n/p} \psi_\mu(\varepsilon_k x)\| = \|\psi_\mu\| ,$$

we have by (3.1)

$$P(\mathrm{D})\varphi_k \to 0 \text{ as } k\to\infty \tag{3.4}$$

in L^p. This shows that $0\in\sigma(P_0)$ (see §4 of ch. 1). To show that it is in $\sigma_e(P_0)$, note that

$$|\varphi_k(x)| = \varepsilon_k^{n/p}\,|\psi(\varepsilon_k x)|\,, \qquad k=1, 2, \dots .$$

Thus if $p\neq\infty$, the $\varphi_k(x)$ converge uniformly to zero. If $p=\infty$ the $\varphi_k(x)$ converge pointwise to zero since $\psi(x)$ vanishes in a neighborhood of the origin. Now if there existed a subsequence of $\{\varphi_k\}$ which converged in L^p to a function u, then $\|u\|=1$ by (3.3). But by Theorem 1.11 of ch. 2 there would be a subsequence of the subsequence which converged to u almost everywhere. Since the φ_k converge to zero everywhere, $u=0$. This contradiction shows that $\{\varphi_k\}$ has no convergent subsequence. We now apply Theorem 4.4 of ch. 1 to conclude that $0\in\sigma_e(P_0)$. This completes the proof.

Theorem 3.2. *In order that λ be in $\rho(P_0)$ it is necessary, and for $p=2$ also sufficient, that $P(\xi)$ be bounded away from λ for ξ real.*

Proof. By Theorem 3.1, if $P(\xi)$ is not bounded away from λ, then $\lambda\in\sigma_e(P_0)$. Consequently λ cannot be in $\rho(P_0)$. It therefore remains only to show that if $p=2$ and $P(\xi)$ is bounded away from λ, then $\lambda\in\rho(P_0)$. Again we may assume that $\lambda=0$.

Suppose that there is a constant $c_0>0$ such that (1.5) holds. Then $S\subset R(P_0)$ by Lemma 1.3. If we can show that

$$\|u\|\leqslant C\|P_0 u\|\,, \qquad u\in D(P_0)\,, \tag{3.5}$$

then it will follow that $0\in\rho(P_0)$ since S is dense in L^p (see §4 of ch. 1). Moreover, since P_0 is the closure in L^p of $P(\mathrm{D})$ on $C_0^\infty\subset S$, inequality (3.5) will be a consequence of

$$\|u\|\leqslant C\|P(\mathrm{D})u\|\,, \qquad u\in S\,. \tag{3.6}$$

It therefore suffices to show that when $p=2$ inequality (1.5) implies (3.6).

Let u be any function in S and set $f=P(\mathrm{D})u$. Then by Theorem 1.2 of ch. 3

$$Fu = Ff/P\,. \tag{3.7}$$

Hence by (1.5)

$$|Fu|\leqslant |Ff|/c_0\,, \tag{3.8}$$

which implies that

$$\|Fu\|\leqslant \|Ff\|/c_0\,. \tag{3.9}$$

Since $p=2$ we can apply Theorem 2.3 of ch. 2 to conclude

$$\|u\| \leqslant \|f\|/c_0, \tag{3.10}$$

which is precisely that we need.

Corollary 3.3. *$\sigma_e(P_{02}) = \sigma(P_{02})$ and consists of the closure of the set of values assumed by $P(\xi)$ for ξ real.*

In general, the non-vanishing of $P(\xi)-\lambda$ for ξ real does not imply that it is bounded away from zero. For instance, if

$$P(\xi) = (\xi_1\xi_2 - 1)^2 + \xi_2^2 + \ldots + \xi_n^2, \tag{3.11}$$

then $P(\xi) \neq 0$ for all real ξ. However, if we take $\xi_2 = 1/\xi_1$ and $\xi_3 = \ldots = \xi_n = 0$, then $P(\xi) \to 0$ as $|\xi_1| \to \infty$. On the other hand, this cannot happen if we assume

$$|P(\xi)| \to \infty \text{ as } |\xi| \to \infty, \qquad \xi \text{ real}. \tag{3.12}$$

For if (3.12) holds and $P(\xi) \neq \lambda$ for real ξ, then there is a constant $c_0 > 0$ such that

$$|P(\xi) - \lambda| \geqslant c_0, \qquad \xi \text{ real}. \tag{3.13}$$

Thus we have

Corollary 3.4. *If $P(\xi)$ satisfies (3.12), then $\sigma(P_{02})$ consists of the set*

$$\{P(\xi),\ \xi \text{ real}\}.$$

In the next section we shall consider sufficient conditions for λ to be in $\rho(P_{0p})$ when $p \neq 2$.

§4. The case $p \neq 2$

If one follows the proof of Theorem 3.1, one sees that the only thing lacking when $p \neq 2$ is inequality (3.6). In this section we shall discuss further assumptions on $P(\xi)$ in order that (1.5) should imply (3.6).

By (3.7) we see that an inequality of the form

$$\|u\| \leqslant C\|f\| \tag{4.1}$$

holds if $1/P$ is a multiplier in L^p (see §3 of ch. 2). We have

Theorem 4.1. *For* $1 \leqslant p < \infty$ *a point* λ *is in* $\rho(P_{0p})$ *if and only if* $1/[P(\xi)-\lambda]$ *is a multiplier in* L^p.

Proof. We may assume $\lambda=0$. Suppose $0 \in \rho(P_0)$. Then there is a constant C such that

$$\|v\| \leqslant C\|P(\mathrm{D})v\|, \qquad v \in S. \tag{4.2}$$

By Lemma 1.4 there is a $c_0>0$ such that (1.5) holds. To show that $1/P(\xi)$ is an L^p multiplier, suppose $f \in S$, $u \in L^p$ and

$$(f, \bar{F}[g/\bar{P}]) = (u, \bar{F}g), \qquad g \in S.$$

Set $w = \bar{F}[Ff/P]$. Then $w \in S$ and $P(\mathrm{D})w=f$ (see the proof of Lemma 1.3). Thus

$$\|w\| \leqslant C\|f\|, \tag{4.3}$$

where C is the constant in (4.2). Furthermore

$$(f, \bar{F}[g/\bar{P}]) = (Ff/P, g) = (w, \bar{F}g), \qquad g \in S$$

by Theorem 2.4 of ch. 2. Thus

$$(u-w, v) = 0, \qquad v \in S.$$

It follows from this that $w=u$ (see the proof of Lemma 1.1). Thus (4.3) shows that $1/P$ is an L^p multiplier.

Conversely, assume that $1/P$ is an L^p multiplier. In particular, it must be bounded (Theorem 3.2 of ch. 2). Hence there is a $c_0>0$ such that (1.5) holds. By Lemma 1.3 for each $f \in S$ there is a unique $u \in S$ such that

$$P(\mathrm{D})u = f. \tag{4.5}$$

Thus for $v \in S$ we have by Theorem 1.2 of ch. 3

$$\begin{aligned}(f, \bar{F}[v/\bar{P}]) &= (P(\mathrm{D})u, \bar{F}[v/\bar{P}]) \\ &= (u, \bar{P}(\mathrm{D})\bar{F}[v/\bar{P}]) = (u, \bar{F}v).\end{aligned}$$

Since $1/P$ is an L^p multiplier, inequality (4.1) holds with the constant C independent of u or f. Since $S \subset R(P_0)$ and S is dense in L^p, we see that $0 \in \rho(P_0)$ and the proof is complete.

In view of Theorem 4.1 we need a criterion for determining when $1/P$ is a multiplier in L^p. In general this is not a simple matter. We give a sufficient condition.

Theorem 4.2. *Assume that* $1 < p < \infty$ *and that* $P(\xi)$ *satisfies*

$$P^{(\mu)}(\xi)/P(\xi) = O(|\xi|^{-a|\mu|}) \text{ as } |\xi| \to \infty\,, \quad |\mu| \leqslant l\,, \tag{4.6}$$

$$1/P(\xi) = O(|\xi|^{-b}) \qquad \text{as } |\xi| \to \infty\,, \tag{4.7}$$

where l *is the smallest integer* $> n|\tfrac{1}{2} - 1/p|$ *and*

$$a \leqslant 1, \quad b > (1-a)n|\tfrac{1}{2} - 1/p|\,. \tag{4.8}$$

Then $\lambda \in \rho(P_0)$ *if and only if* $P(\xi) \neq \lambda$ *for real* ξ.

Proof. We may assume $\lambda = 0$. By Theorem 4.1 it suffices to show that $1/P$ is a multiplier in L^p. Since $a \leqslant 1$, we must have $b > 0$, which shows that $|P(\xi)| \to \infty$ as $|\xi| \to \infty$. Since $P(\xi) \neq 0$ for ξ real, (1.5) must hold. Now for each μ the derivative $D^\mu(1/P)$ consists of a sum of terms of the form

$$\text{constant } P^{(\mu^{(1)})}(\xi) \cdots P^{(\mu^{(k)})}(\xi)/P(\xi)^{k+1}\,, \tag{4.9}$$

where $\mu^{(1)} + \ldots + \mu^{(k)} = \mu$ (this can be verified by a simple induction argument). Thus

$$|D^\mu(1/P)| \leqslant C|\xi|^{-a|\mu| - b}\,, \qquad |\mu| \leqslant l\,. \tag{4.10}$$

We can now apply Theorem 3.4 of ch. 2 to conclude that $1/P$ is an L^p multiplier. This completes the proof.

Corollary 4.3. *Assume that* $1 < p < \infty$ *and that* $P(\xi)$ *satisfies* (4.7) *for some* $b > 0$. *If*

$$|1/p - 1/2| < b/n(m-b)\,, \tag{4.11}$$

where m *is the degree of* $P(\xi)$, *then* $\lambda \in \rho(P_0)$ *if and only if* $P(\xi) \neq \lambda$ *for all* $\xi \in E^n$. *In particular this is true for all* $1 < p < \infty$ *when* $b \geqslant mn/(n+2)$.

Proof. $P^{(\mu)}(\xi)$ is of degree at most $m - |\mu|$. Hence

$$P^{(\mu)}(\xi)/P(\xi) = O(|\xi|^{m - |\mu| - b}) \text{ as } |\xi| \to \infty\,. \tag{4.12}$$

For $|\mu| \geqslant 1$ we have $m - |\mu| - b \leqslant (m - b - 1)|\mu|$. Hence

$$P^{(\mu)}(\xi)/P(\xi) = O(|\xi|^{(m-b-1)|\mu|}) \text{ as } |\xi| \to \infty\,. \tag{4.13}$$

This shows that (4.6) holds with $a = b + 1 - m$. Hence our conclusion holds provided

$$b > (1-a)n|1/p - 1/2| = (m-b)n|1/p - 1/2|\,.$$

This is precisely (4.11), and the proof is complete.

For the case $p = 1$ we have

Theorem 4.4. *Assume that* (4.7) *and*

$$P^{(\mu)}(\xi)/P(\xi) = O(1/|\xi|^{a|\mu|}) \text{ as } |\xi| \to \infty , \qquad |\mu| \leqslant n+1 , \tag{4.14}$$

hold for some $a \leqslant 1$ *and* $b > n+a-an$. *Then we have*

$$\sigma(P_{01}) = \sigma_e(P_{01}) = \{P(\xi),\ \xi \text{ real}\} . \tag{4.15}$$

Proof. Assume $P(\xi) \neq 0$ for ξ real. By (4.7) we see that $1/P$ is in L^p for $pb > n$. Thus $\bar{F}(1/P)$ is in $T^{0,p}$ and

$$(-x)^{\mu} \bar{F}(1/P) = \bar{F}[D^{\mu}(1/P)] \tag{4.16}$$

(Theorems 4.5 and 4.9 of ch. 2). Now by (4.7), (4.9) and (4.14)

$$|D^{\mu}(1/P)| \leqslant C/|\xi|^{a|\mu|+b} , \qquad |\mu| \leqslant n+1 . \tag{4.17}$$

Thus $D^{\mu}(1/P)$ will be in L^1 whenever $a|\mu| + b > n$. By hypothesis this is true for $|\mu| \geqslant n-1$. By Theorem 2.3 of ch. 2

$$|\bar{F}[D^{\mu}(1/P)]| \leqslant \|D^{\mu}(1/P)\|_1 , \qquad |\mu| \geqslant n-1 . \tag{4.18}$$

Combining (4.16) and (4.18) we see that there is a constant C such that

$$|\bar{F}(1/P)| \leqslant C/|x|^{n-1}$$
$$|\bar{F}(1/P)| \leqslant C/|x|^{n+1} .$$

This shows that $\bar{F}(1/P) \in L^1$. Thus if $f \in S$, then

$$F[\bar{F}(1/P) * f] = (1/P) Ff$$

by Theorem 2.6 of ch. 2. Thus if $u = \bar{F}(1/P) * f$, then

$$\|u\|_1 \leqslant \|\bar{F}(1/P)\|_1 \|f\|_1 \tag{4.19}$$

by Theorem 1.9 of ch. 2. But by Lemma 1.3 there is a $v \in S$ such that $P(D)v = f$. Consequently $Fv = Ff/P = Fu$. Thus $u = v$ (Theorem 4.5 of ch. 2) and $P(D)u = f$. Since S is dense in L^1 and (4.19) holds, we see that $0 \in \rho(P_{01})$. The theorem now follows from Theorem 3.1.

Corollary 4.5. *If* $P(\xi)$ *is of degree* m *and satisfies* (4.7) *with* $b > (m-b)\cdot(n-1)+1$, *then* (4.15) *holds.*

Proof. By (4.13) we see that (4.14) holds with $a = b+1-m$. The condition given implies that $b > a+n-an$.

Corollary 4.6. *Let* $P(D)$ *be an elliptic operator of order* $m > 1$. *Then for* $1 \leqslant p < \infty$

$$\sigma(P_0) = \{P(\xi),\ \xi \in E^n\}\ .$$

Proof. If $P(\mathrm{D})$ is elliptic, $a=1$ and $b=m$ (see §5). Apply Theorems 4.2 and 4.4 (or Corollaries 4.3 and 4.5).

§5. Examples

1. $P(\mathrm{D})$ elliptic. If the order of $P(\mathrm{D})$ is m, then $P(\xi)=P_m(\xi)+Q(\xi)$, where $P_m(\xi)$ is a homogeneous polynomial of degree m satisfying

$$|P_m(\xi)| \geqslant c_0|\xi|^m\,, \qquad c_0>0\,, \tag{5.1}$$

and $Q(\xi)$ is of degree less than m (see ch. 3, §3). Thus for $|\xi|$ large

$$\begin{aligned}|P(\xi)| &\geqslant |P_m(\xi)|-|Q(\xi)|\\ &\geqslant c_0|\xi|^m - C|\xi|^{m-1} = |\xi|^{m-1}(c_0|\xi|-C)\,.\end{aligned}$$

For $|\xi|>2C/c_0$ we have

$$|P(\xi)| \geqslant \tfrac{1}{2}c_0|\xi|^m\,. \tag{5.2}$$

Thus $b=m$ in this case. Moreover

$$P^{(\mu)}(\xi) = \mathrm{O}(|\xi|^{m-|\mu|}) \ \text{ as } |\xi|\to\infty\,,$$

so that $a=1$ in view of (5.2). These are the largest values of a and b possible.

2. $P(\xi)=(\xi_1-\mathrm{i})^j(\xi_2^2+\ldots+\xi_n^2+1)^k$. In this case $a=0$. For

$$\partial P/\partial \xi_1 = kP(\xi_1-\mathrm{i})\,.$$

However $b\geqslant \min(j,\,2k)$. To see this note that

$$|P(\xi)|^2 \geqslant (\xi_1^2+1)^j,\quad |P(\xi)|^2 \geqslant (\xi_2^2+\ldots+\xi_n^2+1)^{2k}\,.$$

Thus if $d=\min(j,\,2k)$, then

$$2|P(\xi)|^2 \leqslant \xi_1^{2d}+(\xi_2^2+\ldots+\xi_n^2)^d \geqslant C_1|\xi|^{2d}$$

for some constant $C_1>0$.

For $j=1$, $\sigma(P_{02})$ consists of the half-plane $\operatorname{Im}\lambda \leqslant -1$.

3. The Laplace operator $P(\xi)=-(\xi_1^2+\ldots+\xi_n^2)$. This is elliptic and $\sigma(P_0)$ consists of the negative real axis.

4. The wave operator $P(\xi)=\xi_1^2-(\xi_2^2+\ldots+\xi_n^2)$. Here $\sigma(P_{02})$ consists of

the entire real axis. It does not satisfy the hypotheses of Theorem 4.2. For

$$\partial P/\partial \xi_1 = 2\xi_1 ,$$

and there are real vectors ξ such that $|\xi_1| \to \infty$ and $P(\xi)=0$.

5. The heat operator $P(\xi)=\mathrm{i}\xi_1+(\xi_2^2+\ldots+\xi_n^2)$. This operator is hypoelliptic with $a=\frac{1}{2}$, $b=1$. To see this note that

$$\partial P/\partial \xi_1 = \mathrm{i}, \quad \partial P/\partial \xi_j = 2\xi_j, \qquad j \neq 1 .$$

Now

$$|P(\xi)|^2 = \xi_1^2+(\xi_2^2+\ldots+\xi_n^2)^2 .$$

Thus

$$1/P(\xi) = \mathrm{O}(|\xi|^{-1}) \text{ as } |\xi| \to \infty .$$

Take $j \neq 1$ and set $\omega=(\partial P/\partial \xi_j)/P$. If $\xi_1^2 \leqslant (\xi_2^2+\ldots+\xi_n^2)^2$, then

$$|\omega|^{-2} \geqslant \xi_2^2+\ldots+\xi_n^2 \geqslant |\xi_1| .$$

Thus

$$2|\omega|^{-2} \geqslant |\xi_1|+\xi_2^2+\ldots+\xi_n^2 \geqslant c_2|\xi|$$

for $|\xi|$ large, where $c_2>0$. If $\xi_1^2 > (\xi_2^2+\ldots+\xi_n^2)^2$, then

$$|\omega|^{-2} \geqslant \xi_1^2/\xi_j^2 \geqslant |\xi_1| .$$

Thus

$$2|\omega|^{-2} \geqslant |\xi_1|+\xi_2^2+\ldots+\xi_n^2 \geqslant c_2|\xi| .$$

Hence

$$\omega = \mathrm{O}(|\xi|^{-\frac{1}{2}}) \text{ as } |\xi| \to \infty .$$

In this case $P(\xi)$ satisfies the hypotheses of Theorem 4.2 for $n|\frac{1}{2}-1/p|<2$. When $n \leqslant 4$, this is true for all $1<p<\infty$. $\sigma(P_{02})$ consists of the half-plane $\mathrm{Re}\,\lambda \geqslant 0$.

6. Semi-elliptic operators. Let $e=(e_1, \ldots, e_n)$ be a multi-index of positive integers. For any other multi-index μ of non-negative integers we set

$$(\mu/e) = \sum_1^n \mu_k/e_k .$$

Consider a polynomial of the form

$$P(\xi) = \sum_{(\mu/e) \leqslant 1} a_\mu \xi^\mu , \tag{5.3}$$

where the a_μ are constants. Note that $a_\mu = 0$ if $(\mu/e) > 1$. The operator $P(\mathrm{D})$ corresponding to $P(\xi)$ is called *semi-elliptic* if

$$\hat{P}(\xi) = \sum_{(\mu/e)=1} a_\mu \xi^\mu \neq 0\,, \quad \xi \text{ real}\,. \tag{5.4}$$

An example of such an operator is

$$P(\xi) = \sum_1^n \xi_k^{2m_k}\,,$$

where the m_k are positive integers. Set $m = \min e_k$, $M = \max e_k$ and

$$h(\xi) = \sum_1^n |\xi_k|^{e_k}\,.$$

Then we have

$$|\xi^\mu| \leqslant h(\xi)^{(\mu/e)} \tag{5.5}$$

$$h(\xi) \leqslant C|\hat{P}(\xi)| \tag{5.6}$$

$$h(\xi) \leqslant C(|P(\xi)|+1) \tag{5.7}$$

$$|P^{(\mu)}(\xi)| \leqslant Ch(\xi)^{1-(\mu/e)}\,, \qquad h(\xi) \geqslant 1\,. \tag{5.8}$$

Since $|\mu|/M \leqslant (\mu/e) \leqslant |\mu|/m$, we have

$$|\xi|^m \leqslant h(\xi) \tag{5.9}$$

by (5.5), and

$$P^{(\mu)}(\xi)/P(\xi) = \mathrm{O}(|\xi|^{-m|\mu|/M}) \ \text{ as } |\xi| \to \infty \tag{5.10}$$

by (5.7) and (5.8). Thus we may take $a = m/M$ and $b = m$. Condition (4.8) becomes

$$\frac{mM}{M-m} > n\left|\tfrac{1}{2} - 1/p\right|.$$

To prove (5.5) note that

$$|\xi_k|^{e_k} \leqslant h(\xi)\,,$$

so that

$$|\xi_k^{\mu_k}| \leqslant h(\xi)^{\mu_k/e_k}\,.$$

Taking the product of the $\xi_k^{\mu_k}$ we obtain (5.5). To prove (5.6) let Ω be the set of those real ξ such that $h(\xi)=1$. Then there is a constant C such that

$$|\hat{P}(\xi)| \geqslant 1/C\,, \qquad \xi \in \Omega\,,$$

since $\hat{P}(\xi)$ does not vanish on Ω. If $\xi \neq 0$ is not in Ω, set $t=1/h(\xi)$ and $\eta=(\xi_1 t^{1/e_1}, \ldots, \xi_n t^{1/e_n})$. Then $h(\eta)=th(\xi)=1$. Hence

$$|\hat{P}(\eta)| \geqslant 1/C\,.$$

But $\hat{P}(\eta)=t\hat{P}(\xi)$. This gives (5.6). To obtain (5.7) note that

$$|P(\xi)-\hat{P}(\xi)| \leqslant C \sum_{(\mu/e)<1} |\xi^\mu| \leqslant Ch(\xi)^{1-\delta}$$

for some $\delta>0$. Hence

$$\begin{aligned} |P(\xi)| &\geqslant |\hat{P}(\xi)| - |P(\xi)-\hat{P}(\xi)| \\ &\geqslant h^{1-\delta}(h^\delta - C')/C. \end{aligned}$$

This means that for $h(\xi)^\delta > 2C'$ one has

$$|P(\xi)| \geqslant h(\xi)/2C\,.$$

Since $|h(\xi)| \leqslant C''$ for $h(\xi)^\delta \leqslant 2C'$, we have (5.7). Finally, to prove (5.8) note that $P^{(\mu)}(\xi)$ is the sum of terms of the form $a_\nu \xi^{\nu-\mu}$ with $(\nu/e) \leqslant 1$. Apply (5.5) to obtain (5.8).

§6. Perturbation by a potential

Let $q(x)$ be a measurable function defined on E^n, and let V be the set of those functions $v \in L^p$ such that $qv \in L^p$. We can consider multiplication by q as an operator on L^p with domain V. We let q denote this operator as well, and set $D(q)=V$. We note that

Lemma 6.1. *q is a closed operator.*

Proof. Let $\{v_k\}$ be a sequence of functions in V such that $v_k \to v$ and $qv_k \to w$ in L^p. Then there is a subsequence (also denoted by $\{v_k\}$) such that $v_k \to v$ and $qv_k \to w$ almost everywhere (Theorem 1.11 of ch. 2). Thus $qv_k \to qv$ a.e., so that we must have $qv = w \in L^p$. This shows that $v \in V$.

For any polynomial $P(\xi)$ the operator $P(\mathrm{D})+q$ is defined on $C_0^\infty \cap D(q)$. We now give a condition on q which will guarantee that this set is dense in L^p. We shall call q almost locally in L^p if for each $\varepsilon>0$ and each bounded set Ω in E^n there is a closed set $F \subset \Omega$ such that the measure mF of F is $<\varepsilon$

and $q \in L^p(\Omega \setminus F)$. In particular this is true if the singularities of q have only a finite number of limit points in any bounded region.

Lemma 6.2. *If $1 \leqslant p < \infty$ and q is almost locally in L^p, then $C_0^\infty \cap D(q)$ is dense in L^p.*

Proof. Let u be a function in L^p and let $\varepsilon > 0$ be given. Then there is a function $\varphi \in C_0^\infty$ such that $\|u - \varphi\| < \frac{1}{2}\varepsilon$ (Theorem 1.7 of ch. 2). Let M denote the maximum of $|\varphi(x)|$. By hypothesis there is a closed set F such that $mF < (\varepsilon/4M)^p$ and q is in $L^p(W)$, where $W = (\text{supp } \varphi) \setminus F$. Let $\psi(x)$ be a function in C_0^∞ such that $0 \leqslant \psi \leqslant 1$, $\psi(x) = 1$ for $x \in F$ and $m(\text{supp } \psi) < (\varepsilon/2M)^p$ (see Theorem 1.2 of ch. 2). Now the function $\varphi(1-\psi)$ is in C_0^∞ and vanishes on F. Thus the function $q\varphi(1-\psi)$ is in L^p. This shows that the function $\varphi(1-\psi)$ is in $C_0^\infty \cap D(q)$. But

$$\|u - \varphi(1-\psi)\| \leqslant \|u - \varphi\| + \|\varphi\psi\| < \tfrac{1}{2}\varepsilon + \left(\int_{\text{supp } \psi} |\varphi|^p \mathrm{d}x\right)^{1/p}$$

$$< \tfrac{1}{2}\varepsilon + M\,[m(\text{supp } \psi)]^{1/p} < \varepsilon\,.$$

This completes the proof.

As a consequence of Lemma 6.2 we have

Lemma 6.3. *The operator $P(\mathrm{D}) + q$* on *$C_0^\infty \cap D(q)$ is closable in L^p for $1 < p \leqslant \infty$ if q is almost locally in $L^{p'}$.*

Proof. Suppose $\{\varphi_k\}$ is a sequence of functions in $C_0^\infty \cap D(q)$ such that $\varphi_k \to 0$ and $[P(\mathrm{D}) + q]\varphi_k \to f$ in L^p. Then

$$([P(\mathrm{D}) + q]\varphi_k, v) = (\varphi_k, [\bar{P}(\mathrm{D}) + \bar{q}]v), \qquad k = 1, 2, \dots,$$

for any $v \in C_0^\infty \cap D(q^*)$ (note that $D(q^*) = D(\bar{q})$ in $L^{p'}$). In the limit we have

$$(f, v) = 0\,, \qquad v \in C_0^\infty \cap D(q^*)\,. \tag{6.2}$$

But $C_0^\infty \cap D(q^*)$ is dense in $L^{p'}$ (Lemma 6.2).

Hence

$$(f, w) = 0\,, \qquad w \in L^{p'}\,. \tag{6.3}$$

This implies that $f = 0$ (Theorem 1.13 of ch. 2). The proof is complete.

We shall be concerned with the spectra of extensions of $P(\mathrm{D}) + q$ under various assumptions on $P(\mathrm{D})$ and q. Note that if q is locally in L^p, then $C_0^\infty \subset D(q)$.

Theorem 6.4. *Assume that* $1 \leqslant p < \infty$ *and that there is a sequence* $\{T_k\}$ *of spheres with radii tending to infinity such that* $q \in L^p(T_k)$ *with*

$$\int_{T_k} |q(x)|^p \, dx / |T_k| \to 0 \text{ as } k \to \infty, \tag{6.4}$$

where $|T_k|$ *is the volume of* T_k. *Then any extension* L *of* $P(D)+q$ *on* $C_0^\infty \cap D(q)$ *in* L^p *satisfies*

$$\sigma_e(L) \supset \{P(\xi), \ \xi \text{ real}\}. \tag{6.5}$$

Proof. Let a_k be the center of T_k and γ_k its radius. Let $\psi(x)$ be a function in C_0^∞ which vanishes for $|x| > 1$ and such that

$$\|\psi\| = 1. \tag{6.6}$$

Let ξ be a real vector and set $\lambda = P(\xi)$. We define

$$\varphi_k(x) = e^{i\xi x} \psi[(x - a_k)/\gamma_k] / \gamma_k^{n/p}, \qquad k = 1, 2, \ldots. \tag{6.7}$$

Then

$$\varphi_k \in C_0^\infty$$

and

$$P(D)\varphi_k = e^{i\xi x} \Sigma P^{(\mu)}(\xi) \psi_\mu [(x - a_k)/\gamma_k] / \mu! \, \gamma_k^{|\mu| + (n/p)},$$

where $\psi_\mu(x) = D^\mu \psi(x)$. Thus

$$\|[P(D) - \lambda] \varphi_k\| \leqslant \sum_{\mu \neq 0} |P^{(\mu)}(\xi)| \, \|\psi_\mu\| / \mu! \, \gamma_k^{|\mu|} \to 0 \text{ as } k \to \infty.$$

Moreover,

$$\int |q(x) \varphi_k(x)|^p \, dx \leqslant M^p \int_{|x - a_k| < \gamma_k} |q(x)|^p \, dx / \gamma_k^n \to 0 \text{ as } k \to \infty,$$

where $M = \max |\psi|$. Thus

$$(L - \lambda) \varphi_k \to 0 \text{ as } k \to \infty. \tag{6.8}$$

But by (6.6) we have

$$\|\varphi_k\| = 1, \qquad k = 1, 2, \ldots.$$

This shows that $\lambda \in \sigma(L)$. To prove (6.5) it suffices to show that $\{\varphi_k\}$ does not have a convergent subsequence (Theorem 4.4 of ch. 1). If $\{\varphi_k\}$ had a subsequence converging to a function $u \in L^p$, then we would have

$$\|u\| = 1 \tag{6.10}$$

by (6.9). But then there would be a subsequence of this subsequence which converged to u a.e. (Theorem 1.11 of ch. 2). But $\varphi_k(x) \to 0$ everwhere. Hence $u = 0$ a.e., contradicting (6.10). Hence $\lambda \in \sigma_e(L)$, and the proof is complete.

Remark 6.5. For $\delta > 0$ define the function $q_\delta^*(x)$ by

$$\begin{aligned} q_\delta^*(x) &= |q(x)| \quad \text{for} \quad |q(x)| > \delta\,, \\ &= 0 \qquad\quad \text{for} \quad |q(x)| \leqslant \delta\,. \end{aligned} \tag{6.11}$$

Suppose that for each $\delta > 0$

$$\int_{|x-y|<1} q_\delta^*(x)^p \, dx \to 0 \text{ as } |y| \to \infty\,. \tag{6.12}$$

Then the hypothesis of Theorem 6.4 is satisfied. For let k be any positive integer and take $\delta = 1/k$. By (6.12) there is a point y_k such that

$$\int_{|x-y_k|<k} q_\delta^*(x)^p \, dx < 1\,.$$

Let T_k be the sphere of radius k and center y_k. Then

$$\int_{T_k} |q(x)|^p dx < 1 + (|T_k|/k^p)\,.$$

This shows that (6.4) holds.

Remark 6.6. In particular, we see that the hypotheses of Theorem 6.4 hold if

$$\int_{|x-y|<1} |q(x)|^p dx \to 0 \text{ as } |y| \to \infty\,. \tag{6.13}$$

We can extend Theorem 6.4 to a larger class of perturbations.

Theorem 6.7. *Let $P(\xi), Q_1(\xi), \ldots, Q_N(\xi)$ be polynomials and let $q_1(x), \ldots, q_N(x)$ be measurable functions. Assume that there is a sequence $\{T_k\}$ of spheres with radii $\gamma_k \to \infty$ and non-negative integers $h_1, \ldots, h_N$ such that $q_j \in L^p(T_k)$ and*

$$\int_{T_k} |q_j(x)|^p dx / \gamma_k^{n+ph_j} \to 0 \text{ as } k \to \infty, \qquad 1 \leqslant j \leqslant N\,. \tag{6.14}$$

Let L be any extension of $P(D) + \Sigma\, q_j Q_j(D)$ on C_0^∞. Then $\sigma_e(L)$ contains $P(\xi)$ for each real vector ξ satisfying

$$Q_j^{(\mu)}(\xi) = 0\,, \qquad |\mu| < h_j\,,\ 1 \leqslant j \leqslant N\,. \tag{6.15}$$

In particular, if the h_j all vanish, then (6.5) holds.

Proof. Let $\{\varphi_k\}$ be the sequence defined in the proof of Theorem 6.4. Then

$$\int |q_j(x)\, Q_j(\mathrm{D})\, \varphi_k(x)|^p \,\mathrm{d}x \leqslant C \sum_{\mu} |Q_j^{(\mu)}(\xi)|^p M_\mu^p \int_{T_k} |q_j(x)|^p \,\mathrm{d}x / \gamma_k^{n+p|\mu|} , \tag{6.16}$$

where $M_\mu = \max |\psi_\mu|$. If $|\mu| \geqslant h_j$, then the corresponding term on the right-hand side of (6.16) tends to zero as $k \to \infty$. Thus if $Q_j^{(\mu)}(\xi) = 0$ for $|\mu| < h_j$, the right-hand side of (6.16) tends to zero. Hence

$$[P(\mathrm{D}) - P(\xi) + \Sigma\, q_j Q_j(\mathrm{D})]\, \varphi_k \to 0 \text{ as } k \to \infty ,$$

and the proof is complete.

In a later chapter we shall need the following generalization of Lemma 6.2.

Lemma 6.8. *Let $Q_1(\mathrm{D}), \ldots, Q_N(\mathrm{D})$ be constant coefficient operators and $q_1(x), \ldots, q_N(x)$ functions almost locally in $L^{p'}$. Let V be the set of these functions $\varphi \in C_0^\infty$ such that $q_j Q_j(\mathrm{D}) \varphi \in L^p$ for $1 \leqslant j \leqslant N$. Then V is dense in L^p.*

Proof. We follow the proof of Lemma 6.2. Let $u \in L^p$ and $\varepsilon > 0$ be given, and choose φ as before. For each j there is a closed set U_j with $m(U_j) < (\varepsilon/4M)^p/N$ such that q_j is in L^p on $(\operatorname{supp} \varphi) \backslash U_j$. If we set

$$U = \bigcup_1^N U_j ,$$

then $m(U) < (\varepsilon/4M)^p$ and each q_j is in L^p on $W = (\operatorname{supp} \varphi) \backslash U$. If ψ is defined as before, then $Q_j(\mathrm{D})[\varphi(1-\psi)]$ is in C_0^∞ and vanishes on U for each j. Hence $q_j Q_j(\mathrm{D})\,[\varphi(1-\psi)]$ is in L^p. The remainder of the proof of Lemma 6.2 carries over.

CHAPTER 5

RELATIVE COMPACTNESS

§1. Orientation

Let A be a closed operator on a Banach space X. If B is an A-compact operator on X, then

$$\sigma_e(A+B)=\sigma_e(A) \tag{1.1}$$

(Theorem 4.6 of ch. 1). Thus if $P(\mathrm{D})$ is a constant coefficient partial differential operator and $q(x)$ is a function which is P_0-compact, it follows that

$$\sigma_e(P_0+q)=\sigma_e(P_0)=\sigma(P_0)\,. \tag{1.2}$$

In this chapter and the one that follows we shall consider conditions on $P(\xi)$ and $q(x)$ which will insure that q is P_0-compact, and consequently that (1.2) holds.

In the present chapter we consider operators $P(\mathrm{D})$ which satisfy

$$|P(\xi)|\to\infty \quad \text{as} \quad |\xi|\to\infty\,. \tag{1.3}$$

For each operator satisfying (1.3) there are constants $a\leqslant 1$ and $b>0$ such that

$$P^{(\mu)}(\xi)/P(\xi)=\mathrm{O}(1/|\xi|^{a|\mu|}) \quad \text{as} \quad |\xi|\to\infty, \text{ all } \mu\,, \tag{1.4}$$

and

$$1/P(\xi)=\mathrm{O}(1/|\xi|^{b}) \quad \text{as} \quad |\xi|\to\infty\,. \tag{1.5}$$

(Note that we can always take $a=b+1-m$, where m is the order of $P(\mathrm{D})$.) Our assumptions on $P(\mathrm{D})$ are in terms of a and b.

The conditions on $q(x)$ are expressed in terms of the expression

$$M_{\alpha,p}(q)=\sup_y\int_{|x|<1}|q(x-y)|^p\,|x|^{\alpha-m}\,\mathrm{d}x\,. \tag{1.6}$$

In §2 we give sufficient conditions that $D(P_0)\subset D(q)$. This, of course, is necessary in order for q to be P_0-compact. Then in §3 we show that if

$$\int_{|y|<1} |q(x-y)|^p \mathrm{d}y \to 0 \text{ as } |x| \to \infty \tag{1.7}$$

in addition to the hypothesis of §2, then q is actually P_0-compact. This implies (1.2).

In §4 we discuss the question when $D(P_0) \subset D(Q_0)$ for two operators $P(\mathrm{D})$, $Q(\mathrm{D})$. When $p \neq 2$ we again assume that $P(\xi)$ satisfies (1.3). Assuming $D(P_0) \subset D(Q_0)$ we discuss in §5 conditions on $q(x)$ which insure that $D(P_0) \subset D(qQ_0)$. We then consider operators of the form

$$L(x, \mathrm{D}) = P(\mathrm{D}) + \sum_{j=1}^{r} q_j(x) Q_j(\mathrm{D}) .$$

Conditions of the type mentioned are given on the q_j and Q_j which imply that the closure in L^p of $L(x, \mathrm{D})$ on C_0^∞ has the same essential spectrum as P_0. In §6 we discuss the adjoint of $P_0 + q$ and in §7 we give a variation of our hypotheses on the q_j and Q_j.

§2. P_0-boundedness

In order that q be P_0-compact, one must have $D(P_0) \subset D(q)$ and

$$\|qu\| \leqslant C(\|P_0 u\| + \|u\|), \qquad u \in D(P_0) \tag{2.1}$$

(Theorem 2.12(c) of ch. 1). In particular, we must have $C_0^\infty \subset D(q)$ and

$$\|q\varphi\| \leqslant C(\|P(\mathrm{D})\varphi\| + \|\varphi\|), \qquad \varphi \in C_0^\infty . \tag{2.2}$$

A sufficient condition for (2.2) to hold is given by

Theorem 2.1. *Suppose $P(\xi)$ satisfies*

$$P^{(\mu)}(\xi)/P(\xi) = \mathrm{O}(|\xi|^{-a|\mu|}) \text{ as } |\xi| \to \infty, \quad |\mu| \leqslant n+1 , \tag{2.3}$$

$$1/P(\xi) = \mathrm{O}(|\xi|^{-b}) \text{ as } |\xi| \to \infty \tag{2.4}$$

for real ξ, with $a \leqslant 1$ and $b > a + n - an$. Let k_0 denote the smallest non-negative integer such that $k_0 a > n - b$. Assume that $1 \leqslant p < \infty$ and that $q(x)$ is a function locally in L^p such that $M_{\alpha,p}(q) < \infty$ for some α satisfying

$$0 < \alpha < p(n - k_0) , \tag{2.5}$$

where

$$M_{\alpha,p}(q) = \sup_y \int_{|x|<1} |q(x-y)|^p |x|^{\alpha - n} \mathrm{d}x . \tag{2.6}$$

Assume also that $\rho(P_0)$ is not empty. Then $D(P_0) \subset D(q)$ and (2.1) holds.

In proving Theorem 2.1 we shall make use of

Theorem 2.2. *Let γ be a number satisfying $0 \leqslant \gamma < n$, and let $G(x)$ be a function such that*

$$|G(x)| \leqslant K_1/|x|^{\gamma}, \qquad |x| \leqslant 1 \tag{2.7}$$
$$|G(x)| \leqslant K_2/|x|^{n+1}, \qquad |x| \geqslant 1. \tag{2.8}$$

Suppose $1 \leqslant p < \infty$, $0 < \alpha < p(n-\gamma)$ and that q is a function locally in L^p such that $M_{\alpha,p}(q) < \infty$. Let T be the operator defined by

$$Tf = q[G*f], \qquad f \in S. \tag{2.9}$$

Then T can be extended to a bounded operator on L^p with

$$\|Tf\| \leqslant C(K_1+K_2)[M_{\alpha,p}(q)]^{1/p}\|f\|, \qquad f \in L^p, \tag{2.10}$$

where the constant C depends only on n, γ, α and p.

Lemma 2.3. *Suppose $w \in H^{s,p}$ for some s, p and that $\mathrm{D}^{\mu} w \in L^1$ for $|\mu| = k_0$ and $|\mu| = n+1$, where $k_0 < n$. If*

$$\|\mathrm{D}^{\mu} w\|_1 \leqslant K_3, \qquad |\mu| = k_0 \tag{2.11}$$
$$\|\mathrm{D}^{\mu} w\|_1 \leqslant K_4, \qquad |\mu| = n+1, \tag{2.12}$$

then there is a constant C such that

$$|\bar{F}(w)| \leqslant CK_3/|x|^{k_0} \tag{2.13}$$
$$|\bar{F}(w)| \leqslant CK_4/|x|^{n+1}. \tag{2.14}$$

Before we prove Theorem 2.2 and Lemma 2.3, let us show how they can be used to give the

Proof of Theorem 2.1. It suffices to prove (2.2). For if $u \in D(P_0)$, then there is a sequence $\{\varphi_k\}$ of functions in C_0^{∞} such that $\varphi_k \to u$ and $P(\mathrm{D})\varphi_k \to P_0 u$ in L^p. By (2.2), $\{q\varphi_k\}$ is a Cauchy sequence in L^p and thus converges to a function $v \in L^p$. Since q is a closed operator (Lemma 6.1 of ch. 4) we have $u \in D(q)$ and $qu = v$. Applying (2.2) to each φ_k and taking the limit as $k \to \infty$, we obtain (2.1).

It therefore remains only to prove (2.2). Without loss of generality we assume $0 \in \rho(P_0)$. Let v be any function in S and set $f = P(\mathrm{D})v$. Then by Theorem 1.2 of ch. 3

$$Fv = Ff/P. \tag{2.15}$$

Now by (2.4), $1/P$ is in L^r for $rb > n$. Thus by Theorem 4.10 of ch. 2

$$v = \bar{F}(1/P) * f. \tag{2.16}$$

Now $w = 1/P$ is infinitely differentiable, bounded and satisfies

$$|D^\mu w| \leqslant C/|\xi|^{a|\mu|+b}, \qquad |\mu| \leqslant n+1 \tag{2.17}$$

(see (4.14) of ch. 4). Thus $D^\mu w \in L^1$ whenever $b + a|\mu| > n$. In particular this is true for $|\mu| = k_0$ and $|\mu| = n+1$. (Note that $n-b < a(n-1)$ so that $k_0 < n$.) We now set $G(x) = \bar{F}w$ and apply Lemma 2.3. This shows that G satisfies inequalities of the form (2.7) and (2.8) with $\gamma = k_0$. Since

$$qv = q[G * f] = Tf,$$

where T satisfies the hypotheses of Theorem 2.2, we see that (2.2) holds and the proof is complete.

We now turn to the proofs of Theorem 2.2 and Lemma 2.3. In proving the former we shall use the following lemmas.

Lemma 2.4. *For $\alpha \leqslant \beta$ and $1 \leqslant p < \infty$*

$$M_{\beta,p}(q) \leqslant M_{\alpha,p}(q). \tag{2.18}$$

Proof. Obvious.

Lemma 2.5. *For any α there is a constant C depending only on α and n such that*

$$M_{n,p}(q) \leqslant C M_{\alpha,p}(q). \tag{2.19}$$

Proof. For $\alpha \leqslant n$, (2.19) follows from (2.18). Thus we may assume $\alpha > n$. Let y be any point of E^n and let z be any point such that $|z-y| = \frac{3}{4}$. Then

$$\int_{|x|<\frac{1}{4}} |q(x-y)|^p \mathrm{d}x \leqslant \int_{\frac{1}{2}<|x|<1} |q(x-z)|^p \mathrm{d}x$$

$$\leqslant 2^{\alpha-n} \int_{\frac{1}{2}<|x|<1} |q(x-z)|^p |x|^{\alpha-n} \mathrm{d}x \leqslant 2^{\alpha-n} M_{\alpha,p}(q).$$

Since this is true for any y and there is a number N depending only on n such that every sphere of radius one can be covered by N spheres of radius $\frac{1}{4}$, we have

$$M_{n,p}(q) \leqslant 2^{\alpha-n} N M_{\alpha,p}(q) . \tag{2.20}$$

This gives (2.19), and the proof is complete.

Lemma 2.6. *If* $G(x)$ *satisfies* (2.8), *then*

$$\int_{|x|>1} |q(x-y)|^p |G(x)| \, \mathrm{d}x \leqslant C K_2 M_{n,p}(q) , \tag{2.21}$$

where the constant C *depends only on* n.

Proof. By (2.8)

$$\int_{|x|>1} |q(x-y)|^p |G(x)| \, \mathrm{d}x \leqslant K_2 \sum_{k=1}^{\infty} \int_{k<|x|<k+1} |q(x-y)|^p \mathrm{d}x / k^{n+1} .$$

Now there is a constant C depending only on n such that for each k the shell $k<|z|<k+1$ can be covered by Ck^{n-1} spheres of radius one. Thus

$$\int_{k<|x|<k+1} |q(x-y)|^p \mathrm{d}x \leqslant C k^{n-1} M_{n,p}(q) .$$

This gives

$$\int_{|x|>1} |q(x-y)|^p |G(x)| \, \mathrm{d}x \leqslant C K_2 M_{n,p}(q) \sum_{k=1}^{\infty} k^{-2} , \tag{2.22}$$

which is the desired inequality.

We are now ready for the

Proof of Theorem 2.2. Assume first that $1<p<\infty$. Then for f, $v \in S$ we have

$$\begin{aligned}
(Tf, v) &= \iint q(x) G(x-y) f(y) \overline{v(x)} \mathrm{d}x \, \mathrm{d}y \\
&= \iint q(y+z) \, G(z) f(y) \overline{v(y+z)} \mathrm{d}y \mathrm{d}z . \\
&= \left(\iint_{|z|<1} + \iint_{|z|>1} \right) q(y+z) G(z) f(y) \overline{v(y+z)} \mathrm{d}y \mathrm{d}z .
\end{aligned}$$

Thus by Hölder's inequality (see (1.8) of ch. 2)

$$\begin{aligned}
|(Tf, v)| \leqslant & \left(\iint_{|z|<1} |q(y+z)|^p |G(z)|^{\beta p} |f(y)|^p \mathrm{d}y \mathrm{d}z \right)^{1/p} \\
& \times \left(\iint_{|z|<1} |G(z)|^{(1-\beta)p'} |v(y+z)|^{p'} \mathrm{d}y \mathrm{d}z \right)^{1/p'} \quad +
\end{aligned}$$

$$+\left(\iint_{|z|>1} |q(y+z)|^p\,|G(z)|\,|f(y)|^p\,\mathrm{d}y\,\mathrm{d}z\right)^{1/p}$$

$$\times\left(\iint_{|z|>1} |G(z)|\,|v(y+z)|^{p'}\,\mathrm{d}y\,\mathrm{d}z\right)^{1/p'} \tag{2.23}$$

for any β satisfying $0 \leqslant \beta \leqslant 1$. The idea of the proof is to find a β such that

$$\int_{|z|<1} |q(y+z)|^p\,|G(z)|^{\beta p}\,\mathrm{d}z \leqslant K_1^{\beta p} M_{\alpha,p}(q) \tag{2.24}$$

$$\int_{|z|<1} |G(z)|^{(1-\beta)p'}\,\mathrm{d}z \leqslant C K_1^{(1-\beta)p'}. \tag{2.25}$$

Suppose such a β can be found. Then by Lemmas 2.5 and 2.6

$$\int_{|z|>1} |q(y+z)|^p\,|G(z)|\,\mathrm{d}z \leqslant C K_2 M_{\alpha,p}(q), \tag{2.26}$$

while by (2.8)

$$\int_{|z|>1} |G(z)|\,\mathrm{d}z \leqslant C K_2. \tag{2.27}$$

If we apply inequalities (2.24)–(2.27) to (2.23) we obtain

$$|(Tf, v)| \leqslant C(K_1+K_2)\,[M_{\alpha,p}(q)]^{1/p}\,\|f\|_p\,\|v\|_{p'}.$$

Since S is dense in $L^{p'}$ (Theorem 1.7 of ch. 2), the theorem now follows from Theorem 1.13 of ch. 2.

To prove (2.24) and (2.25) note that we may assume

$$n-\gamma p \leqslant \alpha \leqslant n \tag{2.28}$$

in addition to

$$0<\alpha<p(n-\gamma), \tag{2.29}$$

which was assumed. This follows from Lemmas 2.4 and 2.5. For if $\alpha > n$, then $M_{n,p}(q) < \infty$ by Lemma 2.5, and we may substitute n for α. Moreover, once (2.10) is proved for a particular value of α, it will hold for any smaller value by Lemma 2.4. Note that since $p>1$ there are always values of α satisfying both (2.28) and (2.29).

Take

$$\beta = (n-\alpha)/\gamma p.$$

Then $0 \leqslant \beta \leqslant 1$ by (2.28). Moreover by (2.29)

$$1-(n/\gamma p') < \beta < n/\gamma p .$$

In particular, we see that $(1-\beta)\gamma p' < n$. This implies (2.25) in view of (2.7). This choice of β also gives $\alpha = n - \beta\gamma p$, which implies (2.24) in view of (2.7). Thus the proof for $1 < p < \infty$ is complete.

It remains only to consider the case $p=1$. Now inequality (2.29) becomes

$$0 < \alpha < n-\gamma .$$

Moreover,

$$\|Tf\|_1 \leqslant \iint |q(y+z)|\,|G(z)|\,|f(y)|\,\mathrm{d}y\mathrm{d}z$$

$$= \left(\iint_{|z|<1} + \iint_{|z|>1}\right)|q(y+z)|\,|G(z)|\,|f(y)|\,\mathrm{d}y\mathrm{d}z .$$

Thus by (2.7) and Lemma 2.6

$$\|Tf\|_1 \leqslant [K_1 M_{n-\gamma,1}(q) + CK_2 M_{n,1}(q)]\,\|f\|_1 , \tag{2.30}$$

which implies (2.10) for $p=1$. This completes the proof of Theorem 2.2.

Proof of Lemma 2.3. By Theorems 2.3 and 4.9 of ch. 2 we have

$$|x^\mu \bar{F} w| = |\bar{F}[\mathrm{D}^\mu w]| \leqslant \|\mathrm{D}^\mu w\|_1 .$$

This gives (2.13) and (2.14) for $|\mu| = k_0$ and $|\mu| = n+1$, respectively.

The following consequences of Theorem 2.1 are useful.

Corollary 2.7. *Suppose that $P(\xi)$ is a polynomial of degree m satisfying* (2.4) *with $b > (mn+1-m)/n$. Assume that $1 \leqslant p < \infty$ and that $M_{\alpha,p}(q) < \infty$ for some $\alpha < p(n-k_0)$, where k_0 is the smallest non-negative integer satisfying $(b+1-m)\,k_0 > n-b$. If $\rho(P_0)$ is not empty, then $D(P_0) \subset D(q)$ and* (2.1) *holds.*

Proof. Since $P^{(\mu)}(\xi)$ is of degree at most $m-|\mu|$, we see that (2.3) is satisfied with $a = b+1-m$. Our assumptions imply that $b > a+n-an$. The result follows from Theorem 2.1.

Corollary 2.8. *Suppose that $P(\xi)$ is a polynomial of degree $m > n$ satisfying* (2.4) *with $b > (mn+1-m)/n$. Assume that $1 \leqslant p < \infty$ and that $M_{\alpha,p}(q) < \infty$ for some $\alpha < pn$. If $\rho(P_0)$ is not empty, then $D(P_0) \subset D(q)$ and* (2.1) *holds.*

Proof. If $m > n$, then the assumptions on b imply that $b > n$. Thus $k_0 = 0$. Everything now follows from Corollary 2.7.

§3. P_0-compactness

Now that we have given conditions on q in order that it be P_0-bounded, we can now give further conditions that it be P_0-compact.

Theorem 3.1. *Assume that $1 \leqslant p < \infty$ and that $P(\xi)$ and q satisfy the hypotheses of Theorem 2.1. If, in addition,*

$$\int_{|x-y|<1} |q(x)|^p \, dx \to 0 \ \text{as} \ |y| \to \infty \,, \tag{3.1}$$

then q is P_0-compact. Thus

$$\sigma_e(P_0+q) = \sigma_e(P_0) = \sigma(P_0) = \{P(\xi),\ \xi \in E^n\} \,.$$

In proving Theorem 3.1 we shall use the following three lemmas and a theorem.

Lemma 3.2. *Let φ be a function in C_0^∞, and let Ω be a bounded open set in E^n. Then the operator*

$$Af = \bar{F}(\varphi) * f \tag{3.2}$$

is a compact operator from L^p to $L^\infty(\bar{\Omega})$.

Proof. For $f \in L^p$, Af is continuously differentiable in E^n, and by Theorems 2.1 and 2.5 of ch. 2

$$D_j Af = D_j \bar{F}(\varphi) * f = -\bar{F}(\xi_j \varphi) * f \,.$$

Since $\bar{F}(\varphi)$ and $\bar{F}(\xi_j \varphi)$ are in S, they are surely in $L^{p'}$. Hence by Young's inequality (see Theorem 1.9 of ch. 2)

$$\|Af\|_\infty + \sum_1^n \|D_j Af\|_\infty \leqslant C \|f\|_p \,. \tag{3.3}$$

Now let $\{f_k\}$ be a sequence of functions in L^p satisfying

$$\|f_k\| \leqslant C \,.$$

By (3.3), $\{Af_k\}$ is a uniformly bounded, equicontinuous sequence of functions on $\bar{\Omega}$. Thus it has a subsequence which converges uniformly on $\bar{\Omega}$. Since uniform convergence on $\bar{\Omega}$ is convergence in $L^\infty(\bar{\Omega})$, the proof is complete.

Lemma 3.3. *Suppose $1 \leqslant p < \infty$ and that $q(x)$ is a function locally in L^p such that $M_{\alpha,p}(q) < \infty$ and (3.1) holds. Then for any $\varepsilon > 0$ one has*

$$\int_{|x-y|<1} |q(x)|^p |x-y|^{\alpha-n+\varepsilon} dx \to 0 \quad as \ |y| \to \infty .$$

Proof. The lemma is obvious for $\alpha+\varepsilon \geqslant n$. For $\alpha+\varepsilon < n$ set $s=(n-\alpha)/\varepsilon > 1$. By Hölder's inequality (see (1.8) of ch. 2)

$$\int_{|x-y|<1} |q(x)|^p |x-y|^{\alpha-n+\varepsilon} dx \leqslant \left(\int_{|x-y|<1} |q(x)|^p dx\right)^{1/s} \times \left(\int_{|x-y|<1} |q(x)|^p |x-y|^{\alpha-n} dx\right)^{1/s'}, \tag{3.5}$$

since $(\alpha-n+\varepsilon)s' = \alpha-n$. Since $M_{\alpha,p}(q) < \infty$, (3.4) follows from (3.1).

Lemma 3.4. *Suppose $w(\xi)$ is a function having continuous derivatives up to order N in E^n and such that*

$$|D^\mu w| \leqslant C/|\xi|^{a|\mu|+b}, \qquad |\mu| \leqslant N, \tag{3.6}$$

where $a \leqslant 1$. Let $\psi(\xi)$ be a function in C_0^∞ satisfying

$$0 \leqslant \psi(\xi) \leqslant 1, \qquad \xi \in E^n \tag{3.7}$$

$$\psi(\xi) = 1, \qquad |\xi| < 1 \tag{3.8}$$

$$\psi(\xi) = 0, \qquad |\xi| > 2, \tag{3.9}$$

and for $r > 0$ set

$$\psi_r(\xi) = \psi(\xi/r) \tag{3.10}$$

(see Theorem 1.2 of ch. 2). If μ is such that $a|\mu|+b > n$ and $|\mu| \leqslant N$, then

$$\|D^\mu[(1-\psi_r)w]\|_1 \to 0 \quad as \ r \to \infty .$$

Proof. Note that

$$D^\mu \psi(\xi/r) = \psi_\mu(\xi/r)/r^{|\mu|},$$

where

$$\psi_\mu(\xi) = D^\mu \psi(\xi).$$

Hence

$$|D^\mu \psi_r(\xi)| \leqslant C/r^{|\mu|}, \text{ each } \mu. \tag{3.12}$$

Now for each μ the function

$$g_\mu(\xi) = \mathrm{D}^\mu[(1-\psi_r)w] - (1-\psi_r)\mathrm{D}^\mu w \tag{3.13}$$

consists of a sum of terms of the form

$$\text{constant } \mathrm{D}^{\mu^{(1)}}\psi_r \mathrm{D}^{\mu^{(2)}} w, \; \mu^{(1)} \neq 0, \; \mu^{(1)} + \mu^{(2)} = \mu\,. \tag{3.14}$$

Thus g_μ vanishes outside the region $r \leqslant |\xi| \leqslant 2r$ and satisfies

$$|g_\mu| \leqslant C/r^{a|\mu|+b}$$

in view of (3.6), (3.12) and the fact that $a \leqslant 1$. Thus

$$\|g_\mu\|_1 \leqslant C \int_{r<|\xi|<2r} \mathrm{d}\xi / r^{a|\mu|+b} \leqslant C'/r^{a|\mu|+b-n}\,. \tag{3.15}$$

This tends to zero as $r \to \infty$ provided $a|\mu| + b > n$. Moreover $1-\psi_r$ vanishes for $|\xi| < r$ so that

$$\|(1-\psi_r)\mathrm{D}^\mu w\|_1 \leqslant C \int_{|\xi|>r} \mathrm{d}\xi / |\xi|^{a|\mu|+b}\,, \tag{3.16}$$

which exists and tends to zero as $r \to \infty$ if $a|\mu| + b > n$. The conclusion in (3.11) follows from (3.13) and the last two statements.

Theorem 3. *Let k_0 be an integer satisfying $0 \leqslant k_0 < n$, and let w be a function in C^∞ such that*

$$|\mathrm{D}^\mu w(\xi)| \leqslant C/|\xi|^{a|\mu|+b}, \qquad |\mu| \leqslant n+1\,, \tag{3.17}$$

where $a \leqslant 1$ and $k_0 a > n - b$. Let $q(x)$ be a function satisfying $M_{\alpha,p}(q) < \infty$, where $1 \leqslant p < \infty$ and α satisfies (2.5). *Assume also that* (3.1) *holds. Define the operator T by*

$$Tf = q[\bar{F}(w) * f]\,, \qquad f \in L^p\,. \tag{3.18}$$

Then T is a compact operator on L^p.

Proof. For $R > 0$ set

$$\begin{aligned} q_R(x) &= q(x)\,, \qquad &|x| \leqslant R\,, \\ &= 0\,, \qquad &|x| > R\,. \end{aligned} \tag{3.19}$$

Let $\psi(\xi)$ be a function in C_0^∞ satisfying (3.7)–(3.9), and define $\psi_r(\xi)$ by (3.10). Set $G(x) = \bar{F}(w)$. Since $a|\mu| + b > n$ for $|\mu| \geqslant k_0$, $\bar{F}(w)$ satisfies (2.13) and (2.14) by Lemma 2.3. Thus $G(x)$ satisfies (2.7) and (2.8) with $\gamma = k_0$. Hence T is bounded on L^p by Theorem 2.2. Next note that

$$\begin{aligned} Tf &= q_R[\bar{F}(\psi_r w) * f] + q_R\{\bar{F}[(1-\psi_r)w] * f\} + (q - q_R)[\bar{F}(w) * f] \\ &= T_1 f + T_2 f + T_3 f \end{aligned}$$

for each $r>0$, $R>0$. Now the operator T_1 is compact on L^p. For by Lemma 3.2, $Af=\bar{F}(\psi_r w)*f$ defines a compact operator from L^p to $L^\infty(\bar{\Omega})$, where Ω denotes the sphere $|x|<R$. Since q is locally in L^p, q_R is a bounded operator from $L^\infty(\bar{\Omega})$ to L^p. Thus T_1 is compact on L^p for each r and R.

Next note that in view of Theorem 4.9 of ch. 2

$$|\bar{F}[(1-\psi_r)w]| \leqslant K_3/|x|^k, \qquad k=k_0, n+1,$$

where K_3 is a bound for

$$\|D^\mu[(1-\psi_r)w]\|_1, \qquad |\mu|=k_0, n+1.$$

Since $a|\mu|+b>n$ for $|\mu|\geqslant k_0$, we see by Lemma 3.4 that $K_3\to 0$ as $r\to\infty$. In view of the fact that

$$M_{\alpha,p}(q_R)\leqslant M_{\alpha,p}(q), \tag{3.20}$$

we may apply Theorem 2.2 to conclude that

$$\|T_2\|\to 0 \text{ as } r\to\infty.$$

Turning to T_3, note that by Lemma 3.3 we have

$$\int_{|x|<1} |q(x-y)|^p |x|^{\beta-n} \mathrm{d}x \to 0 \text{ as } |y|\to\infty \tag{3.22}$$

for any β satisfying $\alpha<\beta<p(n-k_0)$. This is equivalent to

$$M_{\beta,p}(q-q_R)\to 0 \text{ as } R\to\infty. \tag{3.23}$$

Moreover by Theorem 2.2

$$\|T_3\|\leqslant CK_4[M_{\beta,p}(q-q_R)]^{1/p}, \tag{3.24}$$

where K_4 is a bound for

$$\|D^\mu w\|_1, \qquad |\mu|=k_0, n+1,$$

and C depends only on n, k_0, β and p. Thus by (3.23) and (3.24) we have

$$\|T_3\|\to 0 \text{ as } R\to\infty.$$

The conclusion of the theorem now follows from Theorem 2.16 of ch. 1.

Now we can give the

Proof of Theorem 3.1. As before we may assume that $0\in\rho(P_0)$. Set $w=1/P$. Then w satisfies (3.17) (see (4.14) of ch. 4). Thus the operator defined by (3.18) is compact on L^p (Theorem 3.5). Now let $\{u_k\}$ be a sequence of functions in $D(P_0)$ such that

$$\|P_0 u_k\| \leqslant C\,, \qquad k=1, 2, \dots\,.$$

Set $f_k = P_0 u_k$. Then $qu_k = Tf_k$. Since T is a compact operator on L^p, $\{qu_k\}$ has a convergent subsequence. Thus q is P_0-compact, and the proof is complete.

We also have

Corollary 3.6. *Assume that* $1 \leqslant p < \infty$ *and that* $P(\xi)$ *and* $q(x)$ *satisfy the hypotheses of either Corollary* 2.7 *or Corollary* 2.8. *If* (3.1) *holds, then* q *is* P_0*-compact.*

§4. Comparison of operators

Let $P(\xi)$ and $Q(\xi)$ be polynomials, and let P_0 and Q_0 denote the minimal operators corresponding to $P(\mathrm{D})$ and $Q(\mathrm{D})$, respectively. In this section we shall give conditions under which one has $D(P_0) \subset D(Q_0)$.

Theorem 4.1. *A necessary and sufficient condition that* $D(P_0) \subset D(Q_0)$ *is that*

$$\|Q(\mathrm{D})\varphi\| \leqslant C(\|P(\mathrm{D})\varphi\| + \|\varphi\|)\,, \qquad \varphi \in C_0^\infty\,. \tag{4.1}$$

Proof. Suppose $D(P_0) \subset D(Q_0)$. Since P_0 and Q_0 are both closed operators, we must have

$$\|Q_0 v\| \leqslant C(\|P_0 v\| + \|v\|)\,, \qquad v \in D(P_0) \tag{4.2}$$

(Lemma 2.8 of ch. 1). This implies (4.1). Conversely, assume (4.1) holds, and let v be any element in $D(P_0)$. Then there is a sequence $\{\varphi_k\}$ of functions in C_0^∞ such that $\varphi_k \to v$, $P(\mathrm{D})\varphi_k \to P_0 v$ in L^p. By (4.1) the sequence $\{Q(\mathrm{D})\varphi_k\}$ converges to some function w in L^p. Since Q_0 is a closed operator, we see that $v \in D(Q_0)$ and $Q_0 v = w$. This completes the proof.

Theorem 4.2. *A necessary, and for* $p=2$ *also sufficient, condition that* $D(P_0) \subset D(Q_0)$ *is that*

$$|Q(\xi)| \leqslant C(|P(\xi)| + 1)\,, \qquad \xi \in E^n\,. \tag{4.3}$$

Proof. Let ξ be a real vector in E^n and let $\psi(x)$ be a function in C_0^∞ such that

$$\|\psi\| = 1\,.$$

Set

$$\varphi_k(x) = e^{i\xi x}\psi(x/k)/k^{n/p}, \qquad k=1, 2, \ldots .$$

Then $\varphi_k \in C_0^\infty$,

$$\|\varphi_k\| = 1, \qquad k=1, 2, \ldots, \tag{4.5}$$

and

$$P(D)\varphi_k(x) = e^{i\xi x}\sum_\mu P^{(\mu)}(\xi)\psi_\mu(x/k)/\mu! \qquad k^{|\mu|+(n/p)}, \tag{4.6}$$

where $\psi_\mu(x) = D^\mu\psi(x)$. Thus

$$\|P(D)\varphi_k\| \to |P(\xi)| \text{ as } k\to\infty . \tag{4.7}$$

Similarly,

$$\|Q(D)\varphi_k\| \to |Q(\xi)| \text{ as } k\to\infty . \tag{4.8}$$

According to Theorem 4.1, inequality (4.1) must hold. Hence

$$\|Q(D)\varphi_k\| \leqslant C(\|P(D)\varphi_k\| + \|\varphi_k\|), \qquad k=1, 2, \ldots .$$

Taking the limit as $k\to\infty$ and noting (4.5), (4.7) and (4.8), we get (4.3) with the same constant.

Now assume that $p=2$ and (4.3) holds. Then there is a constant C such that

$$|Q(\xi)F\varphi|^2 \leqslant C(|P(\xi)F\varphi|^2 + |F\varphi|^2), \qquad \varphi\in C_0^\infty .$$

If we now integrate with respect to ξ and apply Parseval's identity, we obtain (4.1). Hence $D(P_0)\subset D(Q_0)$ by Theorem 4.1. This completes the proof.

Theorem 4.3. *If there is a scalar λ such that $Q(\xi)/[P(\xi)-\lambda]$ is an L^p multiplier, then $D(P_0)\subset D(Q_0)$.* If $\lambda\in\rho(P_0)$, *and $D(P_0)\subset D(Q_0)$, then $Q(\xi)/[P(\xi)-\lambda]$ is an L^p multiplier.*

Proof. We may assume $\lambda=0$. Let v be a function in S and set $f=P(D)v$. Then

$$F[Q(D)v] = (Q/P)Ff. \tag{4.9}$$

If Q/P is an L^p multiplier, then (4.9) implies

$$\|Q(D)v\| \leqslant C\|P(D)v\|, \qquad v\in S . \tag{4.10}$$

This in turn implies that $D(P_0)\subset D(Q_0)$ by Theorem 4.1.

Conversely, assume that $D(P_0)\subset D(Q_0)$. Then (4.2) holds by Lemma 2.8 of ch. 1. Since $0\in\rho(P_0)$, we have

$$\|v\| \leqslant C\|P_0 v\|, \qquad v \in D(P_0), \tag{4.11}$$

which implies

$$\|v\| \leqslant C\|P(\mathrm{D})v\|, \qquad v \in S. \tag{4.12}$$

Moreover, since $S \subset D(P_0)$ (Lemma 1.2 of ch. 4), (4.2) implies

$$\|Q(\mathrm{D})v\| \leqslant C(\|P(\mathrm{D})v\| + \|v\|), \qquad v \in S. \tag{4.13}$$

Inequalities (4.12) and (4.13) imply (4.10). Since (4.9) holds, Q/P must be an L^p multiplier. This completes the proof.

We now give sufficient conditions on P and Q that Q/P be an L^p multiplier.

Theorem 4.4. *Suppose that* $1 < p < \infty$, *and that* $P(\xi)$ *and* $Q(\xi)$ *satisfy*

$$P^{(\mu)}(\xi)/P(\xi) = \mathrm{O}(|\xi|^{-a|\mu|}) \text{ as } |\xi| \to \infty, \text{ each } \mu, \tag{4.14}$$

$$Q(\xi)/P(\xi) = \mathrm{O}(|\xi|^{-c}) \text{ as } |\xi| \to \infty, \tag{4.15}$$

where $a \leqslant 1$ *and*

$$c > (1-a)n|1/p - 1/2|.$$

If $\{P(\xi), \xi \in E^n\}$ *is not the whole complex plane, then* $D(P_0) \subset D(Q_0)$.

The proof of Theorem 4.4 is based on

Lemma 4.5. *Assume that* $P(\xi)$ *and* $Q(\xi)$ *satisfy* (4.14) *and* (4.15) *for some numbers* $a \leqslant 1$ *and* c. *Then*

$$Q^{(\mu)}(\xi)/P(\xi) = \mathrm{O}(|\xi|^{-a|\mu|-c}), \text{ each } \mu. \tag{4.17}$$

Proof. Let μ be any multi-index of length $|\mu| \leqslant l$. Then there are real vectors $\theta^{(1)}, \ldots, \theta^{(r)}$ and coefficients $\gamma_1, \ldots, \gamma_r$ such that

$$t^{|\mu|} Q^{(\mu)}(\xi) = \sum_1^r \gamma_j Q(\xi + t\theta^{(j)}) \tag{4.18}$$

holds for all $\xi \in E^n$ and real t (Theorem 1.3 of ch. 3). Let M be such that

$$|\theta^{(j)}| \leqslant M, \qquad 1 \leqslant j \leqslant r, \tag{4.19}$$

and set $t = |\xi|^a/2M$. Since $a \leqslant 1$, we have for large $|\xi|$

$$|\xi + t\theta^{(j)}| \geqslant |\xi| - \tfrac{1}{2}|\xi|^a \geqslant \tfrac{1}{2}|\xi|.$$

Thus by (4.14), (4.15) and (4.18) we have

$$|\xi|^{a|\mu|}|Q^{(\mu)}(\xi)| \leqslant C\sum_j |P(\xi+t\theta^{(j)}|\,|\xi+t\theta^{(j)}|^{-c}$$
$$\leqslant C\sum_\nu |P^{(\nu)}(\xi)|\,|\xi|^{a|\nu|-c} \leqslant C|P(\xi)|\,|\xi|^{-c}$$

for $|\xi|$ large. This completes the proof.

We can now give the

Proof of Theorem 4.4. Assume that $P(\xi)\neq 0$ for $\xi\in E^n$. By Theorem 4.3 it suffices to show that Q/P is a multiplier in L^p. Now a simple induction shows that $\mathrm{D}^\mu(Q/P)$ is a sum of terms of the form

$$\text{constant } Q^{(\nu)}(\xi)\,P^{(\mu^{(1)})}(\xi)\cdots P^{(\mu^{(k)})}(\xi)/P(\xi)^{k+1}, \tag{4.20}$$

where $\nu+\mu^{(1)}+\ldots+\mu^{(k)}=\mu$. By (4.14) and Lemma 4.5

$$|\mathrm{D}^\mu(Q/P)| \leqslant C|\xi|^{-a|\mu|-c}, \text{ each } \mu. \tag{4.21}$$

We can now apply Theorem 3.4 of ch. 2 to conclude that Q/P is an L^p multiplier. This completes the proof.

For $p=1$ we have

Theorem 4.6. *Assume that* (4.14) *and* (4.15) *hold for* $a\leqslant 1$ *and* $c>(1-a)n+a$. *Then* $D(P_{0p})\subset D(Q_{0p})$ *for* $1\leqslant p<\infty$ *if* $\rho(P_{0p})$ *is not empty.*

Proof. Assume $0\in\rho(P_0)$. By (4.9)

$$Q(\mathrm{D})v=\bar F(Q/P) * P(\mathrm{D})v, \qquad v\in S. \tag{4.22}$$

If $\bar F(Q/P)$ is in L^1, then (4.22) and Young's inequality will imply (4.1), from which the conclusion follows via Theorem 4.1. Now (4.21) holds by Lemma 4.5. Thus $\mathrm{D}^\mu(Q/P)$ will be in L^1 whenever $a|\mu|+c>n$. By hypothesis this is true for $|\mu|\geqslant n-1$. We also have

$$|x|^{|\mu|}|\bar F(Q/P)| = |\bar F[\mathrm{D}^\mu(Q/P)]| \leqslant \|\mathrm{D}^\mu(Q/P)\|_1,$$

from which we see that

$$|\bar F(Q/P)| \leqslant C|x|^{-k}, \qquad k\geqslant n-1. \tag{4.23}$$

This implies that $\bar F(Q/P)$ is in L^1, and the proof is complete.

We have the following corollary to Theorem 4.4.

Theorem 4.7. *Suppose* $1<p<\infty$ *and that* $P(\xi)$, $Q(\xi)$ *are polynomials satisfying* (2.4) *and* (4.15) *with*

$$c>(m-b)n|1/p-1/2|\,, \tag{4.24}$$

where m is the degree of $P(\xi)$. *If* $\{P(\xi),\ \xi\in E^n\}$ *is not the whole complex plane, then* $D(P_0)\subset D(Q_0)$.

Proof. Since $P^{(\mu)}(\xi)$ is of degree at most $m-|\mu|$, (2.4) implies (4.14) with $a=b+1-m$. The result now follows directly from Theorem 4.4.

§5. The operator qQ_0

We now give conditions under which the operator qQ_0 will be P_0-bounded and P_0-compact.

Theorem 5.1. *If* $D(P_0)\subset D(Q_0)$ *and*

$$\|qQ(\mathrm{D})\varphi\|\leqslant C(\|P(\mathrm{D})\varphi\|+\|\varphi\|)\,,\qquad \varphi\in C_0^\infty\,, \tag{5.1}$$

then $D(P_0)\subset D(qQ_0)$.

Proof. Let v be a function in $D(P_0)$. Then there is a sequence $\{\varphi_k\}$ of functions in C_0^∞ such that $\varphi_k\to v$ and $P(\mathrm{D})\varphi_k\to P_0v$ in L^p. By Theorem 4.1, $Q(\mathrm{D})\varphi_k$ converges in L^p to some element w. Since Q_0 is a closed operator, $Q_0v=w$. By (5.1), $qQ(\mathrm{D})\varphi_k$ converges to an element h. Since q is a closed operator, we see that $w\in D(q)$ and $qw=h$. Hence $v\in D(qQ_0)$. This completes the proof.

Theorem 5.2. *Suppose* $P(\xi)$ *and* $Q(\xi)$ *satisfy* (4.14) *and* (4.15) *with* $a\leqslant 1$ *and and* $c>a+n-an$. Let k_0 *denote the smallest non-negative integer such that* $k_0a>n-c$. *Assume that* $1\leqslant p<\infty$, *and that* $q(x)$ *is a function locally in* L^p *such that* $M_{\alpha,p}(q)<\infty$ *for some* α *satisfying* (2.5). *Assume also that* $\rho(P_0)$ *is not empty. Then* $D(P_0)\subset D(qQ_0)$ *and* (4.1) *and* (5.1) *hold.*

Proof. Since the hypotheses of this theorem imply those of Theorem 4.4, we know that $D(P_0)\subset D(Q_0)$. It therefore suffices to prove (5.1). The proof of this inequality is almost identical to that of Theorem 2.1 if we replace b by c, $1/P$ by Q/P and inequality (2.3) by (4.17). We omit the details.

Theorem 5.3. *Assume that* $1\leqslant p<\infty$ *and that* $P(\xi)$, $Q(\xi)$ *and* q *satisfy the hypotheses of Theorem* 5.2. *If, in addition,* q *satisfies* (3.1), *then* qQ_0 *is* P_0-*compact.*

The proof of Theorem 5.3 is the same as that of Theorem 3.1 with $1/P$ replaced by Q/P and b replaced by c.

Next consider the variable coefficient operator of the form

$$L(x, \mathrm{D}) = P(\mathrm{D}) + \sum_{j=1}^{r} q_j(x) Q_j(\mathrm{D}) . \tag{5.2}$$

We have

Theorem 5.4. *Assume that* $P(\xi)$ *satisfies* (4.14) *with* $a \leqslant 1$ *and*

$$Q_j(\xi)/P(\xi) = \mathrm{O}(|\xi|^{-c_j}) \text{ as } |\xi| \to \infty , \qquad 1 \leqslant j \leqslant r , \tag{5.3}$$

for real ξ, *where* $c_j > a + n - an$. *Suppose that* $1 \leqslant p < \infty$ *and that*

$$M_{\alpha_j, p}(q_j) < \infty , \qquad 1 \leqslant j \leqslant r , \tag{5.4}$$

where

$$0 < \alpha_j < p(n - k_j) , \qquad 1 \leqslant j \leqslant r , \tag{5.5}$$

and k_j *is the smallest non-negative integer satisfying*

$$ak_j > n - c_j , \qquad 1 \leqslant j \leqslant r . \tag{5.6}$$

If $\rho\,(P_0)$ *is not empty and*

$$\int_{|x-y|<1} |q_j(x)|^p \,\mathrm{d}x \to 0 \text{ as } |y| \to \infty , \qquad 1 \leqslant j \leqslant r , \tag{5.7}$$

then the operator $L(x, \mathrm{D})$ *is closable in* L^p. *If* L_0 *denotes its closure, then* $D(L_0) = D(P_0)$ *and* $L_0 - P_0$ *is* P_0*-compact. Consequently*

$$\sigma_e(L_0) = \sigma_e(P_0) = \sigma(P_0) = \{P(\xi),\ \xi \in E^n\} . \tag{5.8}$$

Proof. By Theorem 5.3, $D(q_j Q_{j0}) \supset D(P_0)$ for each j, while the operator $q_j Q_{j0}$ is P_0-compact by Theorem 5.3. Thus

$$\Sigma \|q_j Q_j(\mathrm{D})\varphi\| \leqslant C(\|P(\mathrm{D})\varphi\| + \|\varphi\|) , \qquad \varphi \in C_0^\infty .$$

Moreover, by Theorem 2.13 of ch. 1, for any $\varepsilon > 0$ there is a constant K_ε such that

$$\Sigma \|q_j Q_j(\mathrm{D})\varphi\| \leqslant \varepsilon \|P(\mathrm{D})\varphi\| + K_\varepsilon \|\varphi\| , \qquad \varphi \in C_0^\infty . \tag{5.9}$$

Thus

$$\begin{aligned} \|P(\mathrm{D})\varphi\| &\leqslant \|L(x, \mathrm{D})\varphi\| + \Sigma \|q_j Q_j(\mathrm{D})\varphi\| \\ &\leqslant \|L(x, \mathrm{D})\varphi\| + \varepsilon \|P(\mathrm{D})\varphi\| + K_\varepsilon \|\varphi\| . \end{aligned} \tag{5.10}$$

Hence

$$\|P(\mathrm{D})\varphi\| \leqslant C(\|L(x, \mathrm{D})\varphi\| + \|\varphi\|), \qquad \varphi \in C_0^\infty . \tag{5.11}$$

Similarly

$$\|L(x, \mathrm{D})\varphi\| \leqslant C(\|P(\mathrm{D})\varphi\| + \|\varphi\|), \qquad \varphi \in C_0^\infty . \tag{5.12}$$

These inequalities show that $L(x, \mathrm{D})$ is closable and $D(L_0) = D(P_0)$. For if $\varphi_k \to 0$ and $L(x, \mathrm{D})\varphi_k \to f$ in L^p for a sequence $\{\varphi_k\}$ of functions in C_0^∞ then $P(\mathrm{D})\varphi_k$ converges by (5.11). Since P_0 is closed, $P(\mathrm{D})\varphi_k \to 0$. Inequality (5.12) then shows that $f = 0$. A similar argument shows that $D(L_0) = D(P_0)$.

Now on $D(P_0)$

$$L_0 - P_0 = \Sigma\, q_j(x) Q_{j0} ,$$

and each of the operators $q_j Q_{j0}$ is P_0-compact. Hence the same is true of $L_0 - P_0$, and the proof is complete.

§6. The adjoint of $P_0 + q$

We now give a criterion for determining the adjoint of $P_0 + q$.

Theorem 6.1. *Assume that* $1 < p < \infty$, *that* q *is* P_{0p}*-compact and* $\bar{P}_{0p'}$*-compact and that* $\rho(P_{0p})$ *is not empty. Then*

$$(P_{0p} + q)^* = \bar{P}_{0p'} + \bar{q} . \tag{6.1}$$

Proof. Assume $0 \in \rho(P_{0p})$. Since q is P_{0p}-compact, $P_{0p} + q$ is a Fredholm operator on L^p with

$$i(P_{0p} + q) = i(P_{0p}) = 0 \tag{6.2}$$

(Theorem 3.4 of ch. 1). Thus $(P_{0p} + q)^*$ is a Fredholm operator on $L^{p'}$ with

$$i[(P_{0p} + q)^*] = 0 \tag{6.3}$$

(Theorem 3.6 of ch. 1 and Theorem 1.14 of ch. 2). Moreover $\bar{q}$ is $\bar{P}_{0p'}$-compact. Since $0 \in \rho(\bar{P}_{0p'})$, we see that $\bar{P}_{0p'} + \bar{q}$ is a Fredholm operator satisfying

$$i(\bar{P}_{0p'} + \bar{q}) = i(\bar{P}_{0p'}) = 0 .$$

Now by (2.10) of ch. 3

$$(P(\mathrm{D})\varphi + q\varphi, \Psi) = (\varphi, \bar{P}(\mathrm{D})\Psi + \bar{q}\Psi), \qquad \varphi, \Psi \in C_0^\infty . \tag{6.5}$$

This implies

$$(P_{0p}v+qv,\, w) = (v,\, \bar{P}_{0p'}w+\bar{q}w)\,, \qquad v\in D(P_{0p}),\ w\in D(\bar{P}_{0p'})\,. \tag{6.6}$$

Consequently we see that $(P_{0p}+q)^*$ is an extension of $\bar{P}_{0p'}+\bar{q}$. In view of (6.3) and (6.4) we conclude that (6.1) must hold (Theorem 3.4 of ch. 1).

Corollary 6.2. *Assume that q and the coefficients of $P(\xi)$ are real and that q is P_{02}-compact. Then*

(a) *$P_{02}+q$ is self-adjoint*,

(b) *$P_{02}+q$ is the closure in L^2 of $P(\mathrm{D})+q$ on C_0^∞*,

(c) *$P(\mathrm{D})+q$ on C_0^∞ is essentially self-adjoint in L^2.*

Proof. By Corollary 2.3 of ch. 4, $\rho(P_{02})$ is not empty. If we now apply Theorem 6.1 with $p=2$, we obtain (a). To prove (b), note that

$$\|qu\| \leqslant \tfrac{1}{2}\|P_0 u\| + C\|u\|\,, \qquad u\in D(P_0)$$

by Theorem 2.13 of ch. 1. This implies (b) by Theorem 2.9 of that chapter. The conclusion (c) is an immediate consequence of (a) and (b).

§7. Smooth coefficients in L^2

In the preceding sections of this chapter we proved inequalities (2.2) and (5.1) under very weak assumptions on the function $q(x)$. In order to accomplish this we made rather strong assumptions on the operators $P(\mathrm{D})$ and $Q(\mathrm{D})$. In the present section we shall show that when $p=2$ one can relax the assumptions on $P(\mathrm{D})$ and $Q(\mathrm{D})$ provided we strengthened those on $q(x)$. In a sense the results of this section are dual to those of §§2, 5.

Theorem 7.1. *Suppose Fq exists almost everywhere and satisfies*

$$|Fq| \leqslant C/|\xi|^{\gamma}\,, \qquad |\xi|\leqslant 1 \tag{7.1}$$

$$|Fq| \leqslant C/|\xi|^{n+1} \qquad |\xi| > 1 \tag{7.2}$$

where $\gamma<n$. Asssume that there is a scalar λ such that $M_{\alpha,2}\{Q(\xi)/[P(\xi)-\lambda]\}<\infty$ for some $\alpha<2(n-\gamma)$. Then

$$\|q(x)Q(\mathrm{D})v\| \leqslant C\|[P(\mathrm{D})-\lambda]v\|\,, \qquad v\in S\,. \tag{7.3}$$

Proof. Let $v\in S$ be given and set $h(x)=[P(\mathrm{D})-\lambda]v$, $w(\xi)=Q(\xi)/[P(\xi)-\lambda]$. Then by Theorems 2.4 and 2.6 of ch. 2

$$(qQ(\mathrm{D})v, u) = (v, \bar{Q}(\mathrm{D})[\bar{q}u])$$
$$= (Fv, \bar{Q}(\xi)[F\bar{q} * Fu]) = (Fh, \bar{w}\,[F\bar{q} * Fu])$$

for all $u \in S$. If we can show that

$$\|\bar{w}[F\bar{q} * Fu]\| \leqslant C\|u\|\,, \qquad u \in S\,, \tag{7.4}$$

it will follow that

$$|(qQ(\mathrm{D})v, u)| \leqslant C\|h\|\;\|u\|\,, \qquad u \in S\,,$$

and consequently that (7.3) holds. It therefore suffices to prove (7.4). But this inequality follows immediately from Theorem 2.2. The proof is complete.

Corollary 7.2. *If all derivatives of* $q(x)$ *up to order* $n+1$ *are in* L^1 *and*

$$\int_{|\xi-\eta|<1} |Q(\xi)/[P(\xi)-\lambda]|^2\,\mathrm{d}\xi \leqslant C, \qquad \eta \in E^n\,, \tag{7.5}$$

then (7.3) *holds.*

Proof. By Theorem 4.9 of ch. 2, inequalities (7.1) and (7.2) hold with $\gamma = 0$. Moreover, (7.5) implies that $M_{n,2}\{Q(\xi)/[P(\xi)-\lambda]\} < \infty$. The result now follows from Theorem 7.1.

Note that (7.3) does not imply that $qQ(\mathrm{D})$ on C_0^∞ is closable in L^2. However, it does imply that $qQ(\mathrm{D})$ has a unique extension R such that $D(R) = D(P_0)$ and R is P_0-bounded. To construct R, let u be any function in $D(P_0)$. Then there is a sequence $\{\varphi_k\}$ of functions in C_0^∞ such that $\varphi_k \to u$ and $P(\mathrm{D})\varphi_k \to P_0 u$ in L^2. By (7.3) (or (5.1) for that matter), $qQ(\mathrm{D})\varphi_k$ converges in L^2 to a function v which is independent of the sequence $\{\varphi_k\}$. We define Ru to be v. Clearly R is P_0-bounded.

Theorem 7.3. *Suppose* $q(x) \in C^{n+1}$ *and satisfies*

$$|\mathrm{D}^\mu q(x)| \leqslant C|x|^{a|\mu|+b}, \qquad x \in E^n\,, \quad |\mu| \leqslant n+1\,, \tag{7.6}$$

where $a \leqslant 1$ *and there is an integer* k_0 *satisfying* $0 \leqslant k_0 < n$ *and* $k_0 a > n-b$. *Assume that there is a scalar* λ *such that* $M_{\alpha,2}\{Q(\xi)/[P(\xi)-\lambda]\} < \infty$ *for some* $\alpha < 2(n-k_0)$ *and that*

$$\int_{|\xi-\eta|<1} |Q(\xi)/[P(\xi)-\lambda]|^2\,\mathrm{d}\xi \to 0 \text{ as } |\eta| \to \infty\,. \tag{7.7}$$

Then the operator R defined above is P_0-compact. Thus for each $\varepsilon > 0$ there is a constant C such that

$$\|qQ(\mathrm{D})v\| \leqslant \varepsilon \|P(\mathrm{D})v\| + C\|v\|, \qquad v \in S. \tag{7.8}$$

Proof. Let v be any function in S. Then

$$F[qQ(\mathrm{D})v] = Fq * [Q(\xi)Fv] = Fq * [w(\xi)Fh],$$

where $w(\xi) = Q(\xi)/[P(\xi) - \lambda]$ and $h(x) = [P(\mathrm{D}) - \lambda]v$. Thus it suffices to prove that the operator

$$Tg = Fq * [w(\xi)g]$$

is a compact operator on L^2. Moreover, one easily checks that

$$T^* f = \overline{w}[F(\overline{q}) * f].$$

It follows from Theorem 3.5 that T^* is a compact operator on L^2 and consequently from Theorem 2.17 of ch. 1 that T is compact. The second statement follows from Theorem 2.13 of ch. 1. This completes the proof.

Theorem 7.4. *Let* $q_1(x), \ldots, q_N(x)$ *be functions satisfying*

$$|\mathrm{D}^\mu q_j(x)| \leqslant C/|x|^{a_j|\mu| + b_j}, \qquad x \in E^n, \quad |\mu| \leqslant n+1, \quad 1 \leqslant j \leqslant N \tag{7.9}$$

where each $a_j \leqslant 1$ and there is an integer k_j satisfying $0 \leqslant k_j < n$, $k_j a_j > n - b_j$. Let $Q_1(\xi), \ldots, Q_N(\xi)$ be polynomials such that $M_{\alpha_j,2}\{Q_j(\xi)/[P(\xi)-\lambda]\} < \infty$ for some scalar λ with $a_j < 2(n-k_j)$. Assume that

$$\int_{|\xi-\eta|<1} |Q_j(\xi)/[P(\xi)-\lambda]|^2 \,\mathrm{d}\xi \to 0 \ \text{as}\ |\eta| \to \infty, \qquad 1 \leqslant j \leqslant N. \tag{7.10}$$

Let R_j be the extension of $q_j Q_j(\mathrm{D})$ on C_0^∞ defined before Theorem 7.3. Then $L = P_0 + \Sigma R_j$ is the closure in L^2 of the operator $L(x, \mathrm{D}) = P(\mathrm{D}) + \Sigma q_j(x) Q_j(\mathrm{D})$ on C_0^∞ and

$$\sigma_{\mathrm{e}}(L) = \sigma(P_0). \tag{7.11}$$

Proof. By Theorem 7.3 there is a constant C such that

$$\Sigma \|q_j Q_j(\mathrm{D})v\| \leqslant \tfrac{1}{2}\|P(\mathrm{D})v\| + C\|v\|, \qquad v \in S. \tag{7.12}$$

Thus

$$\|P(\mathrm{D})v\| \leqslant 2\,\|L(x, \mathrm{D})v\| + 2C\,\|v\|, \qquad v \in S. \tag{7.13}$$

Now suppose there is a sequence $\{\varphi_k\}$ of functions in C_0^∞ such that $\varphi_k \to u$ and $L(x, \mathrm{D})\varphi_k \to f$ in L^2. By (7.13), $\{P(\mathrm{D})\varphi_k\}$ converges in L^2. Thus $P(\mathrm{D})\varphi_k \to$

P_0u by the definition of P_0. By (7.12), $\{q_jQ_j(\mathrm{D})\varphi_k\}$ converges for each j. By the definition of R_j, $q_jQ_j(\mathrm{D})\varphi_k \to R_ju$ for each j. Thus $L(x, \mathrm{D})\varphi_k \to Lu$. This shows that $L(x, \mathrm{D})$ on C_0^∞ is closable and L is an extension of its closure. Conversely, if $\varphi_k \to u$ and $P(\mathrm{D})\varphi_k \to P_0u$, then $q_jQ_j(\mathrm{D})\varphi_k \to R_ju$ by (7.12). Thus $L(x, \mathrm{D})\varphi_k \to Lu$. This shows that the closure of $L(x, \mathrm{D})$ on C_0^∞ is an extension of L. Hence L is the closure. The relation (7.11) follows from the fact that the R_j are P_0-compact. This completes the proof.

Corollary 7.5. *In addition to the hypotheses of Theorem 7.4, assume that the coefficients of $P(\xi)$ are real and that*

$$(L(x, \mathrm{D})u, v) = (u, L(x, \mathrm{D})v), \qquad u, v \in S. \tag{7.14}$$

Then L is self-adjoint. Thus $L(x, \mathrm{D})$ on C_0^∞ is essentially self-adjoint in L^2.

Proof. By (7.12)

$$\|\Sigma R_ju\| \leqslant \tfrac{1}{2}\|P_0u\| + C\|u\|, \qquad u \in D(P_0).$$

Since P_0 is self-adjoint (Corollary 2.3 of ch. 2), we see that the same is true of L (Lemma 7.3 of ch. 1). Since L is the closure of $L(x, \mathrm{D})$ on C_0^∞, the latter is essentially self-adjoint.

A very important special case of the results of this section is when $Q(\xi) = 1$. We discuss this case in detail.

Theorem 7.6. *Let $P(\xi)$ be a polynomial such that*

$$\sum_\mu |P^{(\mu)}(\xi)| \to \infty \text{ as } |\xi| \to \infty, \qquad \xi \in E^n, \tag{7.15}$$

and $\rho(P_{02})$ is not empty. If $q(x)$ is a continuous function on E^n such that $q(x) \to 0$ as $|x| \to \infty$, then q is P_{02}-compact. Thus $q(P_{02} - \lambda)^{-1}$ is compact for $\lambda \in \rho(P_{02})$ and

$$\sigma_e(P_{02} + q) = \sigma(P_{02}). \tag{7.16}$$

In proving the theorem we shall make use of two lemmas.

Lemma 7.7. *If $q(x)$ is a function in L^∞ and $P(\xi)$ is a polynomial such that $1/P(\xi)$ is in L^∞, then*

$$\|qP_{02}^{-1}\| \leqslant \|q\|_\infty \|1/P\|_\infty. \tag{7.17}$$

This lemma is an immediate consequence of Theorems 1.5 and 2.4 of ch. 2.

Lemma 7.8. *For any polynomial, (7.15) is equivalent to*

$$\int_{|\zeta|<1} \frac{d\zeta}{|P(\xi+\zeta)|^2+1} \to 0 \text{ as } |\xi| \to \infty . \tag{7.18}$$

The proof of this lemma is rather involved. Before we prove it, let us give the

Proof of Theorem 7.6. First assume $q=\psi \in C_0^\infty$. We may assume $0 \in \rho(P_{02})$. Then the operator ψP_{02}^{-1} is the adjoint of the operator given by

$$Tu = \bar{F}\left\{\frac{1}{\bar{P}}[F\bar{\psi} * Fu]\right\}, \qquad u \in L^2$$

(see the proof of Theorem 7.1). By Theorem 2.17 of ch. 1, T^* will be a compact operator on L^2 if T is compact. By Theorem 2.4 of ch. 2, T will be compact if the operator

$$Rw = \frac{1}{\bar{P}}[F\bar{\psi} * w]$$

is compact. But by Theorem 3.5, R will be compact if

$$M_{\alpha,2}[1/P] < \infty \tag{7.19}$$

for some $\alpha < 2n$ and

$$\int_{|\zeta|<1} \frac{d\zeta}{|P(\xi-\zeta)|^2} \to 0 \text{ as } |\xi| \to \infty . \tag{7.20}$$

Since $1/P$ is in L^∞, (7.19) holds for $\alpha = n$. Moreover, for the same reason there is a constant C such that

$$|P(\xi)|^2 + 1 \leqslant C|P(\xi)|^2 , \qquad \xi \in E^n .$$

From this we see that (7.18) implies (7.20). If we now employ Lemma 7.8, we see that (7.15) implies (7.20), and the theorem is proved for $q=\psi \in C_0^\infty$.

Next suppose $q(x)$ is any function continuous in E^n and tending to zero as $|x| \to \infty$. Then there is a sequence $\{\psi_k\}$ of functions in C_0^∞ such that $\|\psi_k - q\|_\infty \to 0$. By the proof just given the operators $\psi_k P_{02}^{-1}$ are compact on L^2. Furthermore by Lemma 7.7

$$\|qP_{02}^{-1} - \psi_k P_{02}^{-1}\| \leqslant \|q-\psi_k\|_\infty \|1/P\|_\infty \to 0 .$$

Thus qP_{02}^{-1} is the limit in norm of compact operators. It follows that it itself is also compact (see Theorem 2.16 of ch. 2).

Our proof of Lemma 7.8 will be based upon

Lemma 7.9. *For each $m \geqslant 1$ and $n \geqslant 1$ there are positive constants $\varepsilon_0(m, n)$, $\gamma(m, n)$ and $\delta(m, n)$ such that the inequalities*

$$|\xi| \leqslant 1, \quad \sum_{|\mu| \leqslant m} a_\mu \xi^\mu \leqslant \varepsilon, \quad \sum_{0<|\mu| \leqslant m} |a_\mu| \geqslant \varepsilon^\gamma, \qquad 0<\varepsilon<\varepsilon_0 \tag{7.21}$$

for complex a_μ and $\xi \in E^n$ imply that ξ is contained in a set of measure $\leqslant \varepsilon^\delta$.

Proof. First assume that $n=1$. Then the lemma claims that

$$|t| \leqslant 1, \quad \left|\sum_{k=0}^{m} a_k t^k\right| \leqslant \varepsilon, \quad \sum_{k=1}^{m} |a_k| \geqslant \varepsilon^\gamma, \qquad 0<\varepsilon<\varepsilon_0 \tag{7.22}$$

implies that t is in a set of measure $\leqslant \varepsilon^\delta$. This is certainly true for $m=1$. In fact we can take $\gamma(1, 1)$ any positive number and note that

$$|a_1(t-\alpha)| \leqslant \varepsilon, \quad |a_1| \geqslant \varepsilon^\gamma$$

imply that $|t-\alpha| \leqslant \varepsilon^{1-\gamma}$. Thus we can take $\delta(1, 1)=1-\gamma(1, 1)$. Now assume that the lemma is proved for $n=1$ and $m-1$. Let $\gamma(m, 1)$ be any positive number less than $\frac{1}{2}\min[\gamma(m-1, 1), 1]$ and $\delta(m, 1)$ any positive number less than $\frac{1}{2}\min[1/m, \delta(m-1, 1)]$. Since

$$\frac{\varepsilon^{\gamma(m, 1)} - \varepsilon^{\frac{1}{2}}}{(\varepsilon+\varepsilon^{\frac{1}{2}})^{\gamma(m-1, 1)}} \to \infty \text{ as } \varepsilon \to 0,$$

we can find a positive number ε_1 such that $\varepsilon_1+\varepsilon_1^{\frac{1}{2}} < \varepsilon_0(m-1, 1)$ and

$$\varepsilon^{\gamma(m, 1)} - \varepsilon^{\frac{1}{2}} \geqslant (\varepsilon+\varepsilon^{\frac{1}{2}})^{\gamma(m-1, 1)}, \qquad 0<\varepsilon<\varepsilon_1 . \tag{7.23}$$

Similarly, there is an $\varepsilon_2>0$ such that

$$\varepsilon^{\delta(m, 1)} \geqslant (\varepsilon+\varepsilon^{\frac{1}{2}})^{\delta(m-1, 1)}, \qquad 0<\varepsilon<\varepsilon_2 . \tag{7.24}$$

We take $\varepsilon_0(m, 1)$ to be $\min(\varepsilon_1, \varepsilon_2)$. Now let $\alpha_1, \ldots, \alpha_m$ be the complex roots of

$$\sum_{k=0}^{m} a_k t^k = 0 .$$

Then (7.22) gives

$$\left|a_m \prod_{k=1}^{m} (t-\alpha_k)\right| \leqslant \varepsilon . \tag{7.25}$$

If $|a_m| \geqslant \varepsilon^{\frac{1}{2}}$, then (7.25) implies that there is a k such that

$$|t-\alpha_k| \leqslant \varepsilon^{\frac{1}{2}m} .$$

Thus t is contained in a set of measure $\leqslant \varepsilon^{\delta(m,1)}$. If $|a_m| < \varepsilon^{\frac{1}{2}}$, then (7.22) implies

$$\left| a_m t^m + \sum_{k=0}^{m-1} a_k t^k \right| \leqslant \varepsilon ,$$

or

$$\left| \sum_{k=0}^{m-1} a_k t^k \right| \leqslant \varepsilon + \varepsilon^{\frac{1}{2}} . \tag{7.26}$$

We also have

$$\sum_{k=1}^{m-1} |a_k| \geqslant \varepsilon^{\gamma(m,1)} - \varepsilon^{\frac{1}{2}} \geqslant (\varepsilon + \varepsilon^{\frac{1}{2}})^{\gamma(m-1,1)} \tag{7.27}$$

by (7.23). Now by the induction hypothesis, inequalities (7.26) and (7.27) imply that t is contained in a set of measure $\leqslant (\varepsilon + \varepsilon^{\frac{1}{2}})^{\delta(m-1,1)}$. If we now apply (7.24) we see that the lemma is proved for $n=1$ and m.

Next assume that the lemma is proved for m and n. We shall prove it for m and $n+1$. Let $\gamma(m, n+1)$ be any positive number less than $\gamma(m, 1)\cdot$ min $[\gamma(m, n), 1]$, and let $\delta(m, n+1)$ be any positive number less than min $[\delta(m, 1), \gamma(m, 1)\, \delta(m, n)]$. Thus there exist $\varepsilon_3 > 0$ and $\varepsilon_4 > 0$ such that

$$\varepsilon^{\gamma(m,n+1)} \geqslant (m+1)^n \varepsilon^{\gamma(m,1)\,\gamma(m,n)} , \qquad 0 < \varepsilon < \varepsilon_3 ; \tag{7.28}$$

$$\varepsilon^{\delta(m,n+1)} \geqslant 2^n \varepsilon^{\delta(m,1)}, \ \varepsilon^{\delta(m,n+1)} \geqslant 2\varepsilon^{\gamma(m,1)\,\delta(m,n)}, \qquad 0 < \varepsilon < \varepsilon_4 . \tag{7.29}$$

We take $\varepsilon_0(m, n+1) = \min [\varepsilon_0(m, 1), \varepsilon_3, \varepsilon_4]$. Now suppose

$$t^2 + |\xi|^2 \leqslant 1 , \quad \left| \sum_{k=0}^{m} Q_k(\xi) t^k \right| \leqslant \varepsilon ,$$

$$\sum_{0 < k + |\mu| \leqslant m} |a_{k\mu}| \geqslant \varepsilon^{\gamma(m,n+1)} , \qquad 0 < \varepsilon < \varepsilon_0(m, n+1) \tag{7.30}$$

where $\xi = (\xi_1, \ldots, \xi_n) \in E^n$, $\mu = (\mu_1, \ldots, \mu_n)$ and

$$Q_k(\xi) = \sum_{|\mu| \leqslant m-k} a_{k\mu} \xi^\mu . \tag{7.31}$$

Now by (7.30) there is a j and a μ such that

$$|a_{j\mu}| \geqslant \varepsilon^{\gamma(m,n+1)} / (m+1)^n \geqslant \varepsilon^{\gamma(m,1)\,\gamma(m,n)} \tag{7.32}$$

(see (7.28)). Moreover, we may assume that $j \neq 0$. For if (7.32) holds only for $j=0$, then there is a $\mu \neq 0$ such that (7.32) holds for $j=0$ and μ. Let k be such that $\mu_k \neq 0$. Then we can interchange the roles of t and ξ_k. All hypotheses remain intact. Now suppose $|Q_j(\xi)| \geqslant \varepsilon^{\gamma(m,1)}$. Then we have

$$\left|\sum_{k=0}^{m} Q_k(\xi)t^k\right| \leqslant \varepsilon, \quad \sum_{k=1}^{m} |Q_k(\xi)| \geqslant \varepsilon^{\gamma(m,1)}.$$

This implies that t is contained in a set of measure $\leqslant \varepsilon^{\delta(m,1)}$. Since $|\xi| \leqslant 1$, (t, ξ) is contained in a set of measure $\leqslant 2^n \varepsilon^{\delta(m,1)}$. By (7.29) this is $\leqslant \varepsilon^{\delta(m,n+1)}$. Next suppose that $|Q_j(\xi)| < \varepsilon^{\gamma(m,1)}$. By (7.32)

$$\sum_{|\mu| \leqslant m-j} |a_{j\mu}| \geqslant \varepsilon^{\gamma(m,1)\gamma(m,n)}.$$

Thus by the inductive hypothesis, ξ is contained in a set of measure $\leqslant \varepsilon^{\gamma(m,1)\delta(m,n)}$. Since $|t| \leqslant 1$, it follows that (t, ξ) is in a set of measure $\leqslant 2\varepsilon^{\gamma(m,1)\delta(m,n)}$. By (7.29) this is $\leqslant \varepsilon^{\delta(m,n+1)}$. This completes the proof of the lemma.

We can now give the

Proof of Lemma 7.8. For any polynomial $P(\xi)$ we have

$$\int_{|\eta|<1} \frac{\mathrm{d}\eta}{|P(\xi+\eta)|^2+1} \geqslant \frac{C}{1+\max\limits_{|\eta|\leqslant 1}|P(\xi+\eta)|^2}.$$

By Taylor's theorem

$$|P(\xi+\eta)| = |\Sigma\, P^{(\mu)}(\xi)\eta^{\mu}/\mu!| \leqslant \Sigma\, |P^{(\mu)}(\xi)|.$$

This proves the implication in one direction. To prove it in the other, suppose that $P(\xi)$ satisfies (7.15). For each ξ set

$$\tilde{P}(\xi) = \Sigma\, |P^{(\mu)}(\xi)| \tag{7.33}$$

$$b_\mu = b_\mu(\xi) = P^{(\mu)}(\xi)/\mu!\, \tilde{P}(\xi)^{\frac{3}{4}}. \tag{7.34}$$

Note that

$$\Sigma\, |b_\mu| \geqslant c_0 \tilde{P}(\xi)^{\frac{1}{4}} \tag{7.35}$$

for some $c_0 > 0$. Let $\Omega_{1\xi}$ be the set of those $\zeta \in E^n$ such that $|\zeta| \leqslant 1$ and $|P(\zeta+\xi)|^2 > \tilde{P}(\xi)$, let $\Omega_{2\xi}$ be the set of those $\zeta \in E^n$ such that $|P(\zeta+\xi)|^2 \leqslant \tilde{P}(\xi)$ and $\Sigma_{\mu\neq 0}|b_\mu| < 1$, and let $\Omega_{3\xi}$ be the remainder of the set $|\zeta| \leqslant 1$. Note that for points in $\Omega_{2\xi}$ and $\Omega_{3\xi}$ we have

$$|\Sigma b_\mu \zeta^\mu| \leqslant 1/\tilde{P}(\xi)^{\frac{1}{4}} \to 0 \text{ as } |\xi| \to \infty.$$

Now

$$\int_{\Omega_{1\xi}} \frac{\mathrm{d}\zeta}{|P(\zeta+\xi)|^2+1} \leqslant \frac{2^n}{\tilde{P}(\xi)+1} \to 0 \text{ as } |\xi| \to \infty. \tag{7.37}$$

For points of $\Omega_{2\xi}$ we have by (7.35)

$$\frac{|P(\xi)|}{\tilde{P}^{\frac{3}{4}}} = \Sigma\,|b_\mu| - \sum_{\mu\neq 0} |b_\mu| \geqslant c_0 \tilde{P}(\xi)^{\frac{1}{4}} - 1 \to \infty \quad \text{as} \quad |\xi| \to \infty \tag{7.38}$$

(recall that $b_\mu = P(\xi)/\tilde{P}(\xi)^{\frac{3}{4}}$ for $\mu = 0$). Thus

$$|\Sigma b_\mu \zeta^\mu| \geqslant \frac{|P(\xi)|}{\tilde{P}^{\frac{3}{4}}} - 1 \to \infty \quad \text{as} \quad |\xi| \to \infty\,. \tag{7.39}$$

Inequalities (7.36) and (7.39) show that $\Omega_{2\xi}$ is empty for $|\xi|$ sufficiently large. Finally we note that the measure of $\Omega_{3\xi}$ tends to zero as $|\xi| \to \infty$ by Lemma 7.9. Thus (7.18) holds, and the proof is complete.

CHAPTER 6

ELLIPTIC OPERATORS

In this chapter we specialize some of our results of ch. 5 to a particular type of partial differential operator which arises very frequently in applications. In the case of elliptic operators we are able to employ special techniques which allow us to obtain better results than are possible in general.

§1. An improvement

As we noted in ch. 4, §5 an elliptic operator $P(\mathrm{D})$ of order m satisfies

$$P^{(\mu)}(\xi)/P(\xi) = \mathrm{O}(1/|\xi|^{|\mu|}) \text{ as } |\xi| \to \infty \tag{1.1}$$

$$1/P(\xi) = \mathrm{O}(1/|\xi|^{m}) \text{ as } |\xi| \to \infty \,. \tag{1.2}$$

Thus the hypotheses of Theorem 2.1 of ch. 5 are satisfied with $a=1$ and $b=m$. In this case we have $k_0=n-m+1$. Thus we have $D(P_0) \subset D(q)$ provided $M_{\alpha,p}(q) < \infty$ for some $\alpha < p(m-1)$ and $\rho(P_0)$ is not empty. In the next few sections we shall show how this statement can be strengthened. We shall show that $D(P_0) \subset D(q)$ provided $\alpha < pm$ without any assumption on $\rho(P_0)$. At first glance this may not appear to be much of a generalization, but the difference is significant in applications.

It should be noted that the stipulation $\alpha < p(m-1)$ was necessitated by the fact that k_0 was required to be an integer. This is turn resulted from the rather crude methods employed in the proof of Theorem 2.1 of ch. 5. The condition $\alpha < pm$ results primarily from the fact that for elliptic operators, at least, one can allow k_0 to take on fractional values.

The basic idea stems from the fact that for $P(\mathrm{D})$ elliptic of order m,

$$D(P_0) = H^{m,p}\,, \tag{1.3}$$

where $H^{s,p}$ are the spaces discussed in ch. 2, §4. Once this is known, the questions of P_0-boundedness and P_0-compactness of q become considerations of when the function q is a bounded operator from $H^{m,p}$ to L^p or a compact operator from $H^{m,p}$ to L^p. These questions are discussed in §§2–4.

It is shown that a sufficient condition for q to be a bounded operator from $H^{s,p}$ to L^p with $s>0$ is that $M_{\alpha,p}(q)<\infty$ for some $\alpha<ps$. If

$$\int_{|x|<1} |q(x-y)|^p \mathrm{d}x \to 0 \text{ as } |y|\to\infty \tag{1.4}$$

as well, then q is a compact operator from $H^{s,p}$ to L^p. The main tools in our investigations are the *Bessel potentials* described in §3. We then apply the results to operators of the form

$$L(x, \mathrm{D}) = P(\mathrm{D}) + \sum_{j=1}^{r} q_j(x) Q_j(\mathrm{D}) . \tag{1.5}$$

In §5, we show how one can handle the case when q is not locally in L^p. Here we make use of *s-extensions* which are special cases of the intermediate extensions discussed in ch. 1, §6. We are able to preserve most of the preceding results provided there is a real value of s such that $0<s<m$ and $q(x)=q_1(x)q_2(x)$, where

$$M_{\alpha,p}(q_1)<\infty\,, \qquad \alpha<ps \tag{1.6}$$

$$M_{\beta,p'}(q_2)<\infty\,, \qquad \beta<p'(m-s) \tag{1.7}$$

$(p'=p/(p-1))$.

§2. A condition for $D(q) \supset H^{s,p}$

We have defined the spaces $H^{s,p}$ for s real and $1\leqslant p\leqslant\infty$ (see §4 of ch. 2). We shall be concerned with the following problem. Let $s>0$ and $1\leqslant p<\infty$ be given. To determine conditions on a function $q(x)$ such that $D(q)\supset H^{s,p}$. Our answer will be given by

Theorem 2.1. *If $M_{\alpha,p}(q)<\infty$ for some real α satisfying*

$$0<\alpha<ps \tag{2.1}$$

then

$$\|q\varphi\|_{0,p} \leqslant C[M_{\alpha,p}(q)]^{1/p}\|\varphi\|_{s,p}\,, \qquad \varphi\in C_0^\infty\,, \tag{2.2}$$

where the constant C depends only on n, s, α and p. Thus $D(q)\supset H^{s,p}$.

Proof. It suffices to prove (2.2). Let φ be a function in C_0^∞, and set

$$f=\bar{F}(1+|\xi|^2)^{\frac{1}{2}s}F\varphi\,.$$

Then $f\in S$ and

$$\|f\|_{0,p} = \|\varphi\|_{s,p}\,. \tag{2.3}$$

Moreover by Theorem 2.2 of ch. 2

$$F\varphi = (1+|\xi|^2)^{-\frac{1}{2}s} F f\,.$$

If we set

$$G_s(x) = \bar{F}\left[(1+|\xi|^2)^{-\frac{1}{2}s}\right], \tag{2.4}$$

then

$$F\varphi = FG_s F f\,,$$

and consequently by Theorem 4.10 of ch. 2

$$\varphi = G_s * f\,. \tag{2.5}$$

We shall show that

$$G_s(x) = \mathrm{O}(|x|^{s-n}) \quad \text{as } |x| \to 0, \quad 0 < s < n\,, \tag{2.6}$$

$$G_n(x) = \mathrm{O}(-\log|x|) \quad \text{as } |x| \to 0\,, \tag{2.7}$$

$$G_s(x) = \mathrm{O}(1) \quad \text{as } |x| \to 0, \quad s > n\,, \tag{2.8}$$

$$G_s(x) = \mathrm{O}(|x|^{-k}) \quad \text{as } |x| \to \infty,\ s > 0, \qquad k = 1, 2, \ldots\,. \tag{2.9}$$

Assume these for the moment. Statement (2.9) with $k = n+1$ implies that

$$\int_{|x-y|>1} |q(x)|^p\, |G_s(x-y)|\, \mathrm{d}x \leqslant CM_{n,p}(q) \tag{2.10}$$

$$\int_{|z|>1} |G_s(z)|\, \mathrm{d}z < \infty\,. \tag{2.11}$$

(The proof of (2.10) is identical to that of (2.27) of ch. 5.) Moreover, there exists a β satisfying $0 \leqslant \beta \leqslant 1$ such that for $p > 1$

$$\int_{|x-y|<1} |q(x)|^p\, |G_s(x-y)|^{\beta p}\, \mathrm{d}x \leqslant CM_{\alpha,p}(q) \tag{2.12}$$

$$\int_{|z|<1} |G_s(z)|^{(1-\beta)p}\, \mathrm{d}x < \infty\,. \tag{2.13}$$

Inequalities (2.10)–(2.13) imply

$$\|q\varphi\|_{0,p} \leqslant C\left[M_{\alpha,p}(q)\right]^{1/p} \|f\|_{0,p} \tag{2.14}$$

(see the proof of Theorem 2.2 of ch. 5). This gives (2.2) in view of (2.3).

To choose β, note that if $ps > n$, we may take $\alpha = n$, $\beta = 0$. Then (2.12) is trivial and (2.13) follows from (2.6) and the fact that $p'(s-n)+n = p'(ps-n)/p > 0$. (If $s \geqslant n$ we use (2.7) or (2.8). Then (2.13) is trivial.) If $ps < n$, we may assume that

$$ps-(p-1)n < \alpha < ps .$$

In this case we take $\beta = (n-\alpha)/p(n-s)$. Clearly $0 < \beta < 1$ and (2.12) holds. Inequality (2.13) follows from the fact that

$$(1-\beta)p'(s-n)+n = (ps-\alpha)/(p-1) > 0 .$$

Thus the proof of Theorem 2.1 will be complete as soon as we prove (2.6)–(2.9). This will be done in the next section.

§3. Bessel potentials

In this section we shall prove

Theorem 3.1. *Let s be a positive real number, and define $G_s(x)$ by* (2.4). *Then $G_s(x)$ satisfies* (2.6)–(2.9).

In order to prove Theorem 3.1 note that the function $(1+|\xi|^2)^{\frac{1}{2}s}$ depends only on $|\xi|$. If $s > n$, then this function is in L^1 nd we can apply Bochner's theorem (Theorem 2.7 of ch. 2) to obtain

$$G_s(x) = (2\pi)^{\frac{1}{2}n}|x|^{\frac{1}{2}(2-n)} \int_0^\infty (1+t^2)^{-\frac{1}{2}s} t^{\frac{1}{2}n} J_{\frac{1}{2}(n-2)}(|x|t)\mathrm{d}t , \tag{3.1}$$

where $J_\lambda(t)$ denotes the Bessel function of the first kind of order λ. Setting $\tau = |x|t$ in (3.1) we get

$$G_s(x) = (2\pi)^{\frac{1}{2}n}|x|^{s-n} \int_0^\infty (|x|^2+\tau^2)^{-\frac{1}{2}s} \tau^{\frac{1}{2}n} J_{\frac{1}{2}(n-2)}(\tau)\mathrm{d}\tau . \tag{3.2}$$

In analyzing (3.2) we shall need

Lemma 3.2. *Let $g(r) = g_{\beta,\lambda}(r)$ be defined by*

$$g(r) = \int_0^\infty (r^2+t^2)^{-\beta} t^\lambda J_{\lambda-1}(t)\mathrm{d}t , \tag{3.3}$$

where $r > 0$, $\lambda > 0$ and $2\beta > \lambda + \frac{1}{2}$. Then the following relationships hold.

$$g(r) = 2\beta \int_0^\infty (r^2+t^2)^{-\beta-1} t^{\lambda+1} J_\lambda(t)\,\mathrm{d}t\,, \tag{3.4}$$

$$g(r) = 2^k \beta(\beta+1)\cdots(\beta+k-1) \int_0^\infty (r^2+t^2)^{-\beta-k} t^{\lambda+k} J_{\lambda+k-1}(t)\,\mathrm{d}t\,, \quad k=1,2,\ldots \tag{3.5}$$

$$g(r) = \mathrm{O}(1) \qquad \textit{as } r\to 0\,, \qquad \beta<\lambda \tag{3.6}$$

$$g(r) = \mathrm{O}(-\log r) \quad \textit{as } r\to 0\,, \qquad \beta=\lambda \tag{3.7}$$

$$g(r) = \mathrm{O}(r^{2(\lambda-\beta)}) \quad \textit{as } r\to 0\,, \qquad \beta>\lambda \tag{3.8}$$

$$g(r) = \mathrm{O}(r^{-k}) \qquad \textit{as } r\to\infty\,, \qquad k=1,2,\ldots\,. \tag{3.9}$$

Proof. In order to prove (3.4) note that

$$\mathrm{d}[t^\lambda J_\lambda(t)]/\mathrm{d}t = t^\lambda J_{\lambda-1}(t)\,, \tag{3.10}$$

$$J_\lambda(t) = \mathrm{O}(t^\lambda) \quad \text{as } t\to 0\,, \tag{3.11}$$

$$J_\lambda(t) = \mathrm{O}(t^{-\frac{1}{2}}) \text{ as } t\to\infty \tag{3.12}$$

(Theorem 2.8 of ch. 2). Thus

$$\begin{aligned} g(r) &= \int_0^\infty (r^2+t^2)^{-\beta}\,\mathrm{d}[t^\lambda J_\lambda(t)] \\ &= -\int_0^\infty t^\lambda J_\lambda(t)\,\mathrm{d}[(r^2+t^2)^{-\beta}] \\ &= 2\beta \int_0^\infty (r^2+t^2)^{-\beta-1} t^{\lambda+1} J_\lambda(t)\,\mathrm{d}t\,. \end{aligned}$$

This gives (3.4). Repeated applications of (3.4) give (3.5). To obtain (3.6) note that

$$\begin{aligned} |g(r)| &\leqslant \int_0^\infty t^{-2\beta+\lambda} |J_{\lambda-1}(t)|\,\mathrm{d}t \\ &\leqslant C\int_0^1 t^{-2\beta+2\lambda-1}\,\mathrm{d}t + C\int_1^\infty t^{-2\beta+\lambda-\frac{1}{2}}\,\mathrm{d}t \end{aligned}$$

by (3.11) and (3.12). Thus $g(r)$ is bounded for $\beta<\lambda$. Relation (3.7) follows from (3.5) and

$$\int_0^\infty (r^2+t^2)^{-\lambda-2}t^{\lambda+2}|J_{\lambda-1}(t)|\,dt$$

$$\leqslant C\int_0^1 (r^2+t^2)^{-1}t\,dt + C\int_1^\infty (r^2+t^2)^{-\frac{1}{2}\lambda-\frac{7}{4}}t\,dt$$

$$\leqslant C'(1-\log r)\,.$$

We merely take $k=2$ in (3.5). In proving (3.8) we note that by (3.4), (3.11) and (3.12)

$$|g(r)| \leqslant C\int_0^1 (r^2+t^2)^{-\beta-1}t^{2\lambda+1}\,dt + C\int_1^\infty (r^2+t^2)^{-\beta-1}t^{\lambda+\frac{1}{2}}\,dt$$

$$\leqslant C'\int_0^\infty (r^2+t^2)^{-\beta-1}t^{2\lambda+1}\,dt \leqslant C''\int_0^\infty (r^2+t^2)^{\lambda-\beta-1}t\,dt$$

$$\leqslant C'''r^{2(\lambda-\beta)}\,.$$

Finally, to obtain (3.9), note that by (3.5)

$$|g(r)| \leqslant C\int_0^\infty (r^2+t^2)^{\frac{1}{2}(\lambda+k)-\beta-k-\frac{3}{4}}t\,dt$$

$$\leqslant C'r^{\lambda-k-2\beta+\frac{1}{2}}\,.$$

Since this is true for each positive k, (3.9) follows. The proof of Lemma 3.2 is complete.

Now we can give the

Proof of Theorem 3.1. First assume that $s>n$. Then $(1+|\xi|^2)^{-\frac{1}{2}s}$ is in L^1 so that Bochner's theorem applies (see Theorem 2.7 of ch. 2). Hence $G_s(x)$ satisfies (3.2). Setting $s=2\beta$ and $n=2\lambda$ in Lemma 3.2 we see by (3.8) and (3.9) that (2.8) and (2.9) hold. Thus the lemma is proved for $s>n$. Note also that for this case (3.5) implies

$$G_s(x) = Q_k(s)\,|x|^{s-n}\int_0^\infty (|x|^2+t^2)^{-\frac{1}{2}s-k}t^{\frac{1}{2}n+k}J_{\frac{1}{2}(n+2k-2)}(t)\,dt\,, \tag{3.13}$$

where

$$Q_k(s) = s(s+2)(s+4)\cdots(s+2k-2)\,. \tag{3.14}$$

Next let us consider the case of arbitrary $s>0$. Let k be an integer $>\frac{1}{2}(n+1)$, and let $H_s(x)$ denote the right-hand side of (3.13). Setting $\beta=\frac{1}{2}(s+2k)$, $\lambda=\frac{1}{2}(n+2k)$ in Lemma 3.2, we see that $H_s(x)$ satisfies

$$H_s(x) = O(|x|^{s-n}) \quad \text{as } |x| \to 0, \quad 0 < s < n \tag{3.15}$$

$$H_n(x) = O(-\log |x|) \quad \text{as } |x| \to 0, \tag{3.16}$$

$$H_s(x) = O(1) \quad \text{as } |x| \to 0, \quad s > n \tag{3.17}$$

$$H_s(x) = O(|x|^{-k}) \quad \text{as } |x| \to \infty, \quad s > 0, \quad k = 1, 2, \ldots . \tag{3.18}$$

Furthermore if we allow s to take on complex values, then for each $x \neq 0$, $H_s(x)$ is an analytic function of s in $\operatorname{Re} s > 0$. If we compute $\mathrm{d}H_s(x)/\mathrm{d}s$ and apply Lemma 3.2 we obtain

$$\begin{aligned} \mathrm{d}H_s(x)/\mathrm{d}s &= O(|x|^{s-n} \log |x|) \quad \text{as } |x| \to 0, \quad 0 < \operatorname{Re} s < n \\ &= O[(\log |x|)^2] \quad \text{as } |x| \to 0, \quad \operatorname{Re} s = n \\ &= O(\log |x|) \quad \text{as } |x| \to 0, \quad \operatorname{Re} s > n \\ &= O(|x|^{-k}) \quad \text{as } |x| \to \infty, \quad \operatorname{Re} s > 0, \quad k = 1, 2, \ldots . \end{aligned}$$

In particular we see that H_s and $\mathrm{d}H_s/\mathrm{d}s$ are both in L^1 for $\operatorname{Re} s > 0$, and for complex z the difference quotient $(H_{s+z} - H_s)/z$ converges in L^1 to $\mathrm{d}H_s/\mathrm{d}s$ as $|z| \to 0$. Thus if w is any function in C_0^∞, the scalar function

$$f(s) = (FH_s, w)$$

is analytic in $\operatorname{Re} s > 0$. Since $H_s(x) = G_s(x)$ for $s > n$ by (3.13), we know that $FH_s = (1 + |\xi|^2)^{-\frac{1}{2}s}$ in this case. But one checks easily that for all complex values of s in $\operatorname{Re} s > 0$ the scalar function

$$h(s) = ([1 + |\xi|^2]^{-\frac{1}{2}s}, w)$$

is analytic. Hence by analytic continuation we must have $h(s) = f(s)$ in $\operatorname{Re} s > 0$. Since both FH_s and $(1 + |\xi|^2)^{-\frac{1}{2}s}$ are continuous functions and w was an arbitrary function in C_0^∞, we see that $FH_s = (1 + |\xi|^2)^{-\frac{1}{2}s}$ for $s > 0$. The desired properties now follow from (3.15)–(3.18), and the proof is complete.

Functions which are of the form $G_s * f$ are called *Bessel potentials* because the functions $G_s(x)$ can be expressed in terms of Bessel functions.

There are two properties of the functions $G_s(x)$ which we have not used. One is that $G_s(x) = O(\exp\{-a|x|\})$ as $|x| \to \infty$ for some $a > 0$. We shall not need this property; the weaker relation (3.9) suffices for our purposes. The second property is that $G_s(x) \geqslant 0$. This fact is not needed for the present chapter. However, in the next chapter we shall want to make use of it. In proving it we shall use

Lemma 3.3. *Let $g(r)$ be given as in Lemma 3.2. Then $g(r)>0$ for $r>0$.*

Proof. We first note that

$$\frac{\Gamma(\beta)}{(r^2+t^2)^\beta} = \int_0^\infty \tau^{\beta-1} \exp\{-\tau(r^2+t^2)\}\,d\tau$$

(Lemma 2.9 of ch. 2). Hence

$$g(r) = \int_0^\infty \tau^{\beta-1} \exp(-\tau r^2) \int_0^\infty t^\lambda J_{\lambda-1}(t) \exp(-\tau t^2)\,dt\,d\tau\,.$$

But by Theorem 2.10 of ch. 2

$$\int_0^\infty t^\lambda J_{\lambda-1}(t) \exp(-\tau t^2) dt = \exp\left(-\frac{1}{4\tau}\right)\Big/(2\tau)^\lambda\,.$$

Hence

$$g(r) = 2^{-\lambda} \int_0^\infty \tau^{\beta-\lambda-1} \exp\left(-\tau r^2 - \frac{1}{4\tau}\right) d\tau\,. \tag{3.19}$$

This clearly shows that $g(r)>0$.

§4. Some consequences

We now show how Theorem 2.1 leads to strengthening of the theorems of ch. 5 for the case of elliptic operators. We first have

Theorem 4.1. *Suppose $1 \leqslant p < \infty$, $s>0$ and that $M_{\alpha,p}(q) < \infty$ for some α satisfying (2.1). If*

$$\int_{|x-y|<1} |q(x)|^p\,dx \to 0 \text{ as } |y|\to\infty\,, \tag{4.1}$$

then q is a compact operator from $H^{s,p}$ to L^p.

Proof. By Lemma 3.3 of ch. 5, we have for any $\varepsilon>0$

$$\int_{|x-y|<1} |q(x)|^p |x-y|^{\alpha+\varepsilon-n}\,dx \to 0 \text{ as } |y|\to\infty\,. \tag{4.2}$$

Let β be any number satisfying

$$\alpha<\beta<ps\,. \tag{4.3}$$

Set $t=\beta/p$. Then by Theorem 2.1 multiplication by q is a bounded operator from $H^{t,p}$ to L^p. Let $\psi(x)$ be a function in C_0^∞ satisfying (see Theorem 1.2 of ch. 2)

$$0 \leqslant \psi(x) \leqslant 1, \qquad x \in E^n \tag{4.4}$$

$$\psi(x) = 1, \qquad |x| \leqslant 1 \tag{4.5}$$

$$\psi(x) = 0, \qquad |x| > 2, \tag{4.6}$$

and set

$$\psi_R(x) = \psi(x/R), \qquad R > 0. \tag{4.7}$$

Since $t < s$ and $\psi_R \in C_0^\infty$, multiplication by ψ_R is a compact operator from $H^{s,p}$ to $H^{t,p}$ (Theorem 4.4 of ch. 2). Hence $q\psi_R$ is a compact operator from $H^{s,p}$ to L^p. Moreover, the norm of the operator $q(1-\psi_R)$ as an operator from $H^{s,p}$ to L^p is bounded by

$$CM_{\beta,p}[q(1-\psi_R)]^{1/p}$$

(Theorem 2.1). By (4.3), (4.4) and (4.5) this expression tends to 0 as $R \to \infty$. Thus q is the limit in norm of the compact operators $q\psi_R$. Consequently q is a compact operator $H^{s,p}$ to L^p. This completes the proof.

A consequence of Theorems 2.1 and 4.1 is

Corollary 4.2. *Let $P(D)$ be an elliptic operator of order m. Suppose that $q(x)$ is a function such that $M_{\alpha,p}(q) < \infty$ for some α satisfying* (1.2) *with $1 < p < \infty$. Then $D(P_0) \subset D(q)$. If, in addition, q satisfies* (4.1), *then q is P_0-compact.*

To derive Corollary 4.2 we merely note the following.

Lemma 4.3. *If $P(D)$ is an elliptic operator of order m, then*

$$D(P) = H^{m,p}. \tag{4.8}$$

The lemma is a simple consequence of the inequality

$$\|\varphi\|_{m,p} \leqslant C(\|P(D)\varphi\|_{0,p} + \|\varphi\|_{0,p}) \leqslant C'\|\varphi\|_{m,p}, \qquad \varphi \in C_0^\infty \tag{4.9}$$

(Theorems 2.4 and 3.1 of ch. 3) and the fact that C_0^∞ is dense in $H^{m,p}$ (Theorem 4.11 of ch. 2).

Another consequence is

Theorem 4.4. *Let r be an integer satisfying $0 \leqslant r < s$. Suppose that $1 \leqslant p < \infty$, that $Q(D)$ is an operator of order r and that $q(x)$ is a function satisfying $M_{\alpha,p}(q) < \infty$ for some α satisfying*

$$0 < \alpha < p(s-r) . \tag{4.10}$$

Then

$$\|qQ(\mathrm{D})\varphi\|_{0,p} \leqslant C\|\varphi\|_{s,p} , \qquad \varphi \in C_0^\infty . \tag{4.11}$$

Thus $H^{s,p} \subset D(qQ_0)$. *If* q *satisfies* (4.1), *then* qQ_0 *is a compact operator from* $H^{s,p}$ *to* L^p.

Proof. By Theorem 2.1

$$\|qQ(\mathrm{D})\varphi\|_{0,p} \leqslant C\|Q(\mathrm{D})\varphi\|_{s-r,p} \leqslant C'\|x\|_{s,p} \tag{4.12}$$

(see Theorem 2.4 of ch. 3). This proves (4.11). To prove the second statement let v be any element in $H^{s,p}$. Then there is a sequence $\{\varphi_k\}$ of functions in C_0^∞ such that $\varphi_k \to v$ in $H^{s,p}$. By the second inequality in (4.12), $Q(\mathrm{D})\varphi_k$ converges in $H^{s-r,p}$ so that $v \in D(Q_0)$. By the first inequality in (4.12),$qQ(\mathrm{D})\varphi_k$ converges. Thus $Q_0 v \in D(q)$. To prove the last statement, note that Q_0 is a bounded operator from $H^{s,p}$ to $H^{s-r,p}$ by (4.12). By Theorem 4.1, q is a compact operator from $H^{s,p}$ to L^p. This completes the proof.

We also have the following result for variable coefficient operators.

Theorem 4.5. *Let*

$$L(x, \mathrm{D}) = P(\mathrm{D}) + \sum_{j=1}^{r} q_j(x) Q_j(\mathrm{D}) , \tag{4.13}$$

where $P(\mathrm{D})$ *is an elliptic operator of order* m *and each* $Q_j(\mathrm{D})$ *is an operator order* $m_j < m$, $1 \leqslant j \leqslant r$. *Suppose that* $1 \leqslant p < \infty$ *and that*

$$M_{\alpha_j,p}(q_j) < \infty , \qquad 1 \leqslant j \leqslant r , \tag{4.14}$$

where

$$0 < \alpha_j < p(m - m_j) , \qquad 1 \leqslant j \leqslant r . \tag{4.15}$$

Assume that

$$\int_{|x-y|<1} |q_j(x)|^p \,\mathrm{d}x \to 0 \text{ as } |y| \to \infty , \qquad 1 \leqslant j \leqslant r . \tag{4.16}$$

Then the operator $L(x, \mathrm{D})$ *is closable in* L^p. *If* L_0 *denotes its closure, then* $D(L_0) = H^{m,p}$ *and* $L_0 - P_0$ *is a compact operator from* $H^{m,p}$ *to* L^p. *Thus*

$$\sigma_e(L_0) = \sigma_e(P_0) = \sigma(P_0) = \{P(\xi), \xi \in E^n\} .$$

The proof of Theorem 4.5 depends on Theorem 4.4 and follows the same reasoning as the proof of Theorem 5.4 of ch. 5.

The following consequence of Theorem 4.1 is often very useful.

Corollary 4.6. *Assume that there is an r satisfying* $1 \leqslant r < \infty$ *such that* $q \in L^r \cap L^\infty$. *Then q is a compact operator from* $H^{s,p}$ *to* L^p *for each* $s > 0$ *and p satisfying* $1 \leqslant p < \infty$.

Proof. Since $q \in L^\infty$, one has $M_{\alpha,p}(q) < \infty$ for any $\alpha > 0$. Furthermore if $r \leqslant p$,

$$\int_{|y|<1} |q(x-y)|^p \mathrm{d}y \leqslant \|q\|_\infty^{p-r} \int_{|y|<1} |q(x-y)|^r \mathrm{d}y \to 0$$

as $|x| \to \infty$. If $r > p$, we have by Hölder's inequality (Theorem 1.5 of ch. 2)

$$\int_{|y|<1} |q(x-y)|^p \mathrm{d}y \leqslant \left(\int_{|y|<1} |q(x-y)|^r \mathrm{d}y \right)^{1/t} \left(\int_{|y|<1} \mathrm{d}y \right)^{1/t'}$$

where $t = r/p > 1$. We now apply Theorem 4.1.

§5. *s*-extensions

All of the theorems of the last chapter and the present chapter up to this point assumed that $q(x)$ was locally in L^p. In this section we shall show how this can be relaxed for $1 < p < \infty$. However, it now becomes unreasonable to work with the closure of $P(\mathrm{D}) + q$ on $D(q) \cap C_0^\infty$ in L^p. We must therefore search for a suitable extension.

Let $P(\mathrm{D})$ be an operator of order m and let s be a real number satisfying $0 < s < m$. Suppose $q(x) = q_1(x) q_2(x)$ where

$$M_{\alpha,p}(q_1) < \infty, \qquad \alpha < ps \tag{5.1}$$

$$M_{\beta,p'}(q_2) < \infty, \qquad \beta < p'(m-s). \tag{5.2}$$

Set

$$b(\varphi, \psi) = (P(\mathrm{D})\varphi, \psi) + (q_1\varphi, \bar{q}_2\psi), \qquad \varphi, \psi \in C_0^\infty. \tag{5.3}$$

By Theorem 2.4 of ch. 3 and Theorem 2.1

$$|b(\varphi, \psi)| \leqslant C\|\varphi\|_{s,p}\|\psi\|_{m-s,p'}, \qquad \varphi, \psi \in C_0^\infty. \tag{5.4}$$

Thus the bilinear form $b(\varphi, \psi)$ can be extended by continuity to $H^{s,p} \times H^{m-s,p'}$. Moreover, if B_0 denotes the operator $P(\mathrm{D}) + q$ acting on $D(q) \cap C_0^\infty$, then

$$(B_0\varphi, v) = b(\varphi, v), \qquad \varphi \in D(B_0), \quad v \in H^{m-s,p'}.$$

Hence by Theorem 6.4 of ch. 1, B_0 has an intermediate extension relative to the spaces $H^{s,p}$ and $H^{m-s,p'}$. We call such an extension an s-extension of $P(\mathrm{D})+q$. We have

Theorem 5.1. *Let $P(\mathrm{D})$ be an elliptic operator of order m, and let s be a real number satisfying $0<s<m$. Assume that $1<p<\infty$ and that $q(x)=q_1(x)\,q_2(x)$, where q_1 and q_2 satisfy* (5.1) *and* (5.2), *respectively. If*

$$\int_{|x-y|<1} |q(x)|\,\mathrm{d}x \to 0 \text{ as } |y|\to\infty\,, \tag{5.5}$$

then $P(\mathrm{D})+q$ acting on $D(q)\cap C_0^\infty$ has an s-extension B such that

$$\sigma_{\mathrm{e}}(B)\subset\sigma(P_0)\,. \tag{5.6}$$

If $\rho(P_0)$ is not empty, then

$$\sigma_{\mathrm{e}}(B)=\sigma(P_0)\,. \tag{5.7}$$

In order to prove the theorem, we shall need several lemmas. In the sequel it will be convenient to let $M_{\alpha,p}$ denote the set of those measurable function q such that $M_{\alpha,p}(q)<\infty$.

Lemma 5.2. *If $1<p<\infty$ and*

$$\|q\varphi\|_{0,p}\leqslant c_0\|\varphi\|_{s,p}\,,\qquad \varphi\in C_0^\infty\,, \tag{5.8}$$

then

$$\|q\varphi\|_{s,p'}\leqslant c_0\|\varphi\|_{0,p'}\qquad \varphi\in C_0^\infty\,. \tag{5.9}$$

Proof. We have for $\varphi,\psi\in C_0^\infty$

$$\begin{aligned}|(q\varphi,\psi)|=|(\varphi,\bar{q}\psi)|&\leqslant\|\varphi\|_{0,p'}\|q\psi\|_{0,p}\\ &\leqslant c_0\|\varphi\|_{0,p'}\|\psi\|_{s,p}\,.\end{aligned}$$

By Theorem 4.12 of ch. 2 this implies (5.9).

Corollary 5.3. *If $1<p<\infty$ and $q\in M_{\alpha,p}$ for $\alpha<ps$, then* (5.9) *holds.*

Lemma 5.4. *If $\rho\geqslant 1$, $\theta\geqslant 0$, $\theta\rho\leqslant p$ and $q\in M_{\alpha,p}$, then $|q|^\theta$ is in $M_{\beta,\rho}$ for each β satisfying $\beta p>\alpha\theta\rho$. If $\theta\rho=p$, then $q\in M_{\alpha,\rho}$.*

Proof. The second statement is trivial. If $\theta\rho<p$, we have by Hölder's inequality (see (1.8) of ch. 2)

$$\int_{|x|<1} |q(x-y)|^{\theta\rho} |x|^{\beta-n} \mathrm{d}x$$

$$\leqslant \left(\int_{|x|<1} |q(x-y)|^{p} |x|^{\alpha-n} \mathrm{d}x\right)^{1/t} \left(\int_{|x|<1} |x|^{\gamma-n} \mathrm{d}x\right)^{1/t'}, \tag{5.10}$$

where $t = p/\theta\rho$ and $\gamma = t'[(\beta-n)-(\alpha-n)/t]+n = t[\beta-(\alpha/t)] = t'[\beta p-\alpha\theta\rho]/p$. By hypothesis $q \in M_{\alpha,p}$ and $\gamma > 0$. Thus both integrals on the right-hand side of (5.10) are uniformly bounded.

Lemma 5.5. *Suppose*

$$\frac{1}{p} = \frac{1}{p_1} + \frac{1}{p_2}, \qquad \frac{\alpha}{p} = \frac{\alpha_1}{p_1} + \frac{\alpha_2}{p_2} \tag{5.11}$$

and

$$q_j \in M_{\alpha_j, p_j}, \qquad j = 1, 2, \tag{5.12}$$

where $p \geqslant 1$, $p_j \geqslant 1$ *and* $\alpha_j > 0$ *for* $j = 1, 2$. *Then* $q_1 q_2 \in M_{\alpha,p}$.

Proof. Set $t = p_1/p > 1$. Then $t' = p_2/p$ and $\alpha = \alpha_1/t + \alpha_2/t'$. Hence by Hölder's inequality

$$\int_{|x|<1} |q_1(x-y)\, q_2(x-y)|^{p} |x|^{\alpha-n} \mathrm{d}x$$

$$\leqslant \left(\int_{|x|<1} |q_1(x-y)|^{p_1} |x|^{\alpha_1-n} \mathrm{d}x\right)^{1/t} \left(\int_{|x|<1} |q_2(x-y)|^{p_2} |x|^{\alpha_2-n} \mathrm{d}x\right)^{1/t'}.$$

This completes the proof.

Lemma 5.6. *Suppose* $1 < p < \infty$ *and* $q(x) = q_1(x)\, q_2(x)$, *where* q_1 *satisfies* (5.1) *and* q_2 *satisfies* (5.2). *If* (5.5) *holds, then* $q(x) = h_1(x)\, h_2(x)$ *where*

$$h_1 \in M_{\gamma,p}, \qquad \gamma < ps \tag{5.13}$$

$$h_2 \in M_{\delta,p'}, \qquad \delta < p'(m-s) \tag{5.14}$$

$$\int_{|x-y|<1} |h_1(x)|^{p} \mathrm{d}x \to 0 \;\; as \;\; |y| \to \infty\,. \tag{5.15}$$

Proof. We take

$$h_1 = |q_1|^{1-\theta_1} |q_2|^{\theta_2}, \qquad h_2 = |q_1|^{\theta_1} |q_2|^{1-\theta_2},$$

where $0 < \theta_1, \theta_2 < 1$ are to be chosen. Since $q_1 \in M_{\alpha,p}$ and $q_2 \in M_{\beta,p'}$, we have $|q_1|^{1-\theta_1} \in M_{\alpha, p/(1-\theta_1)}$ and $|q_2|^{\theta_2} \in M_{\beta p'/\theta_2}$ (Lemma 5.4). Hence $h_1 \in M_{\gamma,r}$, where

$$\frac{1}{r}=\frac{1-\theta_1}{p}+\frac{\theta_2}{p'},\qquad \frac{\gamma}{r}=\frac{\alpha(1-\theta_1)}{p}+\frac{\beta\theta_2}{p'}$$

(Lemma 5.5). We want $r=p$ and $\gamma<ps$. This will happen if

$$\theta_1 p'=\theta_2 p \tag{5.16}$$

and

$$\gamma=\alpha+(\beta-\alpha)\theta_1<ps\,. \tag{5.17}$$

Similarly since $|q_1|^{\theta_1}\in M_{\alpha,p/\theta_1}$ and $|q_2|^{1-\theta_2}\in M_{\beta,p'/(1-\theta_2)}$, we have $h_2\in M_{\delta,p'}$, where

$$\frac{\delta}{p'}=\frac{\gamma\theta_1}{p}+\frac{\beta(1-\theta_2)}{p'}.$$

This means that we want

$$\delta=\beta+(\alpha-\beta)\theta_2<p'(m-s)\,. \tag{5.18}$$

We can satisfy (5.16), (5.17) and (5.18) by taking $\theta_1=\varepsilon/p$, where $\varepsilon<1$ is taken so small that (5.17) and (5.18) hold. Moreover, we will have $\theta_1+\theta_2=\varepsilon<1$. Thus by Hölder's inequality

$$\int_{|x-y|<1}|h_1(x)|^p\,\mathrm{d}x=\int_{|x-y|<1}|q_1(x)|^{p(1-\varepsilon)}|q(x)|^{\varepsilon}\,\mathrm{d}x$$
$$\leqslant\left(\int_{|x-y|<1}|q_1(x)|^p\,\mathrm{d}x\right)^{1-\varepsilon}\left(\int_{|x-y|<1}|q(x)|\,\mathrm{d}x\right)^{\varepsilon}\to 0 \text{ as } |y|\to\infty\,.$$

Thus (5.15) holds and the proof is complete.

Lemma 5.7. *Under the hypotheses of Lemma* 5.6, q *is a compact operator from* $H^{s,p}$ *to* $H^{s-m,p}$ *and from* $H^{m-s,p'}$ *to* $H^{-s,p'}$.

Proof. By Lemma 5.6, $q=h_1h_2$ and (5.13)–(5.15) hold. By (5.13) and (5.15), h_1 is a compact operator from $H^{s,p}$ to L^p (Theorem 4.1). By (5.15), h_2 is a bounded operator from L^p to $H^{s-m,p}$ (Lemma 5.2). Hence q is compact from $H^{s,p}$ to $H^{s-m,p}$. Similar reasoning proves the second assertion. One can also obtain it from the first by duality.

Lemma 5.8. *Let*

$$a(\varphi,\psi)=(P(\mathrm{D})\varphi,\psi)\,,\qquad \varphi,\psi\in S\,. \tag{5.19}$$

By Theorem 2.4 *of ch.* 3 *the bilinear form* $a(\varphi,\psi)$ *can be extended to a bounded* *bilinear form $a(u, v)$ on $H^{s,p} \times H^{m-s,p'}$. Let $\hat{A}$ denote the extended operator associated with $a(u, v)$ (see §6 of ch. 1). If $P(\xi) \neq 0$ for $\xi \in E^n$, then $\hat{A}$ is one–to–one and onto.*

Proof. By definition $u \in D(\hat{A})$ and $\hat{A}u = f$ if $u \in H^{s,p}$, $f \in H^{s-m,p}$ and

$$a(u, w) = (f, w), \qquad w \in H^{m-s,p'}. \tag{5.20}$$

Now there is a sequence $\{\varphi_k\}$ of functions in C_0^∞ which converges to u in $H^{s,p}$. Thus

$$|(P(\mathrm{D})\varphi_k, w)| \leqslant \|\varphi_k - u\|_{s,p} \|w\|_{m-s,p'} + |(f, w)|$$

by Theorem 2.4 of ch. 3. This implies

$$\|P(\mathrm{D})\varphi_k\|_{s-m,p} \leqslant \|\varphi_k - u\|_{s,p} + \|f\|_{s-m,p}$$

(Theorem 4.12 of ch. 2). If we now apply Theorem 3.1 of ch. 3 we obtain

$$\|\varphi_k\|_{s,p} \leqslant C(\|\varphi_k - u\|_{s,p} + \|f\|_{s-m,p}).$$

Letting $k \to \infty$ we have

$$\|u\|_{s,p} \leqslant C\|f\|_{s-m,p}. \tag{5.21}$$

This shows that $\hat{A}$ is one–to–one and that its range is closed (Theorem 2.3 of ch. 1). To complete the proof we must show that it is onto. If this were not so, there would exist a $w \in H^{m-s,p'}$ such that $(\hat{A}u, w) = 0$ for all $u \in D(\hat{A})$. In particular we have $(P(\mathrm{D})v, w) = 0$ for all $v \in S$. But by Lemma 1.3 of ch. 4 for each $g \in S$ there is a $v \in S$ such that $P(\mathrm{D})v = g$. Hence $(g, w) = 0$ for all $g \in S$. This implies that $w = 0$ (Theorem 4.12 of ch. 2). Hence $\hat{A}$ is onto, and the proof is complete.

We can now give the

Proof of Theorem 5.1. We have already shown that B exists. Set $W = H^{s,p}$, $Z = H^{m-s,p'}$ and $X = Y = L^p$. Let A_0 denote the operator $P(\mathrm{D})$ acting on $D(q) \cap C_0^\infty$, and let $a(\varphi, \psi)$ be the bilinear form given by (5.19). Let $\hat{A}$ be defined as in Lemma 5.8. If $\lambda \in \rho(P_0)$, then $P(\xi) \neq \lambda$ for $\xi \in E^n$ (Theorem 3.2 of ch. 4). Hence $\hat{A} - \lambda$ is one–to–one and onto (Lemma 5.8). Let $\hat{C}$ denote the extended operator associated with the bilinear form $c(u, v) = (q_1 u, \bar{q}_2 v)$. By Lemma 5.7, $\hat{C}$ is a compact operator from $H^{s,p}$ to $H^{s-m,p}$. We now apply Theorem 6.3 of ch. 1 to conclude that $B - \lambda \in \Phi(L^p)$ and that $i(B - \lambda) = 0$. Thus λ is not in $\sigma_e(B)$. This proves (5.6).

To prove (5.7) we may assume that $0 \in \rho(P_0)$. By Lemma 5.8, $\hat{A}^{-1}$ exists and is a bounded operator from $H^{s-m,p}$ to $H^{s,p}$ (see Theorem 2.3 of ch. 1). Since $\hat{C}$ is compact, our conclusion follows from Theorem 6.13 of ch. 1.

We now consider perturbation of $P(\mathrm{D})$ by an operator of the form

$$Q(x, \mathrm{D}) = \sum_{\substack{|\mu|<s \\ |\nu|<m-s}} \mathrm{D}^{\nu} g_{\mu\nu}(x) h_{\mu\nu}(x) \mathrm{D}^{\mu}, \tag{5.22}$$

where we assume that $P(\mathrm{D})$ is an elliptic operator of order m. As before, let s be a fixed real number satisfying $0<s<m$.

We assume

$$h_{\mu\nu} \in M_{\alpha-p|\mu|,p}, \qquad \alpha < ps, \tag{5.23}$$

$$g_{\mu\nu} \in M_{\beta-p'|\nu|,p'}, \qquad \beta < p'(m-s), \tag{5.24}$$

and

$$\int_{|x-y|<1} |g_{\mu\nu}(x) h_{\mu\nu}(x)| \, \mathrm{d}x \to 0 \text{ as } |y| \to \infty, \text{ each } \mu, \nu. \tag{5.25}$$

Let V be the set of those $\varphi \in C_0^\infty$ such that $Q(x, \mathrm{D})\varphi \in L^p$. We have

Theorem 5.9. *Under the above hypotheses the operator $P(\mathrm{D})+Q(x, \mathrm{D})$ on V has an s-extension E satisfying*

$$\sigma_e(E) \subset \sigma(P_0). \tag{5.26}$$

Moreover, if $\rho(P_0)$ is not empty, then

$$\sigma_e(E) = \sigma(P_0). \tag{5.27}$$

Proof. Set

$$a(\varphi, \psi) = (P(\mathrm{D})\varphi, \psi), \qquad \varphi, \psi \in C_0^\infty$$

$$c(\varphi, \psi) = \Sigma (h_{\mu\nu} \mathrm{D}^{\mu} \varphi, g_{\mu\nu} \mathrm{D}^{\nu} \psi).$$

By Theorem 2.4 of ch. 3

$$|a(\varphi, \psi)| \leqslant C \|\varphi\|_{s,p} \|\psi\|_{m-s,p'}, \qquad \varphi, \psi \in C_0^\infty, \tag{5.28}$$

while by Theorem 4.4

$$|c(\varphi, \psi)| \leqslant C \|\varphi\|_{s,p} \|\psi\|_{m-s,p'}, \qquad \varphi, \psi \in C_0^\infty. \tag{5.29}$$

Moreover by definition

$$(P(\mathrm{D})\varphi + Q(x, \mathrm{D})\varphi, \psi) = a(\varphi, \psi) + c(\varphi, \psi), \qquad \varphi \in V, \ \psi \in C_0^\infty. \tag{5.30}$$

Thus $P(\mathrm{D})+Q(x, \mathrm{D})$ on V has an s-extension corresponding to the bilinear form $b(\varphi, \psi)=a(\varphi, \psi)+c(\varphi, \psi)$. By Lemma 5.6 we may assume

$$\int_{|x-y|<1} |h_{\mu\nu}(x)|^p \mathrm{d}x \to 0 \text{ as } |y| \to \infty, \text{ each } \mu, \nu . \tag{5.31}$$

Hence the operator $h_{\mu\nu}\mathrm{D}^\mu$ is a compact operator from $H^{s,p}$ to L^p. Since the operator $\mathrm{D}^\nu g_{\mu\nu}$ is bounded from L^p to $H^{s-m,p}$, we see that the extended operator $\hat{C}$ corresponding to $c(u, v)$ is compact from $H^{s,p}$ to $H^{s-m,p}$. We now apply Theorem 6.3 of ch. 1 to obtain (5.26). To prove (5.24) we apply the reasoning of the proof of Theorem 5.1.

CHAPTER 7

OPERATORS BOUNDED FROM BELOW

In this chapter we specialize to the case $p=2$ and show that one can strengthen the theorems of the preceding chapter when $\sigma(P_0)$ is contained in a half-plane.

§1. Introduction

In Theorem 5.1 of ch. 6 we assumed that $q=q_1q_2$, where

$$q_1\in M_{\alpha,p}\,, \qquad \alpha<ps \tag{1.1}$$

$$q_2\in M_{\beta,p'}, \qquad \beta<p'(m-s) \tag{1.2}$$

where s satisfies $0<s<m$. It was shown that

$$\int_{|x|<1}|q(x-y)|\,\mathrm{d}x\to 0 \text{ as } |y|\to\infty \tag{1.3}$$

implies that $P(\mathrm{D})+q$ on $C_0^\infty\cap D(q)$ has an extension B satisfying

$$\sigma_{\mathrm{e}}(B)=\sigma(P_0)\,, \tag{1.4}$$

where $P(\mathrm{D})$ is an elliptic operator of order m for which $\rho(P_0)$ is not empty. Conditions (1.1) and (1.2) imply that there is a $\gamma<m$ such that $q\in M_{\gamma,1}$ (Lemma 5.5 of ch. 6). In this chapter we shall show that this stipulation can be weakened when $p=2$ and $\sigma(P_0)$ is contained in an angle less than π. We first make a transformation to bring $\sigma(P_0)$ into an angle of the form $|\theta|\leqslant\theta_0<\frac{1}{2}\pi$. The fact that L^2 is a Hilbert space allows us to employ regularly accretive extension (see ch. 1, §6). This allows us to conclude under weaker assumptions than (1.1)–(1.3) that $P(\mathrm{D})+q$ on $C_0^\infty\cap D(q)$ has an extension B such that

$$(P_0-\lambda)^{-1}-(B-\lambda)^{-1}$$

is a compact operator on L^2 for some $\lambda\in\rho(P_0)\cap\rho(B)$. We then apply Theorem 4.7 of ch. 1 to obtain (1.4). Details are carried out in §§7–9.

The same method works under suitable hypotheses for operators of the form

$$P(\mathrm{D})+\sum_{j,k=1}^{N}\bar{P}_j(\mathrm{D})\,q_{jk}(x)\,P_k(\mathrm{D}),$$

where the $P_k(\mathrm{D})$ are constant coefficient operators of orders $<m$ (the order of $P(\mathrm{D})$). In §10 we give sufficient conditions for this operator to have a regularly accretive extension B satisfying (1.4) even though the perturbing operator is of order $2m-2$.

The method is also applied to non-elliptic operators using the methods employed at the beginning of ch. 5. In §§2–5 we give conditions for the existence of regularly accretive extensions having the desired properties. As before the results are not as strong as those for elliptic operators. Examples are discussed in §5.

We also consider the problem of determining when $\sigma_e(B)$ is contained in a half-plane. For non-elliptic operators this is discussed in §6; for elliptic operators, in §10.

§2. Regularly accretive extensions

Throughout this chapter we shall assume that $P(\xi)$ satisfies for some constants C, M

$$|\operatorname{Im} P(\xi)| \leqslant C\,(\operatorname{Re} P(\xi)+M), \quad \xi \text{ real}. \tag{2.1}$$

This merely states that the spectrum of P_0 is contained in the region

$$|y| \leqslant C(x+M)$$

of the x–y plane (Corollary 3.3 of ch. 4). Clearly if $\sigma(P_0)$ is contained in an angle less than π, then (2.1) can be achieved by multiplying $P(\xi)$ by an appropriate complex constant. If $P(\xi)=\Sigma\, a_\mu \mathrm{D}^\mu$, we let $\operatorname{Re} P(\xi)$ denote the polynomial $\Sigma\,(\operatorname{Re} a_\mu)\,\xi^\mu$ and $\operatorname{Im} P(\xi)$ the polynomial $\Sigma\,(\operatorname{Im} a_\mu)\,\xi^\mu$.

Lemma 2.1. *If* $P(\xi)$ *satisfies* (2.1), *then*

$$|\operatorname{Im}\,(P(\mathrm{D})v,\, v)| \leqslant \operatorname{Re}(P(\mathrm{D})v,\, v)+M\,\|v\|^2, \qquad v\in S. \tag{2.2}$$

Proof. We have by Theorem 2.4 of ch. 2 and Theorem 1.2 of ch. 3

$$\begin{aligned}\operatorname{Re}\,(P(\mathrm{D})v,\, v) &= \operatorname{Re}\int P(\xi)\,|Fv|^2\,\mathrm{d}\xi = \int \operatorname{Re} P(\xi)\,|Fv|^2\,\mathrm{d}\xi\\ &\geqslant \pm C^{-1}\int \operatorname{Im} P(\xi)\,|Fv|^2\,\mathrm{d}\xi - M\int |Fv|^2\,\mathrm{d}\xi\\ &= \pm C^{-1}\operatorname{Im}\,(P(\mathrm{D})v,\, v)-M\,\|v\|^2.\end{aligned}$$

This gives (2.2).

Now let $q(x)$ be a function locally in L^1. Set $\rho(x)^2 = \max[\mathrm{Re}\, q(x), 0]$, $\sigma(x)^2 = -\min[\mathrm{Re}\, q(x), 0]$. Then

$$\mathrm{Re}\, q = \rho^2 - \sigma^2 , \tag{2.3}$$

and ρ, σ are locally in L^2. Set

$$(\varphi, \psi)_W = \int (\mathrm{Re}\, P(\xi) + M + 1) F\varphi \overline{F\psi}\, d\xi + (\rho\varphi, \rho\psi), \quad \varphi, \psi \in C_0^\infty . \tag{2.4}$$

Then $(\varphi, \psi)_W$ is a scalar product on C_0^∞, and the corresponding norm satisfies

$$\|\varphi\| \leqslant \|\varphi\|_W , \qquad \varphi \in C_0^\infty . \tag{2.5}$$

Thus the completion W of C_0^∞ with respect to this norm is a Hilbert space continuously and densely embedded in L^2. Since q is locally in L^1, $(q\varphi, \psi)$ is defined for $\varphi, \psi \in C_0^\infty$. Set

$$b(\varphi, \psi) = (P(D)\varphi + q\varphi, \psi), \qquad \varphi, \psi \in C_0^\infty . \tag{2.6}$$

If we can show that there are constants C, N such that

$$|b(\varphi, \psi)| \leqslant C \|\varphi\|_W \|\psi\|_W , \qquad \varphi, \psi \in C_0^\infty , \tag{2.7}$$

and

$$\|\varphi\|_W^2 \leqslant C\, [\mathrm{Re}\, b(\varphi) + N \|\varphi\|^2] , \qquad \varphi \in C_0^\infty , \tag{2.8}$$

it will follow via Theorem 6.6 of ch. 1 that $P(D) + q$ on $C_0^\infty \cap D(q)$ has a regularly accretive extension. A sufficient condition is given in

Theorem 2.2. *If there is a constant C such that*

$$|\mathrm{Im}\, q| \leqslant C(\rho^2 + 1) , \qquad \sigma^2 \leqslant C , \tag{2.9}$$

then (2.7) *and* (2.8) *hold. Hence $P(D) + q$ on $C_0^\infty \cap D(q)$ has a regularly accretive extension.*

In case (2.9) does not hold, we have

Theorem 2.3. *Set $\tilde{P}(\xi) = \mathrm{Re}\, P(\xi)$. Assume that there are constants a, b such that $a \leqslant 1$, $b > a + n - an$ and*

$$\tilde{P}^{(\mu)}(\xi)/\tilde{P}(\xi) = O(|\xi|^{-a|\mu|}) \text{ as } |\xi| \to \infty, \text{ all } \mu , \tag{2.10}$$

$$1/\tilde{P}(\xi) = O(|\xi|^{-2b}) \text{ as } |\xi| \to \infty . \tag{2.11}$$

Let k_0 be the smallest non-negative integer such that $k_0 a > n - b$. Assume that $|\mathrm{Im}\, q| = \tau_1^2 + \tau_2^2$ with

$$\tau_1^2 \leqslant C(\rho^2 + 1) \tag{2.12}$$

for some constant C, while τ_2 *and* σ *are in* $M_{\alpha,2}$, *where*

$$0<\alpha<2(n-k_0)\,. \tag{2.13}$$

Then (2.7) *and* (2.8) *hold. In particular,* $P(\mathrm{D})+q$ *on* $C_0^\infty \cap D(q)$ *has a regularly accretive extension.*

In proving Theorem 2.3 we shall make use of

Lemma 2.4. *Let* k_0 *be an integer satisfying* $0\leqslant k_0<n$, *and let* w *be a function in* $C_B^\infty \cap H^{s,p}$ *for some* s, p *and such that* $\mathrm{D}^\mu w\in L^1$ *for* $|\mu|=k_0, n+1$ *with*

$$\|\mathrm{D}^\mu w\|_1 \leqslant K_1\,, \qquad |\mu| = k_0$$
$$\|\mathrm{D}^\mu w\|_1 \leqslant K_2\,, \qquad |\mu| = n+1\,.$$

Let σ *be a function in* $M_{\alpha,2}$, *where* α *satisfies* (2.13). *Then*

$$\|\sigma u\|^2 \leqslant C(K_1+K_2)^2 M_{\alpha,2}(\sigma) \int |w(\xi)|^{-2}|Fu|^2 \mathrm{d}\xi\,, \qquad u\in S\,, \tag{2.14}$$

where the constant C *depends only on* k_0, n *and* α.

Proof. Inequality (2.14) requires proof only for those u such that $Fu/w\in L^2$. In this case there is a $v\in L^2$ such that

$$Fv = Fu/w$$

(Theorem 4.5 of ch. 2). Thus by Theorem 4.6 of ch. 2

$$u=\bar{F}(w) * v\,.$$

Hence (2.14) will follow from

$$\|\sigma(\bar{F}(w) * v)\|^2 \leqslant C(K_1+K_2)^2 M_{\alpha,2}(\sigma)\|v\|^2, \qquad v\in L^2\,. \tag{2.15}$$

But (2.15) is a consequence of Theorem 2.2 and Lemma 2.3 of ch. 5. This completes the proof.

We now give the

Proof of Theorem 2.3. By Lemma 6.10 of ch. 1, it suffices to prove

$$|b(\varphi)| \leqslant C\|\varphi\|_W^2\,, \qquad \varphi\in C_0^\infty \tag{2.16}$$

and (2.8). By Lemma 2.1, inequality (2.16) will follow from

$$\|\tau_1\varphi\|^2 \leqslant C(\|\rho\varphi\|^2+\|\varphi\|^2), \qquad \varphi\in C_0^\infty\,, \tag{2.17}$$

and

$$\|\sigma\varphi\|^2+\|\tau_2\varphi\|^2 \leqslant C\int (\tilde{P}(\xi)+M+1)|F\varphi|^2 \mathrm{d}\xi, \qquad \varphi\in C_0^\infty\,. \tag{2.18}$$

Now (2.17) is an obvious consequence of (2.12). To prove (2.16), set $w = 1/(\tilde{P}(\xi)+M+1)^{\frac{1}{2}}$. Then $w \in C^\infty$ and $D^\mu w$ consists of a sum of terms of the form

$$\text{constant } \tilde{P}^{(\mu^{(1)})}(\xi)\cdots\tilde{P}^{(\mu^{(k)})}(\xi)/(\tilde{P}(\xi)+M+1)^{k+\frac{1}{2}}, \; \mu^{(1)}+\ldots\mu^{(k)}=\mu.$$

Hence by (2.10) and (2.11)

$$|D^\mu w| \leqslant C|\xi|^{-a|\mu|-b}, \text{ each } \mu. \tag{2.19}$$

Thus $D^\mu w \in L^1$ for $a|\mu|+b>n$, i.e., for $|\mu| \geqslant k_0$. By hypothesis $k_0<n$, so that we can now apply Lemma 2.4. Since σ and τ_2 are in $M_{\alpha,2}$ with α satisfying (2.13), we obtain (2.18).

Inequality (2.8) is a bit more delicate. By definition

$$\|\varphi\|_W^2 = \operatorname{Re} b(\varphi)+\|\sigma\varphi\|^2+(M+1)\|\varphi\|^2. \tag{2.20}$$

Thus (2.8) will follow if we can show that there are an $\varepsilon>0$ and a constant C such that

$$\|\sigma\varphi\|^2 \leqslant (1-\varepsilon)\|\varphi\|_W^2+C\|\varphi\|^2. \tag{2.21}$$

Inequality (2.21) will follow if we can prove that

$$\|\sigma\varphi\|^2 \leqslant (1-\varepsilon)\int(\tilde{P}(\xi)+M+1)|F\varphi|^2\,d\xi+C\|\varphi\|^2. \tag{2.22}$$

We apply Lemma 2.4. Let $\psi(\xi)$ be a function in C_0^∞ satisfying (3.7)–(3.9) of ch. 5, and define the function $\psi_r(\xi)$ by (3.10) of that chapter. Let $w(\xi)$ be the function defined above and set $w_1=(1-\psi_r)w$, $w_2=\psi_r w$. Then by Lemma 3.4 of ch. 5

$$\|D^\mu w_1\|_1 \to 0 \text{ as } r\to\infty, \qquad |\mu| \geqslant k_0. \tag{2.23}$$

Hence if $\varepsilon>0$ is given, we have by Lemma 2.4

$$2\|\sigma u\|^2 \leqslant (1-\varepsilon)\int|w_1(\xi)|^{-2}|Fu|^2\,d\xi, \qquad u\in S \tag{2.24}$$

for r sufficiently large. Fix r so that (2.24) holds, and let u_1 and u_2 be defined by

$$Fu_1=(1-\psi_r)F\varphi, \qquad Fu_2=\psi_r F\varphi.$$

Then by (2.24)

$$2\|\sigma u_1\|^2 \leqslant (1-\varepsilon)\int|w(\xi)|^{-2}|F\varphi|^2\,d\xi. \tag{2.25}$$

Moreover, by Theorem 2.1 of ch. 6

$$\|\sigma u_2\|^2 \leqslant C\|u_2\|_{j,2}^2,$$

where $j=n-k_0$. But

$$\|u_2\|_{j,2}^2 = \int (1+|\xi|^2)^j |Fu_2|^2 \,\mathrm{d}\xi$$
$$= \int |\psi_r|^2 (1+|\xi|^2)^j |F\varphi|^2 \,\mathrm{d}\xi \leqslant C\|\varphi\|^2 ,$$

since the function $(1+|\xi|^2)^j \psi_r$ is in C_0^∞ and hence is bounded for each r. Thus

$$\|\sigma u_2\|^2 \leqslant C\|\varphi\|^2 . \tag{2.26}$$

We now merely note that $\varphi = u_1 + u_2$. Thus (2.22) is a consequence of (2.25) and (2.26). This completes the proof.

The proof of Theorem 2.2 is similar to, but much simpler than that of Theorem 2.3. In fact we can merely set $\tau_2 = 0$ and ignore the hypotheses (2.10) and (2.11) in Theorem 2.3. Then we note that

$$\|\sigma\varphi\| \leqslant C\|\varphi\| .$$

§3. Invariance of the essential spectrum

Now that we have sufficient conditions for $P(\mathrm{D})+q$ on $C_0^\infty \cap D(q)$ to have a regularly accretive extension, the next question we ask is under what conditions will it have a regularly accretive extension B satisfying

$$\sigma_e(B) = \sigma(P_0) . \tag{3.1}$$

One answer is given by

Theorem 3.1. *Let $P(\xi)$ and $q(x)$ satisfy all of the hypotheses of Theorem 2.3. Let j_0 be the smallest non-negative integer such that $j_0 a > n-2b$. Assume that $\rho \in M_{\beta,2}$ for some β satisfying*

$$0 < \beta < 2(n-j_0) , \tag{3.2}$$

and that

$$\int_{|x-y|<1} |q(x)| \,\mathrm{d}x \to 0 \quad as \quad |y| \to \infty . \tag{3.3}$$

Assume also that

$$\tau_1(x) = \mathrm{O}(\rho(x)) \quad as \quad |x| \to \infty . \tag{3.4}$$

Then $P(\mathrm{D})+q$ *on* $C_0^\infty \cap D(q)$ *has a regularly accretive extension B, satisfying* (3.1).

Proof. Consider the norms given by

$$\|v\|_Z^2 = \int (|\tilde{P}(\xi)|^2+1)\,|Fv|^2 \mathrm{d}\xi\,, \qquad v \in S \tag{3.5}$$

$$\|v\|_{\tilde{Z}}^2 = \int (|\tilde{P}(\xi)|^2+1)^{-1}|Fv|^2 \mathrm{d}\xi, \quad v \in S\,, \tag{3.6}$$

and let Z and $\tilde{Z}$ denote the Hilbert spaces obtained by completing S with respect to these norms. By (2.1) there is a constant C such that

$$C^{-1}\|v\|_Z \leqslant \|P(\mathrm{D})v\| + \|v\| \leqslant C\|v\|_Z, \qquad v \in S\,, \tag{3.7}$$

so that $Z = D(P_0)$.

Now for $u \in L^2$ and $v \in Z$ we have

$$|(u, \bar{P}_0 v)| \leqslant C\|u\|\,\|v\|_Z\,.$$

Thus for each $u \in L^2$ there is an $f \in \tilde{Z}$ such that

$$(u, \bar{P}_0 v) = (f, v)\,, \qquad v \in Z\,. \tag{3.8}$$

We define an operator $\tilde{P}$ from L^2 to $\tilde{Z}$ by $\tilde{P}u = f$. Clearly the operator $\tilde{P}$ is an extension of P_0. I claim that if $\lambda \in \rho(P_0)$, then $\tilde{P}-\lambda$ is one–to–one and onto. For then $\bar{\lambda}$ is in $\rho(\bar{P}_0)$ and hence there is a $v \in Z$ such that $(\bar{P}_0 - \bar{\lambda})v = u$. Hence by (3.8)

$$(u, u) = (\tilde{P}u - \lambda u, v)\,,$$

and consequently

$$\|u\|^2 \leqslant \|(\tilde{P}-\lambda)u\|_{\tilde{Z}}\,\|v\|_Z \leqslant C\|(\tilde{P}-\lambda)u\|_{\tilde{Z}}\,\|u\|\,.$$

Thus

$$\|u\| \leqslant C\|(\tilde{P}-\lambda)u\|_{\tilde{Z}}\,, \qquad u \in L^2\,, \tag{3.9}$$

showing that $\tilde{P}-\lambda$ is one–to–one and has closed range. To show that it is onto, let f be any element in $\tilde{Z}$. Then there is a sequence $\{f_k\}$ of functions in S converging to f in $\tilde{Z}$. Since $\lambda \in \rho(P_0)$, for each k there is a $u_k \in Z$ such that $(P_0 - \lambda)u_k = f_k$. Thus

$$(u_k, (\bar{P}_0 - \bar{\lambda})v) = (f_k, v)\,, \qquad v \in Z\,. \tag{3.10}$$

By (3.9), the sequence $\{u_k\}$ converges in L^2 to some function u. Taking the limit in (3.10) we obtain

$$(u, (\bar{P}_0 - \bar{\lambda})v) = (f, v)\,, \qquad v \in Z\,,$$

which shows that $(\tilde{P}-\lambda)u = f$. Thus $\tilde{P}-\lambda$ is onto $\tilde{Z}$.

Let B be the regularly accretive extension constructed in the proof of Theorem 2.3. Note that

$$(B-\tilde{P})u = qu\,, \qquad u \in D(B)\,. \tag{3.11}$$

For we have

$$(u, (\bar{P}(\mathrm{D})+\bar{q})v) = (Bu, v)\,, \qquad u \in D(B)\,,\ v \in S\,.$$

If we compare this with (3.8), we obtain (3.11).

Now I claim that $\rho, \sigma, \tau_1, \tau_2$ are all compact operators from Z to L^2. For ρ, σ, τ_2 this follows from Theorem 3.1 of ch. 5 if we note that

$$\int_{|x-y|<1} (\rho^2+\sigma^2+\tau_2^2)\mathrm{d}x \leqslant C \int_{|x-y|<1} |q(x)|\,\mathrm{d}x \to 0 \text{ as } |y| \to \infty\,. \tag{3.12}$$

Moreover, by (2.12) and (3.4) there is a bounded domain Ω such that

$$\|\tau_1 v\|^2 \leqslant C(\|\rho v\|^2+\|v\|_\Omega^2)\,, \qquad v \in D(\rho)\,, \tag{3.13}$$

where $\|v\|_\Omega$ denotes the $L^2(\Omega)$ norm of v. By (2.11)

$$\|v\|_{2b,2} \leqslant C\|v\|_Z\,, \qquad v \in Z\,.$$

Since $b>0$, this shows that every sequence which is bounded in the Z norm has a subsequence converging in $L^2(\Omega)$ (Theorem 4.4 of ch. 2). Since ρ is a compact operator from Z to L^2, inequality (3.13) shows that the same is true of τ_1.

By the same token, $\rho, \sigma, \tau_1, \tau_2$ are compact operators from L^2 to $\tilde{Z}$. For they are the adjoints of their complex conjugates considered as operators from Z to L^2 (see Theorem 2.17 of ch. 1).

Since B is regularly accretive, $(B-\lambda)^{-1}$ exists for $\operatorname{Re}\lambda+N \leqslant 0$ (Theorem 6.7 of ch. 1). Note that it is a bounded operator from L^2 to W. In fact by (2.8)

$$\begin{aligned}\|u\|_W^2/C &\leqslant \operatorname{Re} b(u)+N\|u\|^2 \leqslant \operatorname{Re}((B-\lambda)u, u)\\ &\leqslant \|(B-\lambda)u\|\,\|u\|\,, \qquad u \in D(B)\,.\end{aligned}$$

By (2.4), (2.17) and (2.18) $\rho, \sigma, \tau_1, \tau_2$ are all bounded operators from W to L^2.

We now apply Theorem 4.7 of ch. 1. Since $\rho(P_0)$ and $\rho(B)$ both contain half-planes of the form $\operatorname{Re}\lambda+C \geqslant 0$, there is a $\lambda \in \rho(P_0) \cap \rho(B)$. Thus we have

$$\begin{aligned}(P_0-\lambda)^{-1}-(B-\lambda)^{-1} &= (P_0-\lambda)^{-1}(B-\lambda)(B-\lambda)^{-1}\\ &\qquad -(\tilde{P}-\lambda)^{-1}(\tilde{P}-\lambda)(B-\lambda)^{-1}\\ &= (\tilde{P}-\lambda)^{-1}(B-\tilde{P})(B-\lambda)^{-1} = (\tilde{P}-\lambda)^{-1}q(B-\lambda)^{-1}\\ &= (\tilde{P}-\lambda)^{-1}[\rho^2-\sigma^2+\mathrm{i}(\tau_1^2+\tau_2^2)\operatorname{sgn}\operatorname{Im} q](B-\lambda)^{-1}\,,\end{aligned}$$

where $\operatorname{sgn}\gamma=\gamma/|\gamma|$ for $\gamma\neq 0$ and $\operatorname{sgn} 0=0$. We made use of the fact that $\tilde{P}$ is an extension of P_0. Now $\rho(B-\lambda)^{-1}$ is a bounded operator on L^2 since $(B-\lambda)^{-1}$ is bounded from L^2 to W and ρ is bounded from W to L^2. Moreover, ρ is compact from L^2 to $\tilde{Z}$ while $(\tilde{P}-\lambda)^{-1}$ is bounded from $\tilde{Z}$ to L^2. Hence the operator $(\tilde{P}-\lambda)^{-1}\rho^2(B-\lambda)^{-1}$ is a compact operator on L^2. The same reasoning applies to the rest. In handling $\operatorname{Im} q$, note that $\tau_1(B-\lambda)^{-1}$ is bounded on L^2 and the same is true of $\operatorname{sgn}\operatorname{Im} q$. Since $(\tilde{P}-\lambda)^{-1}\tau_1$ is compact on L^2, the entire expression is likewise. This completes the proof.

§4. Perturbation by an operator

We can extend the results of the preceding sections to cover the case of a perturbing operator. Theorems 2.3 and 3.1 will become special cases. Set

$$Q(x, \mathrm{D}) = \sum_{j,k=1}^{N} \bar{P}_j(\mathrm{D})\, q_{jk}(x)\, P_k(\mathrm{D}), \tag{4.1}$$

where the $P_k(\mathrm{D})$ are constant coefficient operators and the $q_{jk}(x)$ are functions defined on E^n. In addition to (2.10) we assume

$$P_k(\xi)/|\tilde{P}(\xi)|^{\frac{1}{2}} = \mathrm{O}(|\xi|^{-b_k}) \text{ as } |\xi|\to\infty, \qquad 1\leqslant k\leqslant N, \tag{4.2}$$

where $\tilde{P}(\xi)=\operatorname{Re} P(\xi)$ and the b_k are greater than $a+n-an$. Set

$$e(\varphi, \psi) = \Sigma(q_{jk} P_k(\mathrm{D})\varphi, P_j(\mathrm{D})\psi), \qquad \varphi, \psi\in C_0^{\infty}, \tag{4.3}$$

$$b(\varphi, \psi) = (P(\mathrm{D})\varphi, \psi) + e(\varphi, \psi), \qquad \varphi, \psi\in C_0^{\infty}, \tag{4.4}$$

$$\rho_k(x)^2 = \max\,[\operatorname{Re} q_{kk}(x), 0],\ \sigma_k(x)^2 = \rho_k(x)^2 - \operatorname{Re} q_{kk}(x), \qquad 1\leqslant k\leqslant N \tag{4.5}$$

$$(\varphi, \psi)_W = \int (\tilde{P}(\xi)+M+1)\, F\varphi\, \overline{F\psi}\, \mathrm{d}\xi + \sum_{k=1}^{N} (\rho_k P_k(\mathrm{D})\varphi, \rho_k P_k(\mathrm{D})\psi), \qquad \varphi, \psi\in C_0^{\infty}. \tag{4.6}$$

Let s_k be the smallest non-negative integer such that $s_k a > n-b_k$, $1\leqslant k\leqslant N$. Assume that ρ_k are locally in L^2 and that $\sigma_k\in M_{\alpha_k,2}$, where $\alpha_k<2(n-s_k)$. For $j\neq k$ assume that $q_{jk}=f_{jk}h_{jk}$ with $f_{jk}\in M_{\alpha_j,2}$ and $h_{jk}\in M_{\alpha_k,2}$. Finally, assume that

$$|\operatorname{Im} q_{kk}| = \tau_{k1}^2 + \tau_{k2}^2, \qquad 1\leqslant k\leqslant N,$$

where

$$\tau_{k1}^2 \leqslant C(\rho_k^2+1), \qquad 1 \leqslant k \leqslant N, \tag{4.7}$$

and τ_{k2} is in $M_{\alpha_k,2}$. Let V be the set of those $\varphi \in C_0^\infty$ such that $Q(x, \mathrm{D})\varphi$ is in L^2. We have

Theorem 4.1. *Under the assumption stated above, inequalities* (2.7) *and* (2.8) *hold. Hence* $P(\mathrm{D})+Q(x, \mathrm{D})$ *on* V *has a regularly accretive extension.*

Proof. We use the following inequalities.

$$\|\tau_{k1} P_k(\mathrm{D})\varphi\|^2 \leqslant C(\|\rho_k P_k(\mathrm{D})\varphi\|^2 + \|P_k(\mathrm{D})\varphi\|^2), \qquad \varphi \in C_0^\infty, \quad 1 \leqslant k \leqslant N. \tag{4.8}$$

For each $\varepsilon > 0$ there is a constant C such that

$$\begin{aligned} &\Sigma(\|\sigma_k P_k(\mathrm{D})\varphi\|^2 + \|\tau_{k2} P_k(\mathrm{D})\varphi\|^2 + \|P_k(\mathrm{D})\sigma\|^2) \\ &\qquad + \Sigma(\|h_{jk} P_k(\mathrm{D})\varphi\|^2 + \|f_{jk} P_j(\mathrm{D})\varphi\|^2) \\ &\leqslant \varepsilon \int (\tilde{P}(\xi)+M+1)|F\varphi|^2 \mathrm{d}\xi + C\|\varphi\|^2, \qquad \varphi \in C_0^\infty. \end{aligned} \tag{4.9}$$

These inequalities clearly imply (2.16), which in turn implies (2.7) (Lemma 6.10 of ch. 1). Moreover,

$$\begin{aligned} \|\varphi\|_W^2 = \operatorname{Re} b(\varphi) + \Sigma\|\sigma_k P_k(\mathrm{D})\varphi\|^2 + (M+1)\|\varphi\|^2 \\ - \operatorname{Re}\Sigma'(q_{jk} P_k(\mathrm{D})\varphi, P_j(\mathrm{D})\varphi), \end{aligned}$$

where Σ' denotes summation over j, k for which $j \neq k$. Thus (4.9) implies

$$\|\varphi\|_W^2 \leqslant \operatorname{Re} b(\varphi) + \varepsilon \int (\tilde{P}(\xi)+M+1)|F\varphi|^2 \mathrm{d}\xi + C'\|\varphi\|^2,$$

where C' depends on ε. If we take $\varepsilon < 1$, we obtain (2.8). Now inequality (4.8) follows immediately from (4.7). In proving (4.9) we shall make use of

$$P_k^{(\mu)}(\xi)/|\tilde{P}(\xi)|^{\frac{1}{2}} = \mathrm{O}(|\xi|^{-a|\mu|-b_k}) \text{ as } |\xi| \to \infty, \qquad 1 \leqslant k \leqslant N. \tag{4.10}$$

Assume this for the moment, and set $w_k = P_k(\xi)/(\tilde{P}(\xi)+M+1)^{\frac{1}{2}}$. Then $w_k \in C^\infty$ and $\mathrm{D}^\mu w_k$ consists of a sum of terms of the form

$$\text{constant } P_k^{(\nu)}(\xi)\tilde{P}^{(\mu^{(1)})}(\xi)\cdots\tilde{P}^{(\mu^{(i)})}(\xi)/(\tilde{P}(\xi)+M+1)^{i+\frac{1}{2}},$$

where $\nu+\mu^{(1)}+\ldots\mu^{(i)} = \mu$. Thus

$$|\mathrm{D}^\mu w_k| \leqslant C/|\xi|^{a|\mu|+b_k}, \text{ each } \mu, \quad 1 \leqslant k \leqslant N. \tag{4.11}$$

Define ψ_R as in the proof of Theorem 2.3. For $\varphi \in C_0^\infty$, define u_1 and u_2 by

$$Fu_1 = (1-\psi_R)F[P_k(\mathrm{D})\varphi], \qquad Fu_2 = \psi_R F[P_k(\mathrm{D})\varphi].$$

Then following the proof of Theorem 2.3, for any $\delta > 0$ we can take R so large that

$$2\|\sigma_k u_1\|^2 \leqslant \delta \int |w_k(\xi)|^{-2} |P_k(\xi) F\varphi|^2 \, d\xi$$

$$= \delta \int (\tilde{P}(\xi) + M + 1) |F\varphi|^2 \, d\xi \, .$$

Moreover

$$\|\sigma_k u_2\| \leqslant C \|u_2\|_{i,2} \, ,$$

where $i = n - s_k$. But

$$\|u_2\|_{i,2}^2 = \int (1 + |\xi|^2)^i |Fu_2|^2 \, d\xi$$

$$= \int |\psi_R(\xi) \, P_k(\xi)|^2 (1 + |\xi|^2)^i |F\varphi|^2 \, d\xi \leqslant C \|\varphi\|^2 \, ,$$

since $|\psi_R P_k(\xi)|^2 (1 + |\xi|^2)^i$ is in C_0^∞ for each R. This shows that

$$\|\sigma_k P_k(\mathrm{D}) \varphi\|^2 \leqslant \delta \int (\tilde{P}(\xi) + M + 1) |F\varphi|^2 \, d\xi + C \|\varphi\|^2 \, .$$

The same reasoning applies to each of the remaining terms in (4.9).

It remains only to prove (4.10). These relations follow immediately from

Lemma 4.2. *Let $P(\xi)$, $Q(\xi)$ be polynomials satisfying for ξ real*

$$P^{(\mu)}(\xi) / P(\xi) = \mathrm{O}(|\xi|^{-a|\mu|}) \quad \textit{as} \quad |\xi| \to \infty \, , \quad \textit{each } \mu \tag{4.12}$$

$$Q(\xi) / |P(\xi)|^{\frac{1}{2}} = \mathrm{O}(|\xi|^{-c}) \quad \textit{as} \quad |\xi| \to \infty \tag{4.13}$$

with $a \leqslant 1$. Then

$$Q^{(\mu)}(\xi) / |P(\xi)|^{\frac{1}{2}} = \mathrm{O}(|\xi|^{-a|\mu| - c}) \quad \textit{as} \quad |\xi| \to \infty, \quad \textit{each } \mu.$$

Proof. We follow the proof of Lemma 4.5 of ch. 5. By Theorem 1.3 of ch. 3, we have

$$t^{|\mu|} |Q^{(\mu)}(\xi)| \leqslant C \Sigma |Q(\xi + t\theta^{(j)})|$$

for real ξ and $t \geqslant 1$, where $\theta^{(1)}, \ldots, \theta^{(r)}$ are fixed real vectors. If

$$|\theta^{(j)}| \leqslant M \, , \qquad 1 \leqslant j \leqslant r \, ,$$

we take $t = |\xi|^a / 2M$. Then for $|\xi|$ large

$$|\xi + t\theta^{(j)}| \geqslant |\xi| - t|\theta^{(j)}| \geqslant \tfrac{1}{2}|\xi| \, .$$

Thus

$$|\xi|^{a|\mu|}|Q^{(\mu)}(\xi)| \leqslant C\Sigma|P(\xi+t\theta^{(j)})|^{\frac{1}{2}}|\xi+t\theta^{(j)}|^{-c}.$$

But

$$|P(\xi+t\theta^{(j)})| \leqslant C\Sigma|P^{(\nu)}(\xi)|\,|\xi|^{a|\nu|} \leqslant C'|P(\xi)|.$$

This gives the desired conclusion.

We now turn to a generalization of Theorem 3.1.

Theorem 4.3. *Assume that all of the hypotheses of Theorem* 4.1 *hold. Let* l_k *be the smallest non-negative integer such that* $l_k a > n-2b_k$, $1 \leqslant k \leqslant N$, *and assume that* $\rho_k \in M_{\delta_k, 2}$, *where*

$$0 < \delta_k < 2(n-l_k), \qquad 1 \leqslant k \leqslant N. \tag{4.15}$$

Assume further that

$$\tau_{k1}(x) = O(\rho_k(x)) \text{ as } |x| \to \infty, \qquad 1 \leqslant k \leqslant N, \tag{4.16}$$

and that

$$\int_{|x|<1} \Sigma\, |q_{jk}(x-y)|\,dx \to 0 \text{ as } |y| \to \infty. \tag{4.17}$$

Then $P(D)+Q(x, D)$ *on* V *has a regularly accretive extension* B *satisfying* (3.1).

Proof. We follow the proof of Theorem 3.1. In place of (3.11) we use

$$(B-\tilde{P})u = \Sigma\, \tilde{\bar{P}}_k(\rho_k^2-\sigma_k^2+i(\tau_{k1}^2+\tau_{k2}^2) \text{ sgn Im } q_k)P_{k0}+\Sigma'\, \bar{P}_j f_{jk} h_{jk} P_{k0} u. \tag{4.18}$$

In deriving (4.18) note that

$$(u, Q(x, D)^* v) = ([B-\tilde{P}]u, v), \qquad u \in D(B), \quad v \in S.$$

Moreover, by (4.8) and (4.9), W is contained in the domains of each of the operators $\sigma_k P_{k0}$, $\tau_{k1} P_{k0}$, $\tau_{k2} P_{k0}$, $\rho_k P_{k0}$, $f_{jk} P_{j0}$, $\bar{h}_{jk} P_{k0}$ for each j and k. In addition, the hypotheses guarantee that each of these operators is compact from Z to L^2. (Note that all we need is that either $f_{jk} P_{j0}$ or $\bar{h}_{jk} P_{k0}$ be compact from Z to L^2 for each j, k. For this we use Lemma 5.6 of ch. 6.) This not only gives (4.18) but also shows that $B-\hat{P}$ is a compact operator from W to $\tilde{Z}$. This is all that is needed to obtain the desired conclusion.

§5. An illustration

To illustrate the theory of the last few sections, we consider a class of operators which satisfy the hypotheses. Suppose

$$\tilde{P}(\xi) = \operatorname{Re} P(\xi) = \sum_{(\mu/e)=1} a_\mu \xi^\mu + 1$$

$$\operatorname{Im} P(\xi) = \sum_{(\mu/e)\leqslant 1} b_\mu \xi^\mu$$

where the a_μ, b_μ are real and e and (μ/e) are defined as in ch. 4, §5. Assume that $\tilde{P}(\xi)$ is semi-elliptic and, for definiteness that

$$\tilde{P}(\xi) \geqslant 1\,, \quad \xi \text{ real}\,.$$

Then by (5.5) of ch. 4

$$|\operatorname{Im} P(\xi)| \leqslant C \sum_{(\mu/e)\leqslant 1} |\xi^\mu| \leqslant C h(\xi)$$

when $h(\xi) \geqslant 1$. Otherwise we have

$$|\operatorname{Im} P(\xi)| \leqslant C\,.$$

Hence

$$|\operatorname{Im} P(\xi)| \leqslant C\tilde{P}(\xi)\,,$$

and (2.1) is satisfied. If m and M are defined as in ch. 4, §5, we may take $a = m/M$ in (2.10) and $b = \frac{1}{2}m$ in (2.11). The requirement $b > a + n - an$ now becomes $mM > 2(m + nM - mn)$. The integer k_0 of Theorem 2.3 now is the smallest non-negative integer greater than $n/m - M/2$.

Let $Q(\xi)$ be a polynomial defined by

$$Q(\xi) = \sum_{(\mu/e)\leqslant\delta} c_\mu \xi^\mu\,,$$

where $\delta < 1$. Then by (5.5) and (5.9) of ch. 4

$$|Q(\xi)/\tilde{P}(\xi)| \leqslant C h(\xi)^{\delta-1} \leqslant C' |\xi|^{m(\delta-1)}$$

for $|\xi|$ large. Thus if

$$P_k(\xi) = \Sigma\, d_{k\mu} \xi^\mu\,, \qquad (\mu/e) \leqslant 1 - 2b_k/m$$

$$Q_j(\xi) = \Sigma\, f_{j\mu} \xi^\mu\,, \qquad (\mu/e) \leqslant 1 - 2c_j/m$$

$$R_j(\xi) = \Sigma\, g_{j\mu} \xi^\mu\,, \qquad (\mu/e) \leqslant 1 - 2d_j/m\,,$$

where b_k, c_j, d_j are all greater than $(m + nM + mn)/M$, then the hypotheses of Theorem 4.1 are satisfied.

§6. Essential spectrum bounded from below

If $P(\xi)$ satisfies (2.1), then $\sigma(P_0)$ is contained in the half-plane $\operatorname{Re}\lambda+M\geqslant 0$. Theorem 3.1 gives conditions on q which guarantee that $P(\mathrm{D})+q$ on $C_0^\infty\cap D(q)$ has a regularly accretive extension B satisfying (3.1). In particular, $\sigma_e(B)$ will be contained in $\operatorname{Re}\lambda+M\geqslant 0$.

The hypotheses of Theorem 3.1 contain assumptions on $\rho(x)$. The purpose of this section is to show that one can conclude that $\sigma_e(B)$ is contained in $\operatorname{Re}\lambda+M\geqslant 0$ without making any assumptions on ρ.

Theorem 6.1. *Let $P(\mathrm{D})$ be an operator satisfying* (2.1) *and the hypotheses of Theorem* 2.3. *Assume that*

$$\int_{|x-y|<1}\sigma(x)^2\,\mathrm{d}x\to 0 \quad as \quad |y|\to\infty\,. \tag{6.1}$$

Then $P(\mathrm{D})+q$ on $C_0^\infty\cap D(q)$ has a regularly accretive extension B such that $\sigma_e(B)$ is contained in the half-plane $\operatorname{Re}\lambda+M\geqslant 0$.

In proving Theorem 6.1 we shall make use of

Lemma 6.2. *Let k_0 be an integer satisfying $0\leqslant k_0<n$, and let w be a function in $C^\infty(E^n)$ such that*

$$|\mathrm{D}^\mu w(\xi)|\leqslant C|\xi|^{-a|\mu|-b},\qquad |\mu|\leqslant n+1\,, \tag{6.2}$$

where $a\leqslant 1$ and $k_0 a>n-b$. Let $\sigma(x)$ be a function in $M_{\alpha,2}$, where α satisfies (2.13), *and suppose* (6.1) *holds. Let Y be the set of those $u\in L^2$ such that $Fu/w\in L^2$. Then σ is a compact operator from Y to L^2.*

We postpone the proof of Lemma 6.2 until the end of this section. We now use it to give the

Proof of Theorem 6.1. First assume that $\sigma=0$. Set

$$a(\varphi,\psi)=(P(\mathrm{D})\varphi,\psi)+(\rho\varphi,\rho\psi)+\mathrm{i}((\operatorname{Im} q)\varphi,\psi),\qquad \varphi,\psi\in C_0^\infty \tag{6.3}$$

$$(\varphi,\psi)_W=\int(\tilde{P}(\xi)+M+\varepsilon)F\varphi\overline{F\psi}\,\mathrm{d}\xi+(\rho\varphi,\rho\psi)\,,\qquad \varphi,\psi\in C_0^\infty\,, \tag{6.4}$$

where $\varepsilon>0$ is fixed. We let W be the completion of C_0^∞ with respect to the corresponding norm and extend $a(\varphi,\psi)$ to $W\times W$. Clearly a satisfies (see (2.17) and (2.18))

$$|a(u, v)| \leqslant C\|u\|_W \|v\|_W, \qquad u, v \in W \tag{6.5}$$
$$\|u\|_W^2 = \operatorname{Re} a(u) + (M+\varepsilon)\|u\|^2, \qquad u \in W. \tag{6.6}$$

Thus the operator A associated with $a(u, v)$ is regularly accretive and $\sigma(A)$ is contained in the half-plane $\operatorname{Re} \lambda + M + \varepsilon > 0$ (Theorem 6.7 of ch. 1). Since ε was arbitrary, we see that $\sigma(A)$ is contained in $\operatorname{Re} \lambda + M \geqslant 0$. Since A is an extension of $P(\mathrm{D}) + q$, the theorem is proved for $\sigma = 0$.

In the general case, set

$$\begin{aligned} e(\varphi, \psi) &= -(\sigma\varphi, \sigma\psi), \qquad \varphi, \psi \in C_0^\infty \\ c(\varphi, \psi) &= (\sigma\varphi, \sigma\psi), \qquad \varphi, \psi \in C_0^\infty . \end{aligned} \tag{6.7}$$

By (2.18)

$$|e(\varphi)| \leqslant c(\varphi) \leqslant C\|\varphi\|_W^2, \qquad \varphi \in C_0^\infty .$$

Hence e and c can be extended to be bounded on $W \times W$. Next set $w = 1/(\tilde{P}(\xi) + M + \varepsilon)^{\frac{1}{2}}$. Then by (2.19) we see that w satisifies the hypotheses of Lemma 6.2. Moreover, the set of those $u \in L^2$ such that $Fu/w \in L^2$ contains W with continuous inclusion. Thus if $\{u_k\}$ is a sequence of functions in W such that $\|u_k\|_W \leqslant C$, then there is a subsequence (also denoted by $\{u_k\}$) such that $c(u_j - u_k) \to 0$. Thus conditions a) and b) of Theorem 6.11 of ch. 1 are satisfied. To verify c), suppose $u_k \to 0$ in L^2 while $c(u_j - u_k) \to 0$. Then σu_k converges in L^2 to some function h. If $\psi \in C_0^\infty \cap D(\sigma)$, then

$$(h, \psi) = \lim\, (\sigma u_k, \psi) = \lim (u_k, \sigma\psi) = 0 .$$

Since $C_0^\infty \cap D(\sigma)$ is dense in L^2 (Lemma 6.2 of ch. 4), we have $h = 0$. Thus $c(u_k) \to 0$. Thus all of the hypotheses of Theorem 6.11 of ch. 1 are satisfied. Hence the operator B associated with $b(u, v) = a(u, v) + e(u, v)$ is regularly accretive and satisfies $\sigma_e(B) = \sigma_e(A)$. Since $\sigma_e(A)$ is contained in $\operatorname{Re} \lambda + M \geqslant 0$, the same is true of $\sigma_e(B)$. The proof is complete.

A similar generalization for the operator (4.1) is

Theorem 6.3. *Assume in addition to the hypotheses of Theorem* 4.1 *that*

$$\int_{|x|<1} \Sigma\, \sigma_k(x-y)^2\, \mathrm{d}x \to 0 \quad as \quad |y| \to \infty . \tag{6.8}$$

Then $P(\mathrm{D}) + Q(x, \mathrm{D})$ *on* V *has a regularly accretive extension* B *such that* $\sigma_e(B)$ *is contained in* $\operatorname{Re} \lambda + M \geqslant 0$.

The proof is similar to that of Theorem 6.1 and is omitted.

We now give the simple

Proof of Lemma 6.2. Let u be any function in Y and define f by $Ff = Fu/w$. Thus

$$\sigma u = Tf = \sigma[\bar{F}(w) * f].$$

By Theorem 3.5 of ch. 5 the operator T is compact on L^2. This implies that σ is a compact operator from Y to L^2. The proof is complete.

§7. Strongly elliptic operators

As might be expected, one can improve the theorems of the preceding section when $P(\mathrm{D})$ is elliptic, provided that $\sigma(P_0)$ is contained in an angle less than π. The subclass of elliptic operators having this property is well known in the theory of partial differential operators.

An operator $P(\mathrm{D})$ is said to be *strongly elliptic* if there is a complex constant γ such that $\mathrm{Re}\,[\gamma P(\xi)]$ is elliptic (see §3 of ch. 3). If $P(\mathrm{D})$ is of order m and $P_m(\mathrm{D})$ is its principal part, this means that there is a constant $c_0 > 0$ such that

$$|\mathrm{Re}\,[\gamma P_m(\xi)]| \geqslant c_0 |\gamma|\, |\xi|^m, \qquad \xi \in E^n. \tag{7.1}$$

Since all of the points $\xi \neq 0$ of E^n are connected for $n > 1$, $\mathrm{Re}\,[\gamma P_m(\xi)]$ does not change sign in E^n. Hence we may find a γ such that $|\gamma| = 1$ and

$$\mathrm{Re}\,[\gamma P_m(\xi)] \geqslant c_0 |\xi|^m, \qquad \xi \in E^n. \tag{7.2}$$

Note that if $P(\mathrm{D})$ is elliptic and the coefficients of $P_m(\xi)$ are real, then $P(\mathrm{D})$ is strongly elliptic.

Lemma 7.1. *If $P(\mathrm{D})$ is strongly elliptic, then $\sigma(P_0)$ is contained in an angle less than π.*

Proof. By definition we may find a γ such that $|\gamma| = 1$ and (7.2) holds. Since

$$|P(\xi) - P_m(\xi)| \leqslant C|\xi|^{m-1}, \qquad |\xi| > 1,$$

we have

$$\begin{aligned} \mathrm{Re}\,[\gamma P(\xi)] &\geqslant \mathrm{Re}\,[\gamma P_m(\xi)] - |P(\xi) - P_m(\xi)| \\ &\geqslant |\xi|^{m-1}(c_0|\xi| - C) \geqslant \tfrac{1}{2} c_0 |\xi|^m \end{aligned}$$

for $|\xi| > \max\,[1, 2C/c_0]$. Moreover, for $|\xi| \leqslant \max\,[1, 2C/c_0]$, $P(\xi)$ is bound-

Thus there are constants K_0, M_0 such that

$$(|\xi|^2+1)^r/K_0 \leqslant \operatorname{Re}[\gamma P(\xi)]+M_0 \leqslant K_0(|\xi|^2+1)^r, \tag{7.3}$$

where $r=\frac{1}{2}m$. Since

$$|P(\xi)| \leqslant C(|\xi|^2+1)^r, \tag{7.4}$$

we see that

$$|\operatorname{Im}[\gamma P(\xi)]| \leqslant K_1\{\operatorname{Re}[\gamma P(\xi)]+M_0\}. \tag{7.5}$$

This completes the proof.

It should be noted that the order m of a strongly elliptic operator $P(\mathrm{D})$ is always even. For if m were odd, we would have $\operatorname{Re}[\gamma P_m(-\xi)] = -\operatorname{Re}[\gamma P_m(\xi)]$, which clearly contradicts (7.2). Thus $r=\frac{1}{2}m$ is an integer.

Since every strongly elliptic operator satisfies (7.3) and (7.5) for some constant γ satisfying $|\gamma|=1$, we can consider the operator $\gamma P(\mathrm{D})$ in place of $P(\mathrm{D})$ and assume that

$$(|\xi|^2+1)^r/K_0 \leqslant \operatorname{Re} P(\xi)+M_0 \leqslant K_0(|\xi|^2+1)^r, \qquad \xi\in E^n, \tag{7.6}$$

and

$$|\operatorname{Im} P(\xi)| \leqslant K_1[\operatorname{Re} P(\xi)+M_0], \qquad \xi\in E^n, \tag{7.7}$$

hold. Since this entails no loss of generality, we shall assume that (7.6) and (7.7) hold throughout the remainder of this chapter.

In stating our theorems it will be convenient to introduce some new notation. For $\alpha>0$ set

$$\begin{aligned}\omega_\alpha(x)=\omega_\alpha^{(n)}(x) &= |x|^{\alpha-n}, && 0<\alpha<n\\ &=1-\log|x|, && \alpha=n\\ &=1, && \alpha>n\end{aligned}$$

$$N_{\alpha,\delta,y}(q) = \int_{|x|<\delta} |q(x-y)|^2\,\omega_\alpha(x)\,\mathrm{d}x \tag{7.9}$$

$$N_{\alpha,\delta}(q) = \sup_y N_{\alpha,\delta,y}(q), \quad N_{\alpha,y}(q)=N_{\alpha,1,y}(q) \tag{7.10}$$

$$N_\alpha(q) = N_{\alpha,1}(q). \tag{7.11}$$

We let N_α denote the set of functions q which are locally in L^2 and such that $N_\alpha(q)<\infty$. Note that $N_\alpha=M_{\alpha,2}$ for $\alpha\neq n$.

Our first theorem is a counterpart of Theorem 2.3.

Theorem 7.2. *Let $P(\mathrm{D})$ be a strongly elliptic operator of order m. Let $q(x)$ be a function locally in L^1, and define the functions $\rho(x)$ and $\sigma(x)$ as in §2. Assume that $|\mathrm{Im}\, q| = \tau_1^2 + \tau_2^2$, where τ_1 satisfies (2.12) and $\tau_2 \in N_m$. Assume that $\sigma \in N_m$, and if $m \leqslant n$, assume also that*

$$N_{m,\delta}(\sigma) \to 0 \quad as \quad \delta \to 0\,. \tag{7.12}$$

Then $P(\mathrm{D}) + q$ on $C_0^\infty \cap D(q)$ has a regularly accretive extension.

In proving Theorem 7.2 we shall make use of the following.

Theorem 7.3. *Let s be a positive real number. Then there is a constant K_2 depending only on s and n such that*

$$\|qv\|^2 \leqslant K_2 N_{2s,\delta}(q)\|v\|_{s,2}^2 + C_\delta N_{2s}(q)\|v\|^2, \quad q \in N_{2s},\, v \in S,\, \delta > 0, \tag{7.13}$$

where C_δ depends only on s, n and δ. Thus

$$\|qv\|^2 \leqslant C N_{2s}(q)\|v\|_{s,2}^2, \qquad q \in N_{2s}\,,\ v \in S\,, \tag{7.14}$$

where C depends only on s and n. We also have

$$\|qv\|_{-s,2}^2 \leqslant C N_{2s}(q)\|v\|^2\,, \qquad q \in N_{2s}\,,\ v \in S\,, \tag{7.15}$$

with the same constant C. In particular $H^{s,2} \subset D(q)$ for $q \in N_{2s}$.

We postpone the proof of Theorem 7.3 until the next section. We now show how it can be employed in the

Proof of Theorem 7.2. Set

$$\|\varphi\|_W^2 = \|\varphi\|_{r,2}^2 + \|\rho\varphi\|^2\,, \qquad \varphi \in C_0^\infty\,, \tag{7.16}$$

where $r = \frac{1}{2}m$. We let W be the completion of C_0^∞ with respect to this norm. Clearly W is a Hilbert space continuously and densely embedded in L^2. Let $b(\varphi, \psi)$ be given by (2.6). We shall show that (2.7) and (2.8) hold. Our result will thus follow from Theorem 6.6 of ch. 1. Now by Lemma 6.12 of ch. 1, (2.8) is a consequence of (2.16). To prove (2.16) we note that

$$|(P(\mathrm{D})\, v, v)| \leqslant C\,\|v\|_{r,2}^2\,, \qquad v \in S \tag{7.17}$$

$$\|\tau_1 \varphi\|^2 \leqslant C\,(\|\rho\varphi\|^2 + \|\varphi\|^2)\,, \qquad \varphi \in C_0^\infty \tag{7.18}$$

and

$$\|\sigma v\|^2 + \|\tau_2 v\|^2 \leqslant C\,\|v\|_{r,2}^2\,, \qquad v \in S\,. \tag{7.19}$$

Inequality (7.17) follows from Theorem 2.4 of ch. 3, (7.18) follows from (2.12), and (7.19) follows from (7.14) with $s=r$. As for (2.8), note that

$$\operatorname{Re} b(\varphi)=\operatorname{Re}(P(\mathrm{D})\,\varphi,\varphi)+\|\rho\varphi\|^2-\|\sigma\varphi\|^2 . \tag{7.20}$$

By (7.6) we have

$$\operatorname{Re}(P(\mathrm{D})\varphi,\varphi)+M_0\|\varphi\|^2\geqslant\|\varphi\|_{r,2}^2/K_0\,,\qquad \varphi\in C_0^\infty\,, \tag{7.21}$$

where we multiplied (7.6) by $|F\varphi|^2$, integrated and applied Theorem 2.4 of ch. 2. Thus

$$\operatorname{Re} b(\varphi)+M_0\|\varphi\|^2+\|\sigma\varphi\|^2\geqslant\|\varphi\|_W^2/K_0\,,\qquad \varphi\in C_0^\infty\,, \tag{7.22}$$

since $K_0\geqslant 1$. Inequality (2.8) will follow if we can show that for each $\varepsilon>0$ there is a constant C such that

$$\|\sigma\varphi\|^2\leqslant\varepsilon\|\varphi\|_{r,2}^2+C\|\varphi\|^2\,,\qquad \varphi\in C_0^\infty\,. \tag{7.23}$$

This is indeed the case. If $m\leqslant n$, we have by (7.13)

$$\|\sigma\varphi\|^2\leqslant K_2N_{m,\delta}(\sigma)\|\varphi\|_{r,2}^2+C_\delta N_m(\sigma)\|\varphi\|^2\,,\qquad \varphi\in C_0^\infty\,,$$

while by (7.12) we can take φ so small that $K_2N_{m,\delta}(q)<\varepsilon$. This proves (7.23) for $m\leqslant n$. If $m>n$, note that $\sigma\in N_\alpha$ for $n<\alpha<m$. Set $t=\frac{1}{2}\alpha$. Then $t<r$ and

$$\|\sigma\varphi\|^2\leqslant CN_\alpha(q)\|\varphi\|_{t,2}^2\,,\qquad \varphi\in C_0^\infty\,.$$

If we can show that for any ε there is a constant C such that

$$\|\varphi\|_{t,2}\leqslant\varepsilon\,\|\varphi\|_{r,2}+C\,\|\varphi\|\,,\qquad \varphi\in C_0^\infty\,, \tag{7.24}$$

then (7.23) will be proved. Inequality (7.24) is a consequence of the following lemma which will be proved in the next section.

Lemma 7.4. *Let s, t be real numbers satisfying $0<t<s$. Then for each $\varepsilon>0$ there is a constant C such that*

$$\|\varphi\|_{t,2}\leqslant\varepsilon\|\varphi\|_{s,2}+C\|\varphi\|\,,\qquad \varphi\in C_0^\infty\,. \tag{7.25}$$

Remark. If $\alpha<m\leqslant n$ and $\sigma\in N_\alpha$, then (7.12) holds. For let β be such that $\alpha<\beta<m$. Then

$$\int_{|x|<\delta}\sigma(x-y)^2|x|^{\beta-n}\mathrm{d}x\leqslant\delta^{\beta-\alpha}\int_{|x|<\delta}\sigma(x-y)^2|x|^{\alpha-n}\mathrm{d}x\leqslant\delta^{\beta-\alpha}N_\alpha(\sigma)$$

for $\delta\leqslant 1$. Thus

$$N_{m,\delta}(\sigma) \leqslant N_{\beta,\delta}(\sigma) \leqslant \delta^{\beta-\alpha} N_\alpha(\sigma) \to 0 \text{ as } \delta \to 0\,. \tag{7.26}$$

Note also that (7.12) is not actually necessary in the proof of Theorem 7.2. All that is needed is $K_0 K_2 N_{m,\delta}(q) < 1$ for some $\delta > 0$.

We note the important consequence of Theorem 7.3 and Lemma 7.4.

Lemma 7.5. *Let s be a positive number and suppose* $q \in N_{2s}$. *If* $2s \leqslant n$ *assume also that*

$$N_{2s,\delta}(q) \to 0 \quad as \quad \delta \to 0\,. \tag{7.27}$$

Then for each $\varepsilon > 0$ *there is a constant C such that*

$$\|q\varphi\| \leqslant \varepsilon \|\varphi\|_{s,2} + C\|\varphi\|\,, \qquad \varphi \in C_0^\infty\,. \tag{7.28}$$

Proof. If $2s \leqslant n$ we take δ so small that $K_2 N_{2s,\delta}(q) < \varepsilon$. Then (7.28) follows from (7.13).

If $2s > n$ note by (7.8) that $q \in N_\alpha$ for some α satisfying $n < \alpha < 2s$. Set $t = \frac{1}{2}\alpha$. Then $t < s$ and by (7.14)

$$\|q\varphi\| \leqslant C\|\varphi\|_{t,2}\,, \qquad \varphi \in C_0^\infty\,.$$

If we now apply Lemma 7.4 we obtain (7.28). This completes the proof.

§8. The remaining proofs

In this section we give the proofs of Theorem 7.3 and Lemma 7.4. The proof of the latter is simple. It follows directly from the inequality

$$\begin{aligned}(1+|\xi|^2)^t &= (1+|\xi|^2)^{\frac{1}{2}s}(1+|\xi|^2)^{t-\frac{1}{2}s} \\ &\leqslant \varepsilon(1+|\xi|^2)^s + \frac{1}{4\varepsilon}(1+|\xi|^2)^{2t-s}\end{aligned} \tag{8.1}$$

applied repeatedly. Each time this inequality is applied the exponent of the second term is reduced by $s-t$. Thus it will eventually be reduced to zero.

In proving Theorem 7.3 we shall make use of several simple lemmas.

Lemma 8.1. *For* $s > 0$, $G_s(x)$ *is infinitely differentiable for* $x \neq 0$ *and*

$$D^\mu G_s(x) = O(|x|^{-k}) \quad as \quad |x| \to \infty, \qquad k = 1, 2, \ldots, \; all \; \mu\,. \tag{8.2}$$

Proof. By (3.13) of ch. 6 it suffices to show that

$$d^j g(r)/dr^j = O(r^{-k}) \text{ as } r \to \infty, \qquad j, k = 0, 1, 2, \dots, \tag{8.3}$$

where $g(r)$ is given by (3.3) of that chapter. But $d^j g(r)/dr^j$ is a sum of terms each of which is a product of a power of r and a function of the same form as $g(r)$. Hence (8.3) follows from (3.9) of that chapter.

Corollary 8.2. *If $v \in C_B^\infty$ and vanishes near $x=0$, then vG_s is in S.*

Lemma 8.3. *For $s, t > 0$*

$$G_s * G_t = G_{s+t}\,. \tag{8.4}$$

Proof. Taking Fourier transforms we have by Theorem 4.10 of ch. 2

$$F(G_s * G_t) = FG_s \cdot FG_t = (1+|\xi|^2)^{-\frac{1}{2}(s+t)}$$

$$= FG_{s+t}\,.$$

Lemma 8.4. $G_s(x) > 0$ *for $s > 0$ and $x \neq 0$.*

Proof. Let k be an integer $> \frac{1}{2}(n+1)$. By Lemma 3.2 of ch. 6 for $s > \frac{1}{2}(n+1)$

$$G_s(x) = Q_k(s)|x|^{s-n} \int_0^\infty (|x|^2 + t^2)^{-\frac{1}{2}s-k} t^{\frac{1}{2}n+k} J_{\frac{1}{2}(n+2k-2)}(t)\,dt\,,$$

where

$$Q_k(s) = s(s+2)(s+4)\cdots(s+2k-2)\,.$$

By analytic continuation, this holds for all $s > 0$ (see the proof of Theorem 3.1 of ch. 6). Setting $\beta = \frac{1}{2}(s+2k)$ and $\lambda = \frac{1}{2}(n+2k)$ in Lemma 3.3 of ch. 6, we obtain the desired result.

Lemma 8.5. *For each $s > 0$ there is a constant C such that*

$$G_s(x) \leqslant C\omega_s(x)\,, \qquad 0 < |x| < 1\,, \tag{8.5}$$

where $\omega_s(x)$ is defined by (7.8).

Proof. See (2.6)–(2.8) of ch. 6.

Lemma 8.6. *If $s > 0$ and*

$$\int |q(x-y)|^2 G_s(x)\,dx \leqslant C_0\,, \qquad y \in E^n\,, \tag{8.6}$$

then

$$\|\bar{q}G_s * (qv)\| \leqslant C_0 \|v\|\,, \qquad v \in S\,. \tag{8.7}$$

Proof. For $u, v \in S$

$$(\bar{q}G_s * (qv), u) = \int\int \overline{q(y)}\, G_s(y-x)\, q(x)\, v(x)\, \overline{u(y)}\, dx\, dy .$$

Hence by Schwarz's inequality

$$\begin{aligned} |(\bar{q}G_s * (qv), u)|^2 &\leqslant \left(\int \int |q(y)|^2 G_s(y-x)\, |v(x)|^2\, dx\, dy\right) \\ &\quad \times \left(\int \int |q(x)|^2 G_s(y-x)\, |u(y)|^2\, dx\, dy\right) \\ &\leqslant C_0^2 \|v\|^2 \|u\|^2 . \end{aligned}$$

Since this is true for all $u, v \in S$, we can apply Theorems 1.7 and 1.13 of ch. 2 to obtain (8.7).

Lemma 8.7. *If* (8.7) *holds and* $t = \frac{1}{2}s$, *then*

$$\|qv\|_{-t,2}^2 \leqslant C_0 \|v\|^2 , \qquad v \in S . \tag{8.8}$$

Proof. By definition and (8.7)

$$\begin{aligned} \|qv\|_{-t,2}^2 &= \int (1+|\xi|^2)^{-s} |F(qv)|^2\, d\xi \\ &= (G_s * (qv), qv) = (\bar{q}G_s * (qv), v) \\ &\leqslant \|\bar{q}G_s * (qv)\|\, \|v\| \leqslant C_0 \|v\|^2 . \end{aligned}$$

This gives (8.8).

Lemma 8.8. *If* (8.8) *holds, then*

$$\|qv\|^2 \leqslant C_0 \|v\|_{t,2}^2 , \qquad v \in S . \tag{8.9}$$

Proof. For $u, v \in S$ we have

$$\begin{aligned} |(\bar{q}v, u)| &= |(v, qu)| \leqslant \|v\|_{t,2} \|qu\|_{-t,2} \\ &\leqslant \|v\|_{t,2}\, C_0^{\frac{1}{2}} \|u\| . \end{aligned}$$

Since this is true for all $u, v \in S$, it implies

$$\|\bar{q}v\|^2 \leqslant C_0 \|v\|_{t,2}^2 , \qquad v \in S .$$

Since $\|qv\| = \|\bar{q}v\|$, we obtain (8.9).

Lemma 8.9. *For each* $s > 0$ *there is a constant* C *depending only on* s *and* n *such that*

$$\int |q(x-y)|^2 G_s(x)\, dx \leqslant C N_s(q), \qquad q \in N_s,\ y \in E^n . \tag{8.10}$$

Proof. By Lemma 8.5 and the definition of N_s

$$\int_{|x|<1} |q(x-y)|^2\, G_s(x)\mathrm{d}x \leqslant CN_s(q)\,.$$

Moreover by (2.9) of ch. 6, $G_s(x)$ satisfies (2.8) of ch. 5. Thus by Lemma 2.6

$$\int_{|x|>1} |q(x-y)|^2\, G_s(x)\mathrm{d}x \leqslant CM_{n,2}(q)\,.$$

The last two inequalities give (8.10).

We are now ready for the

Proof of Theorem 7.3. We first note that (7.14) and (7.15) are immediate consequences of Lemmas 8.6–8.9. Note also that (7.13) is equivalent to

$$\|q[G_s * u]\|^2 \leqslant K_2 N_{2s,\delta}(q)\|u\|^2 + C_\delta N_{2s}(q)\|G_s * u\|^2, \quad u \in S\,. \tag{8.11}$$

Let $\psi(x)$ be a function in C_0^∞ satisfying

$$\begin{aligned} 0 \leqslant \psi(x) &\leqslant 1\,, && x \in E^n \\ \psi(x) &= 0\,, && |x| > \tfrac{1}{2} \\ \psi(x) &= 1\,, && |x| < \tfrac{1}{4} \end{aligned}$$

(see Theorem 1.2 of ch. 2), and set $\psi_\delta(x) = \psi(x/\delta)$.

Define

$$\begin{aligned} G_{s,\delta}(x) &= \psi_\delta(x)\, G_s(x) \\ \tilde{G}_{s,\delta}(x) &= [1-\psi_\delta(x)]\, G_s(x) \\ H(x) \quad &= G_{s,\delta} * G_{s,\delta}\,. \end{aligned}$$

Then

$$\begin{aligned} H(x) &= \int \psi_\delta(x-y)\, G_s(x-y)\, \psi_\delta(y)\, G_s(y)\mathrm{d}y \\ &\leqslant \int G_s(x-y)\, G_s(y)\mathrm{d}y = G_{2s}(x) \end{aligned}$$

by Lemmas 8.3 and 8.4. Moreover if $|x| > \delta$, then either $|y| > \frac{1}{2}\delta$ or $|x-y| > \frac{1}{2}\delta$. Thus $\psi_\delta(x-y)\, \psi_\delta(y) = 0$ for $|x| > \delta$. Hence by Lemma 8.5

$$\begin{aligned} H(x) &\leqslant K\omega_{2s}(x)\,, && |x| < \delta\,, \\ H(x) &= 0\,, && |x| \geqslant \delta\,, \end{aligned} \tag{8.12}$$

where K depends only on s and n. Now for $v \in S$

$$\begin{aligned}\|G_{s,\delta} * (qv)\|^4 &= (G_{s,\delta} * (qv),\, G_{s,\delta} * (qv))^2 \\ &= (H * (qv),\, qv)^2 \\ &= (\textstyle\int\int H(x-y)\, q(y)\, v(y)\, \overline{q(x)\, v(x)}\, \mathrm{d}x\, \mathrm{d}y)^2 \\ &\leqslant (\textstyle\int\int H(x-y) |q(x)|^2 |v(y)|^2 \mathrm{d}x\, \mathrm{d}y) \\ &\quad \times (\textstyle\int\int H(x-y) |q(y)|^2 |v(y)|^2 \mathrm{d}x\, \mathrm{d}y)\,.\end{aligned}$$

By (8.12)

$$\int |q(x)|^2 H(x-y) \mathrm{d}x \leqslant K N_{2s,\delta}(q)\,.$$

Hence

$$\|G_{s,\delta} * (qv)\|^2 \leqslant K N_{2s,\delta}(q) \|v\|^2\,, \qquad v \in S\,. \tag{8.13}$$

Thus for $u, v \in S$

$$\begin{aligned}|(q[G_{s,\delta} * u],\, v)|^2 &= |(u,\, G_{s,\delta} * (\bar{q}v))|^2 \\ &\leqslant \|u\|^2 \|G_{s,\delta} * (\bar{q}v)\|^2 \leqslant K N_{2s,\delta}(q) \|u\|^2 \|v\|^2\,,\end{aligned}$$

since $N_{\alpha,\delta}(\bar{q}) = N_{\alpha,\delta}(q)$ in general. Since this is true for all $u, v \in S$, we have

$$\|q[G_{s,\delta} * u]\|^2 \leqslant K N_{2s,\delta}(q) \|u\|^2, \qquad u \in S\,. \tag{8.14}$$

Next note that $\tilde{G}_{s,\delta}$ is in S by Corollary 8.2. In particular, there is a constant C_δ depending only on s, n and δ such that

$$(1+|\xi|^2)^{2s} |F\tilde{G}_{s,\delta}|^2 \leqslant C_\delta\,. \tag{8.15}$$

Thus by (7.14)

$$\begin{aligned}\|q[\tilde{G}_{s,\delta} * u]\|^2 &\leqslant C N_{2s}(q) \|\tilde{G}_{s,\delta} * u\|_{s,2}^2 \\ &= C N_{2s}(q) \textstyle\int (1+|\xi|^2)^s\, |F\tilde{G}_{s,\delta}|^2 |Fu|^2 \mathrm{d}\xi \\ &\leqslant C C_\delta N_{2s}(q) \textstyle\int (1+|\xi|^2)^{-s} |Fu|^2 \mathrm{d}\xi \\ &= C C_\delta N_{2s}(q) \|G_s * u\|^2\,.\end{aligned} \tag{8.16}$$

Since $G_{2s} = G_{2s,\delta} + \tilde{G}_{2s,\delta}$, (8.11) follows from (8.14) and (8.16). Thus the proof of (7.13) is complete.

§9. Perturbation by a potential. Elliptic case

We now show how to strengthen Theorem 3.1 when $P(\mathrm{D})$ is strongly elliptic. As before we assume that $P(\xi)$ satisfies (7.6) and (7.7).

Theorem 9.1. *Let* $P(\mathrm{D})$ *and* $q(x)$ *satisfy the hypotheses of Theorem* 7.2. *Assume* (3.4) *holds and that* $\rho \in N_{2m}$ *with*

$$N_{2m,x}(\rho) + N_{2m,x}(\sigma) + N_{2m,x}(\tau_2) \to 0 \quad as \ |x| \to \infty \,. \tag{9.1}$$

Then $P(\mathrm{D}) + q$ *on* $C_0^\infty \cap D(q)$ *has a regularly accretive extension* B *such that*

$$\sigma_e(B) = \sigma_e(P_0) = \{P(\xi), \ \xi \in E^n\} \,. \tag{9.2}$$

We reduce the proof of Theorem 9.1 to a series of lemmas. All of the hypotheses of the theorem are assumed.

Lemma 9.2. *If* Z *and* $\tilde{Z}$ *are defined by* (3.5) *and* (3.6), *respectively, then* $Z = H^{m,2}$ *and* $\tilde{Z} = H^{-m,2}$.

Proof. This follows from (7.3) and (7.4).

Lemma 9.3. *Let* $\tilde{P}$ *be defined as in the proof of Theorem* 3.1. *Then* $\tilde{P}$ *is a bounded linear operator from* L^2 *to* $H^{-m,2}$. *If* Re λ *is sufficiently large* $\tilde{P} + \lambda$ *is onto.*

Proof. See the proof of Theorem 3.1 and note Lemma 9.2.

Lemma 9.4. *Let* B *be the regularly accretive extension of* $P(\mathrm{D}) + q$ *constructed in the proof of Theorem* 7.2 *and let* W *be given by* (7.16). *For* Re λ *large* $(B + \lambda)^{-1}$ *is a bounded operator from* L^2 *to* W.

Proof. See Theorem 6.15 of ch. 1.

Lemma 9.5. ρ *is a bounded operator from* W *to* L^2.

Proof. By (7.16)

$$\|\rho\sigma\| \leqslant \|\varphi\|_W \,, \qquad \varphi \in C_0^\infty \,. \tag{9.3}$$

Corollary 9.6. *For* Re λ *large,* $\rho(B + \lambda)^{-1}$ *is a bounded operator on* L^2.

Lemma 9.7. *If* $\varphi \in C_0^\infty$, *then* φ *is a compact operator from* W *to* L^2.

Proof. We know that φ is a compact operator from $H^{r,2}$ to L^2 (Theorem 4.4 of ch. 2). By (7.16) every sequence bounded in W is also bounded in $H^{r,2}$. It follows therefore that φ is a compact operator from W to L^2.

Corollary 9.8. *If* $\varphi \in C_0^\infty$, *then* φ *is a compact operator from* L^2 *to* $\hat{W}'$.

Proof. Note that φ from L^2 to $\hat{W}'$ is the adjoint of $\bar{\varphi}$ from W to L^2. Apply Theorem 2.17 of ch. 1.

Lemma 9.9 *$H^{m,2} \subset W \subset H^{r,2} \subset L^2 \subset H^{-r,2} \subset \hat{W}' \subset H^{-m,2}$ with a continuous and dense embedding for each inclusion.*

Proof. Apply Lemma 6.5 of ch. 1, Theorem 4.1 of ch. 2 and Theorem 7.3.

Lemma 9.10. *For $\varphi \in C_0^\infty$ and $s \geqslant 0$, $\varphi\tilde{P} - \tilde{P}\varphi$ is a compact operator from $H^{s,2}$ to $H^{s-m,2}$.*

Proof. For $v \in S$ we have

$$[\varphi P(\mathrm{D}) - P(\mathrm{D})\varphi]v = \sum_{|\mu|<m} c_\mu(x)\mathrm{D}^\mu v\,, \tag{9.4}$$

where the coefficients $c_\mu(x)$ are in C_0^∞. Now for $|\mu| < m$, D^μ is a bounded operator from $H^{s,2}$ to $H^{s-m+1,2}$ (Lemma 4.8 of ch. 2) while each $c_\mu(x)$ is a compact operator from $H^{s-m+1,2}$ to $H^{s-m,2}$ (Theorem 4.4 of ch. 2). Thus the right-hand side of (9.4) is a compact operator from $H^{s,2}$ to $H^{s-m,2}$. Moreover since

$$[\varphi\tilde{P} - \tilde{P}\varphi]v = [\varphi P(\mathrm{D}) - P(\mathrm{D})\varphi]v, \qquad v \in S\,,$$

we have

$$[\varphi\tilde{P} - \tilde{P}\varphi]v = \sum_{|\mu|<m} c_\mu(x)\mathrm{D}^\mu v\,, \qquad v \in H^{s,2}\,. \tag{9.5}$$

This gives the desired result.

Lemma 9.11. *For* Re λ *large and $\varphi \in C_0^\infty$, $\rho(B+\lambda)^{-1}\varphi$ is a compact operator on L^2.*

Proof. By Corollary 9.8, φ is a compact operator from L^2 to $\hat{W}'$. By Theorem 6.16 of ch. 1, $(\hat{B}+\lambda)^{-1}$ is bounded from $\hat{W}'$ to W. And by Lemma 9.5, ρ is a bounded operator from W to L^2. Thus $\rho(\hat{B}+\lambda)^{-1}\varphi$ is a compact operator on L^2. But on L^2, $\rho(B+\lambda)^{-1}\varphi = \rho(\hat{B}+\lambda)^{-1}\varphi$. This completes the proof.

Lemma 9.12. *For $\varphi \in C_0^\infty$ and* Re λ *large, $\varphi\rho(B+\lambda)^{-1}$ is a compact operator on L^2.*

Proof. We have

$$\begin{aligned}\varphi\rho(B+\lambda)^{-1} - \rho(B+\lambda)^{-1}\varphi &= \rho[\varphi(B+\lambda)^{-1} - (B+\lambda)^{-1}\varphi] \\ &= \rho(B+\lambda)^{-1}[(B+\lambda)\varphi - \varphi(B+\lambda)](B+\lambda)^{-1} \\ &= \rho(\hat{B}+\lambda)^{-1}[\tilde{P}\varphi - \varphi\tilde{P}](B+\lambda)^{-1} \end{aligned} \tag{9.6}$$

by (3.11). Now by Lemma 9.4, $(B+\lambda)^{-1}$ is a bounded operator from L^2 to W. Since W is continuously embedded in $H^{r,2}$, $(B+\lambda)^{-1}$ is continuous from L^2 to $H^{r,2}$. We also know that $(\hat{B}+\lambda)^{-1}$ is bounded from $\hat{W}'$ to W (Theorem 6.16 of ch. 1). Moreover, since $H^{-r,2}$ is continuously embedded in $\hat{W}'$ (Lemma 9.9), the restriction of $(\hat{B}+\lambda)^{-1}$ to $H^{-r,2}$ is bounded from $H^{-r,2}$ to W. Finally ρ is bounded from W to L^2. Putting all of these pieces of information together we see that the right-hand side of (9.6) is compact on L^2. Since $\rho(B+\lambda)^{-1}\varphi$ was already shown to be compact (Lemma 9.11), the proof is complete.

Proof of Theorem 9.1. By (7.6), $(P_0+\lambda)^{-1}$ exists for Re λ large. Since B is regularly accretive, the same is true of $(B+\lambda)^{-1}$ (Theorem 6.7 of ch. 1). Without loss of generality we may assume that this is true for $\lambda=0$. Now

$$\begin{aligned} P_0^{-1}-B^{-1} &= \tilde{P}^{-1}(B-\tilde{P})B^{-1} = \tilde{P}^{-1}qB^{-1} \\ &= \tilde{P}^{-1}[\rho^2-\sigma^2+\mathrm{i}\,(\operatorname{sgn}\operatorname{Im} q)(\tau_1^2+\tau_2^2)]B^{-1}. \end{aligned} \tag{9.7}$$

Our result will be obtained by showing that the right-hand side of (9.7) is compact on L^2 and then applying Theorem 4.7 of ch. 1. It suffices to consider in detail only the term $\tilde{P}^{-1}\rho^2B^{-1}$ since it presents the most difficulty. Let φ be a function in C_0^∞ such that $\varphi(x)=1$ for $|x|<\frac{1}{2}$, $\varphi(x)=0$ for $|x|>1$ and $0\leqslant\varphi\leqslant 1$ everywhere. Set $\varphi_R(x)=\varphi(x/R)$. Now

$$\tilde{P}^{-1}\rho^2B^{-1} = \tilde{P}^{-1}\rho[\varphi_R\rho B^{-1}]+[\tilde{P}^{-1}\rho(1-\varphi_R)]\rho B^{-1}. \tag{9.8}$$

For each $R>0$ the operator $\varphi_R\rho B^{-1}$ is a compact operator on L^2 (Lemma 9.12). Since $\rho\in N_{2m}$, it is a bounded operator from L^2 to $H^{-m,2}$ (Theorem 7.3). Since $\tilde{P}^{-1}$ is bounded from $H^{-m,2}$ to L^2, we see that the first operator on the right-hand side of (9.8) is compact for each $R>0$. Now the operator ρB^{-1} is bounded on L^2 (Corollary 9.6). Moreover $\rho(1-\varphi_R)$ is a bounded operator from L^2 to $H^{-m,2}$ with bound

$$\text{constant } N_{2m}[\rho(1-\varphi_R)]$$

(Theorem 7.3). But by (9.1)

$$N_{2m}[\rho(1-\varphi_R)]\to 0 \text{ as } R\to\infty .$$

Since $\tilde{P}^{-1}$ is bounded from $H^{-m,2}$ to L^2, we see that the second operator on the right-hand side of (9.8) is a bounded operator on L^2 with norm tending to zero as $R\to\infty$. We see therefore that $\tilde{P}^{-1}\rho^2B^{-1}$ is the limit in $B(L^2)$ of compact operators. By Theorem 2.16 of ch. 1, it is compact as well. The proof is complete.

§10. Perturbation by an operator. Elliptic case

We now show how Theorems 4.1 and 4.3 can be strengthened when $P(\mathrm{D})$ is strongly elliptic.

Consider the operator

$$Q(x, \mathrm{D}) = \sum_{j,k=1}^{N} \bar{P}_j(\mathrm{D})\, q_{jk}(x)\, P_k(\mathrm{D})\,, \tag{10.1}$$

where the $P_j(\mathrm{D})$ are constant coefficient operators and the q_{jk} are functions locally in L^1. In stating our hypotheses it will be convenient to extend the definition of the space N_α to $\alpha \leqslant 0$. We let N_0 denote L^∞ with

$$N_0(q) = N_{0,\delta}(q) = \|q\|_\infty\,.$$

For $\alpha < 0$ we let N_α denote the set of those functions which are zero almost everywhere.

Let m_j be the order of $P_j(\mathrm{D})$, and set

$$\rho_k(x)^2 = \max\, \mathrm{Re}[q_{kk}(x), 0], \quad \sigma_k(x)^2 = \rho_k(x)^2 - q_{kk}(x)\,, \quad 1 \leqslant k \leqslant N\,. \tag{10.2}$$

We assume the following:

H1. $$q_{jk} - \bar{q}_{kj} = (f_{1jk} + g_{1jk})(f_{2jk} + g_{2jk})\,,$$

where

$$f_{1jk} \in N_{m-2m_j}\,, \quad f_{2jk} \in N_{m-2m_k} \tag{10.3}$$

$$|g_{1jk}| \leqslant C\rho_j\,, \quad |g_{2jk}| \leqslant C\rho_k\,. \tag{10.4}$$

H2. For $j < k$

$$q_{jk} + \bar{q}_{kj} = h_{1jk}\, h_{2jk}\,,$$

where

$$h_{1jk} \in N_{m-2m_j}, \quad h_{2jk} \in N_{m-2m_k}\,. \tag{10.5}$$

If $m \leqslant n + 2m_j$, we also assume

$$N_{m-2m_j,\delta}(h_{1jk}) \to 0 \quad \text{as} \quad \delta \to 0\,. \tag{10.6}$$

H3. $\sigma_k \in N_{m-2m_k}$, and if $m \leqslant n + 2m_k$ we assume

$$N_{m-2m_k,\delta}(\sigma_k) \to 0 \quad \text{as} \quad \delta \to 0. \tag{10.7}$$

Theorem 10.1. *Let $P(\mathrm{D})$ be a strongly elliptic operator of order m, and let V denote the set of these $\varphi \in C_0^\infty$ such that $Q(x, \mathrm{D})\varphi \in L^2$. If hypotheses H1–H3 are satisfied, then $P(\mathrm{D})+Q(x, \mathrm{D})$ on V has a regularly accretive extension.*

Proof. Set

$$\|\varphi\|_W^2 = \|\varphi\|_{r,2}^2 + \Sigma \|\rho_k P_k(\mathrm{D})\varphi\|^2 , \qquad \varphi \in C_0^\infty , \tag{10.8}$$

and let W be the completion with respect to this norm. Set

$$q(\varphi, \psi) = \Sigma (q_{jk} P_k(\mathrm{D})\varphi, P_j(\mathrm{D})\psi) \tag{10.9}$$

$$b(\varphi, \psi) = (P(\mathrm{D})\varphi, \psi) + q(\varphi, \psi) . \tag{10.10}$$

It suffices to show that

$$|b(\varphi)| \leqslant C \|\varphi\|_W^2, \qquad \varphi \in C_0^\infty \tag{10.11}$$

and

$$\|\varphi\|_W^2 \leqslant C[\mathrm{Re}\, b(\varphi) + M\|\varphi\|^2] , \qquad \varphi \in C_0^\infty \tag{10.12}$$

(Theorem 6.6 of ch. 1). To this end note that for $j \neq k$

$$2q_{jk} = (f_{1jk} + g_{1jk})(f_{2jk} + g_{2jk}) + h_{1jk} h_{2jk} . \tag{10.13}$$

Now by (10.3) and Theorem 7.3

$$\begin{aligned} |(f_{2jk} P_k(\mathrm{D})\varphi, \bar{f}_{1jk} P_j(\mathrm{D})\varphi)| &\leqslant C \|P_k(\mathrm{D})\varphi\|_{r-m_k,2} \|P_j(\mathrm{D})\varphi\|_{r-m_j,2} \\ &\leqslant C' \|\varphi\|_{r,2}^2 . \end{aligned} \tag{10.14}$$

(Note that $f_{1jk}=0$ if $m_j > r$ and $f_{2jk}=0$ if $m_k > r$.)

Moreover by (10.4)

$$|(g_{2jk} P_k(\mathrm{D})\varphi, \bar{f}_{1jk} P_j(\mathrm{D})\varphi)| \leqslant C \|\rho_k P_k(\mathrm{D})\varphi\| \, \|\varphi\|_{r,2} \tag{10.15}$$

$$|(f_{2jk} P_k(\mathrm{D})\varphi, \bar{g}_{1jk} P_j(\mathrm{D})\varphi)| \leqslant C \|\varphi\|_{r,2} \, \|\rho_j P_j(\mathrm{D})\varphi\| \tag{10.16}$$

$$|(g_{2jk} P_k(\mathrm{D})\varphi, \bar{g}_{1jk} P_j(\mathrm{D})\varphi)| \leqslant C \|\rho_k P_k(\mathrm{D})\varphi\| \, \|\rho_j P_j(\mathrm{D})\varphi\| . \tag{10.17}$$

Let $\varepsilon > 0$ be given. By (10.5) and (10.6) we have in view of Lemma 7.5 that there is a constant C such that

$$\begin{aligned} |(h_{2jk} P_k(\mathrm{D})\varphi, \bar{h}_{1jk} P_j(\mathrm{D})\varphi)| &\leqslant C \|\varphi\|_{r,2} (\varepsilon \|\varphi\|_{r,2} + C' \|\varphi\|) \\ &\leqslant 2\varepsilon \|\varphi\|_{r,2}^2 + C'' \|\varphi\|^2 , \qquad \varphi \in C_0^\infty . \end{aligned} \tag{10.18}$$

A similar argument gives

$$\|\sigma_k P_k(\mathrm{D})\varphi\|^2 \leqslant \varepsilon \|\varphi\|_{r,2}^2 + C\|\varphi\|^2 , \qquad \varphi \in C_0^\infty . \tag{10.19}$$

Inequality (10.11) follows directly from (10.13)–(10.19). To prove (10.12) note that

$$
\begin{aligned}
2\,\mathrm{Re}\, q(\varphi) &= \Sigma\,(q_{jk} P_k(\mathrm{D})\varphi,\, P_j(\mathrm{D})\varphi) + \Sigma\,(\bar{q}_{kj} P_k(\mathrm{D})\varphi,\, P_j(\mathrm{D})\varphi) \\
&= \Sigma'\,(h_{2jk} P_k(\mathrm{D})\varphi,\, \bar{h}_{1jk} P_j(\mathrm{D})\varphi) + 2\Sigma\, \|\rho_k P_k(\mathrm{D})\varphi\|^2 \\
&\quad - 2\Sigma\, \|\sigma_k P_k(\mathrm{D})\varphi\|^2 ,
\end{aligned} \tag{10.20}
$$

where Σ' denotes summation over all j, k such that $j \neq k$. Now by (10.18) and (10.19), for each $\varepsilon > 0$ there is a constant C such that

$$
\mathrm{Re}\, q(\varphi) \geqslant \Sigma\, \|\rho_k P_k(\mathrm{D})\varphi\|^2 - \varepsilon \|\varphi\|_{r,2}^2 - C\|\varphi\|^2 , \qquad \varphi \in C_0^\infty . \tag{10.21}
$$

This inequality in conjunction with (7.21) gives (10.12). This completes the proof.

Theorem 10.2. *If, in addition to the hypotheses of Theorem* 10.1, *we have*

H4. $\quad \rho_j \in N_{2m-2m_j}$ *and*

$$
N_{2m-2m_j,x}(\rho_j + \sigma_j + |f_{1jk}| + |h_{1jk}|) \to 0 \;\; as \;\; |x| \to \infty , \tag{10.22}
$$

then $P(\mathrm{D}) + Q(x, \mathrm{D})$ *on* V *has a regularly accretive extension* B *such that*

$$
\sigma_e(B) = \sigma_e(P_0) = \{P(\xi), \quad \xi \in E^n\} . \tag{10.23}
$$

Proof. Essentially all of the ingredients of the proof are found in the proof of Theorem 9.1. We omit the details. Let B be the regularly accretive extension constructed in the proof of Theorem 10.1. We must show that for $\mathrm{Re}\,\lambda$ sufficiently large the operator $(P_0 + \lambda)^{-1} - (B + \lambda)^{-1}$ is compact. Assuming $\lambda = 0$ we have

$$
\begin{aligned}
P_0^{-1} - B^{-1} = \tilde{P}^{-1} [& \Sigma\, \tilde{P}_k \rho_k^2 P_{k0} - \Sigma\, \tilde{P}_k \sigma_k^2 P_{k0} \\
& + \tfrac{1}{2}\Sigma\, \tilde{P}_j (f_{1jk} + g_{1jk})(f_{2jk} + g_{2jk}) P_{k0} + \tfrac{1}{2}\Sigma'\, \tilde{P}_j h_{1jk} h_{2jk} P_{k0}] B^{-1} .
\end{aligned} \tag{10.24}
$$

Each of the terms is handled in the same way as in the proof of Theorem 9.1. For instance

$$
\begin{aligned}
\tilde{P}^{-1} \tilde{P}_k \rho_k^2 P_{k0} B^{-1} &= \tilde{P}^{-1} \tilde{P}_k \rho_k [\varphi_R \rho_k P_{k0} B^{-1}] \\
&\quad + [\tilde{P}^{-1} \tilde{P}_k \rho_k (1 - \varphi_R)]\, \rho_k P_{k0} B^{-1} .
\end{aligned}
$$

The same reasoning applies here as in (9.8).

Remark. The surprising aspect of Theorem 10.2 is that the operator $Q(x, \mathrm{D})$ can be of order $2m-2$. Thus (10.23) holds even though the unperturbed operator $P(\mathrm{D})$ is of lower order than the perturbing operator $Q(x, \mathrm{D})$.

In ending this chapter we mention the counterparts of Theorems 6.1 and 6.3 for strongly elliptic operators. Because of their similarity to the former, we only sketch the proof of the first.

Theorem 10.6. *Let* $P(\mathrm{D})$ *and* $q(x)$ *satisfy the hypotheses of Theorem* 7.2. *In particular* (7.6) *and* (7.7) *are to hold. If*

$$N_{m,x}(\sigma) \to 0 \quad as \quad |x| \to \infty ,$$

then $p(\mathrm{D})+q$ *on* $C_0^\infty \cap D(q)$ *has a regularly accretive extension* B *such that* $\sigma_e(B)$ *is contained in* $\mathrm{Re}\ \lambda + M_0 \geqslant 0$.

Proof. Set $\tilde{q}=q+\sigma^2=\rho^2+\mathrm{i}\ \mathrm{Im}\ q$. By Theorem 7.2 and (7.22) $P(\mathrm{D})+\tilde{q}$ on $C_0^\infty \cap D(\tilde{q})$ has a regularly accretive extension $\tilde{B}$ such that $\sigma_e(\tilde{B})$ is contained in $\mathrm{Re}\ \lambda + M_0 \geqslant 0$. Now $C_0^\infty \cap D(q) \cap D(\tilde{q})$ is dense in L^2 (Lemma 6.8 of ch. 4). Furthermore on this set

$$(B+\lambda)^{-1} - (\tilde{B}+\lambda)^{-1} = (B+\lambda)^{-1}\sigma \cdot \sigma(\tilde{B}+\lambda)^{-1} . \tag{10.25}$$

If φ_R is as in the proof of Theorem 9.1, the operator $\varphi_R \sigma(\tilde{B}+\lambda)^{-1}$ is a compact operator on L^2. In addition

$$\|(1-\varphi_R)\sigma(\hat{B}+\lambda)^{-1}\| \to 0 \text{ as } R \to \infty .$$

This shows that the left-hand side of (10.25) is compact, and the proof is complete.

Theorem 10.7. *Assume that all of the hypotheses of Theorem* 10.1 *hold. If*

$$N_{m-2m_k,x}(\sigma_k) \to 0 \quad as \quad |x| \to \infty, \qquad 1 \leqslant k \leqslant N , \tag{10.26}$$

then $P(\mathrm{D})+Q(x,\mathrm{D})$ *on* V *has a regularly accretive extension* B *such that* $\sigma_e(B)$ *is contained in the set* $\mathrm{Re}\ \lambda + M_0 \geqslant 0$.

CHAPTER 8

SELF-ADJOINT EXTENSIONS

Suppose $q(x)$ is a real valued function locally in L^2 and $P(\xi)$ is a polynomial with real coefficients. Then the expression

$$(P(\mathrm{D})\varphi+q\varphi,\varphi) = \int P(\xi)|F\varphi|^2\,\mathrm{d}\xi+\int q(x)|\varphi|^2\,\mathrm{d}x$$

is real for each $\varphi\in C_0^\infty\cap D(q)$. By Lemma 6.2 of ch. 4 the set of functions $C_0^\infty\cap D(q)$ is dense in L^2. Hence the operator $P(\mathrm{D})+q$ on $C_0^\infty\cap D(q)$ is symmetric in L^2. In this chapter we shall be concerned with self-adjoint extensions of such an operator. In §1 we present several theorems giving sufficient conditions for a self-adjoint extension to exist. In some instances the extension will be bounded from below. In this case Theorem 7.7 of ch. 1 shows that the extension is regularly accretive.

In §2 we consider the question when $P(\mathrm{D})+q$ on $C_0^\infty\cap D(q)$ will have a self-adjoint extension B such that $\sigma_e(B)=\sigma(P_0)$.

If the conditions of §2 are not met, it is possible that points of $\sigma_e(B)$ are removed from those of $\sigma(P_0)$. An interesting problem is to estimate the distance such points can move. Some results are given in §3.

In §4 we give conditions which insure that an operator of the form

$$L(x,\mathrm{D}) = P(\mathrm{D})+\sum_{j=1}^{r} q_j(x)\,Q_j(\mathrm{D})$$

is essentially self-adjoint (i.e., is such that its closure is self-adjoint). Extensions with finite negative spectrum are discussed in §5.

§1. Existence

We now give sufficient conditions for $P(\mathrm{D})+q$ on $C_0^\infty\cap D(q)$ to have a self-adjoint extension in L^2. In each of the following theorems it is assumed that q and the coefficients of $P(\xi)$ are real and that q is locally in L^2.

Theorem 1.1. *If only even order derivatives appear in* $P(\mathrm{D})$, *then* $P(\mathrm{D})+q$ *on* $C_0^\infty\cap D(q)$ *has a self-adjoint extension.*

Proof. Let J be the conjugation operator, i.e., $Jv(x)=\overline{v(x)}$. Note that $J\varphi\in C_0^\infty\cap D(q)$ whenever φ is in this set. Now

$$J\mathrm{D}^\mu\varphi = J(-\mathrm{i})^{|\mu|}\partial^{|\mu|}\varphi/(\partial x)^\mu = (-1)^{|\mu|}\mathrm{D}^\mu J\varphi\,.$$

Since only derivatives of even order enter into $P(\mathrm{D})$, we have

$$JP(\mathrm{D})\varphi = P(\mathrm{D})J\varphi\,.$$

This shows that all of the hypotheses of Theorem 7.18 of ch. 1 are fulfilled. The conclusion follows immediately.

Theorem 1.2. *If $q(-x)=q(x)$, then $P(\mathrm{D})+q$ in $C_0^\infty\cap D(q)$ has a self-adjoint extension.*

Proof. In this case set $J\varphi(x)=\overline{\varphi(-x)}$. Note that

$$(J\varphi,J\psi)=\int\overline{\varphi(-x)}\psi(-x)\mathrm{d}x=(\psi,\varphi)\,.$$

Moreover $J\varphi\in C_0^\infty\cap D(q)$ whenever φ is in this set. We also have

$$J\mathrm{D}^\mu\varphi = \mathrm{i}^{|\mu|}\partial^{|\mu|}\,\overline{\varphi(-x)}/(-\partial x)^\mu = \mathrm{D}^\mu J\varphi\,.$$

Hence

$$J[P(\mathrm{D})+q]\varphi = [P(\mathrm{D})+q]J\varphi\,.$$

Again we apply Theorem 7.18 of ch. 1 to obtain the desired conclusion.

Theorem 1.3. *Assume that q is locally in L^1 and that*

$$P(\xi)+M\geqslant 0\,,\qquad \xi \text{ real}\,, \tag{1.2}$$

holds for some real number M. Set

$$\rho(x)^2=\max\,[q(x),0]\,,\ \sigma(x)^2=-\min\,[q(x),0]\,. \tag{1.3}$$

If there is a constant C such that

$$\sigma^2\leqslant C\,, \tag{1.4}$$

then $P(\mathrm{D})+q$ on $C_0^\infty\cap D(q)$ has a regularly accretive self-adjoint extension.

Proof. Set

$$b(\varphi,\psi)=(P(\mathrm{D})\varphi+q\varphi,\psi),\qquad \varphi,\psi\in C_0^\infty\,. \tag{1.5}$$

$$(\varphi,\psi)_W=\int(P(\xi)+M+1)F\varphi\overline{F\psi}\,\mathrm{d}\xi+(\rho\varphi,\rho\psi)\,. \tag{1.6}$$

By Theorem 2.2 of ch. 7, $P(\mathrm{D})+q$ has a regularly accretive extension with the closure of $b(\varphi,\psi)$ in W as the corresponding bilinear form. Since b is symmetric, the same is true of its closure. We now apply Theorem 7.1 of ch. 1.

Theorem 1.4. *Assume that $P(\xi)$ satisfies* (1.2) *for some real number M and*

$$P^{(\mu)}(\xi)/P(\xi) = \mathrm{O}(|\xi|^{-a|\mu|}) \text{ as } |\xi| \to \infty, \text{ all } \mu, \tag{1.7}$$

$$1/P(\xi) = \mathrm{O}(|\xi|^{-2b}) \text{ as } |\xi| \to \infty, \tag{1.8}$$

where $b > a+n-am$. Let k_0 be the smallest non-negative integer such that $k_0 a > n-b$. Assume that q is locally in L^1 and that $\sigma \in M_{\alpha,2}$ with

$$0 < \alpha < 2(n-k_0). \tag{1.9}$$

Then $P(\mathrm{D})+q$ on $C_0^\infty \cap D(q)$ has a regularly accretive self-adjoint extension.

Proof. By Theorem 2.3 of ch. 7, $P(\mathrm{D})+q$ has a regularly accretive extension with corresponding bilinear form satisfying (1.5). Since C_0^∞ is dense in W, the bilinear form is symmetric. By Theorem 7.6 of ch. 1 the regularly accretive extension is self-adjoint.

Theorem 1.5. *Let*

$$Q(x, \mathrm{D}) = \sum_{j,k=1}^{N} \bar{P}_j(\mathrm{D})\, q_{jk}(x)\, P_k(\mathrm{D}), \tag{1.10}$$

where the $P_j(\mathrm{D})$ are constant coefficient operators and the functions q_{jk} satisfy $q_{jk}(x) = \overline{q_{kj}(x)}$ for each j, k. Assume that the polynomial $P(\xi)$ with real coefficients satisfies (1.2), (1.7) *and*

$$P_k(\xi)/|P(\xi)|^{\frac{1}{2}} = \mathrm{O}(|\xi|^{-b_k}) \text{ as } |\xi| \to \infty, \qquad 1 \leqslant k \leqslant N, \tag{1.11}$$

with each $b_k > a+n-an$. For each k let s_k be the smallest non-negative integer such that $s_k a > n-b_k$. Set

$$\rho_k(x)^2 = \max[q_{kk}(x), 0], \; \sigma_k(x)^2 = \rho_k(x)^2 - q_{kk}(x), \; 1 \leqslant k \leqslant N. \tag{1.12}$$

Assume that the ρ_k are locally in L^2 and that $\sigma_k \in M_{\alpha_k,2}$, where $\alpha_k < 2(n-s_k)$. Assume also that for $j \neq k$, $q_{jk} = f_{jk} h_{jk}$ with $f_{jk} \in M_{\alpha_j,2}$ and $h_{jk} \in M_{\alpha_k,2}$. Let V be the set of those $\varphi \in C_0^\infty$ such that $Q(x, \mathrm{D})\varphi$ is in L^2. Then $P(\mathrm{D})+Q(x, \mathrm{D})$ on V has a regularly accretive self-adjoint extension.

Proof. By Theorem 4.1 of ch. 7 there is a regularly accretive extension with the corresponding bilinear form $b(u, v)$ satisfying

$$b(\varphi, \psi) = (P(\mathrm{D})\varphi, \psi) + \Sigma(q_{jk}(x) P_k(\mathrm{D})\varphi, P_j(\mathrm{D})\psi), \qquad \varphi, \psi \in C_0^\infty. \tag{1.13}$$

The assumptions on the q_{jk} imply that this bilinear form is symmetric. We now apply Theorem 7.6 of ch. 1.

Theorem 1.6. *Let $P(\mathrm{D})$ be an elliptic operator of order m with real coefficients. Let $q(x)$ be a real valued function locally in L^1, and let $\rho(x)$, $\sigma(x)$ be given by* (1.3). *Assume that $\sigma \in N_m$ and if $m \leqslant n$ also that*

$$N_{m,\delta}(\sigma) \to 0 \quad as \quad \delta \to 0 .$$

Then $P(\mathrm{D})+q$ on *$C_0^\infty \cap D(q)$ has a regularly accretive self-adjoint extension.*

Proof. We apply Theorem 7.2 of ch. 7 and follow the proof of Theorem 1.3.

Theorem 1.7. *Let $P(\mathrm{D})$ be an elliptic operator of order $m=2r$ with real coefficients. Consider the operator $Q(x, \mathrm{D})$ given by* (1.10), *where the functions $q_{jk}(x)$ are locally in L^1 and the operators $P_j(\mathrm{D})$ are of orders m_j, respectively. Let ρ_k and σ_k be defined by* (1.12). *Assume that $\sigma_k \in N_{m-2m_k}$, and if $m \leqslant n+2m_k$ also that*

$$N_{m-2m_k,\delta}(\sigma_k) \to 0 \quad as \quad \delta \to 0, \qquad 1 \leqslant k \leqslant N . \tag{1.15}$$

For $j<k$ assume that $\bar{q}_{kj}=q_{jk}=h_{1jk}h_{2jk}$ with $h_{1jk} \in N_{m-2m_j}$ and $h_{2jk} \in N_{m-2m_k}$. If $m \leqslant n+2m_j$, assume also that

$$N_{m-2m_j,\delta}(h_{1jk}) \to 0 \quad as \quad \delta \to 0 . \tag{1.16}$$

Let V be the set of those $\varphi \in C_0^\infty$ such that $Q(x, \mathrm{D})\varphi \in L^2$. Then $P(\mathrm{D})+Q(x, \mathrm{D})$ on V has a regularly accretive self-adjoint extension.

§2. Extensions with special properties

The next question we consider is when will $P(\mathrm{D})+q$ on $C_0^\infty \cap D(q)$ have a self-adjoint extension B such that

$$\sigma_e(B) = \sigma(P_0) . \tag{2.1}$$

Sufficient conditions are easily obtained by means of Theorems 3.1, 4.3, 9.1 and 10.2 of ch. 7. We state the results here.

Theorem 2.1. *Assume that $P(\xi)$ and $q(x)$ satisfy the hypotheses of Theorem 1.4. Let j_0 be the smallest non-negative integer such that $j_0 a > n-2b$. If $\rho \in M_{\beta,2}$ for some β satisfying $0<\beta<2(n-j_0)$ and*

$$\int_{|x-y|<1} |q(x)| \,\mathrm{d}x \to 0 \quad as \quad |y| \to \infty , \tag{2.2}$$

then $P(\mathrm{D})+q$ on *$C_0^\infty \cap D(q)$ has a regularly accretive self-adjoint extension B satisfying* (2.1).

Theorem 2.2. *Suppose that all of the hypotheses of Theorem 1.5 are satisfied. Let l_k be the smallest non-negative integer such that*

$$l_k a > n - 2b_k\,, \qquad 1 \leqslant k \leqslant N\,, \tag{2.3}$$

and assume that $\rho_k \in M_{\delta_k,2}$, where

$$0 < \delta_k < 2(n - l_k)\,, \qquad 1 \leqslant k \leqslant N\,. \tag{2.4}$$

If

$$\int_{|x-y|<1} \Sigma\, |q_{jk}(x)|\, \mathrm{d}x \to 0 \ \text{ as } \ |y| \to \infty\,, \tag{2.5}$$

then $P(\mathrm{D}) + Q(x, \mathrm{D})$ on V has a regularly accretive self-adjoint extension B satisfying (2.1).

Theorem 2.3. *Let the hypotheses of Theorem 1.6 be satisfied. Assume in addition that $\rho \in N_{2m}$ with*

$$N_{2m,x}(\rho) + N_{2m,x}(\sigma) \to 0 \ \text{ as } \ |x| \to \infty\,. \tag{2.6}$$

Then $P(\mathrm{D}) + q$ on *$C_0^\infty \cap D(q)$ has a regularly accretive self-adjoint extension B satisfying* (2.1).

Theorem 2.4. *Assume that the hypotheses of Theorem 1.7 are satisfied. Assume in addition that $\rho_k \in N_{2m-2m_k}$ and that*

$$N_{2m-2m_j,x}(\rho_j + \sigma_j + |h_{1jk}|) \to 0 \ \text{ as } \ |x| \to \infty\,, \qquad 1 \leqslant j, k \leqslant N\,.$$

Then $P(\mathrm{D}) + Q(x, \mathrm{D})$ on V has a regularly accretive self-adjoint extension B satisfying (2.1).

If $P(\xi)$ satisfies (1.2), then $\sigma(P_0)$ is contained in the half-line $\lambda + M \geqslant 0$. Thus (2.1) guarantees that $\sigma_e(B)$ is contained in this set. If we are satisfied to only draw this conclusion instead of (2.1), we can reduce the hypothesis of the preceding theorems. To do this we make use of the results of ch. 7, §6.

Theorem 2.5. *Assume that $P(\xi)$ and $q(x)$ satisfy the hypotheses of Theorem* 1.4. *If*

$$\int_{|x-y|<1} \sigma(x)^2\, \mathrm{d}x \to 0 \ \text{ as } \ |y| \to \infty\,, \tag{2.7}$$

then $P(\mathrm{D}) + q$ on $C_0^\infty \cap D(q)$ has a regularly accretive self-adjoint extension B such that $\sigma_e(B)$ is contained in the half-line $\lambda + M \geqslant 0$.

Theorem 2.6. *Under the hypotheses of Theorem* 1.5, *if*

$$\int_{|x-y|<1} \Sigma\, \sigma_k(x)^2\, dx \to 0 \quad as \quad |y| \to \infty\,, \tag{2.8}$$

then $P(D)+Q(x, D)$ *on* V *has a regularly accretive self-adjoint extension* B *with* $\sigma_e(B)$ *contained in* $\lambda + M \geqslant 0$.

Similarly we use Theorems 10.6 and 10.7 of ch. 7 to obtain

Theorem 2.7. *The conclusion of Theorem* 2.5 *holds if the hypotheses of Theorem* 1.6 *are satisfied and*

$$N_{m,x}(\sigma) \to 0 \quad as \quad |x| \to \infty\,. \tag{2.9}$$

Theorem 2.8 *Under the hypotheses of Theorem* 1.7 *assume that*

$$N_{2m-2m_k,x}(\sigma_k) \to 0 \quad as \quad |x| \to \infty\,, \quad 1 \leqslant k \leqslant N\,.$$

Then $P(D)+Q(x, D)$ *on* V *has a regularly accretive self-adjoint extension* B *satisfying* (2.1).

§3. Intervals containing the essential spectrum

In ch. 4, §6 we gave criteria for $\sigma_e(L)$ to contain $\sigma(P_0)$ for every closed extension L of $P(D)+q$ on $C_0^\infty \cap D(q)$. If the coefficients of $P(\xi)$ are real and $q(x)$ is real valued, then the assumptions given there for $p=2$ guarantee the same for self-adjoint extensions. However, when one is interested only in self-adjoint extensions one can estimate under weaker assumptions how far points of the essential spectrum can move under the perturbation. We present some of these results in this section.

Theorem 3.1. *Let* $P(\xi)$ *be a polynomial with real coefficients, and let* $q(x)$ *be a real valued function. Assume that there is a sequence* $\{T_k\}$ *of non-intersecting spheres with radii* $r_k \to \infty$ *such that* $q \in L^2(T_k)$ *and*

$$\frac{1}{|T_k|}\int_{T_k} q(x)^2\, dx \to M^2 \quad as \quad k \to \infty\,, \tag{3.1}$$

where $|T_k|$ *is the volume of* T_k. *Then for each* $\lambda_0 \in \sigma(P_0)$ *and each self-adjoint extension* L *of* $P(D)+q$ *on* $C_0^\infty \cap D(q)$ *the interval* $[\lambda_0 - M, \lambda_0 + M]$ *contains a point of* $\sigma_e(L)$.

Proof. Let a_k denote the center of T_k and set $c_n = \int_{|x|<1} \mathrm{d}x$. Let ε satisfying $0 < \varepsilon \leqslant 1$ be given. Then there exists a $\psi \in C_0^\infty$ such that $\psi = 0$ for $|x| \geqslant 1$, $\|\psi\| = 1$ and

$$0 \leqslant \psi \leqslant [(1+\varepsilon)/c_n]^{\frac{1}{2}} . \tag{3.2}$$

To see this, set $\delta = (1+\varepsilon)^{-1/n}$. Then $0 < \delta < 1$ and $\int_{|x|<\delta} \mathrm{d}x = c_n \delta^n = c_n/(1+\varepsilon)$. Now there is a function $\psi_0 \in C_0^\infty$ such that $\psi_0 = 1$ in $|x| < \delta$, $\psi_0 = 0$ in $|x| \geqslant 1$, and $0 \leqslant \psi_0 \leqslant 1$ (see Theorem 1.2 of ch. 2). Thus $\|\psi_0\|^2 \geqslant c_n/(1+\varepsilon)$. If we now take $\psi = \psi_0/\|\psi_0\|$, we see that ψ satisfies the desired conditions.

Assume first that there is a $\xi \in E^n$ such that $P(\xi) = \lambda_0$, and set

$$\varphi_k(x) = \mathrm{e}^{\mathrm{i}\xi x} \psi[(x-a_k)/r_k]/r_k^{\frac{1}{2}n} .$$

Then

$$P(\mathrm{D})\varphi_k = \mathrm{e}^{\mathrm{i}\xi x} \Sigma\, P^{(\mu)}(\xi) \psi_\mu[(x-a_k)/r_k]/\mu!\, r_k^{\frac{1}{2}n+|\mu|} ,$$

where $\psi_\mu(x) = \mathrm{D}^\mu \psi(x)$. Therefore

$$\|[P(\mathrm{D}) - \lambda_0]\varphi_k\| \leqslant \sum_{\mu \neq 0} |P^{(\mu)}(\xi)|\, \|\psi_\mu\|/\mu!\, r_k^{|\mu|} \to 0 \text{ as } k \to \infty . \tag{3.3}$$

Since $\|\psi\| = 1$, we have $\|\varphi_k\| = 1$ for each k. Moreover, the supports of the $\{\varphi_k\}$ do not intersect. Hence

$$(\varphi_j, \varphi_k) = 0 , \quad (P(\mathrm{D})\varphi_j, \varphi_k) = 0 ,$$
$$(P(\mathrm{D})\varphi_j, P(\mathrm{D})\varphi_k) = 0 , \qquad j \neq k .$$

We also have

$$|\varphi_k| \leqslant [(1+\varepsilon)/c_n r_k^n]^{\frac{1}{2}} = [(1+\varepsilon)/|T_k|]^{\frac{1}{2}} .$$

Hence

$$\|q\varphi_k\|^2 = \int_{T_k} |q(x)\varphi_k(x)|^2 \mathrm{d}x \leqslant \frac{1+\varepsilon}{|T_k|} \int_{T_k} |q(x)|^2 \mathrm{d}x .$$

By (3.1) there is an integer N such that

$$\frac{1}{|T_k|} \int_{T_k} |q(x)|^2 \mathrm{d}x \leqslant M^2(1+\varepsilon), \qquad k > N .$$

Thus

$$\|q\varphi_k\| \leqslant M(1+\varepsilon) , \qquad k > N .$$

Combining this with (3.3) we get

$$\|[P(\mathrm{D}) + q - \lambda_0]\varphi_k\| \leqslant M(1+2\varepsilon)$$

for k sufficiently large. We now apply Theorem 7.16 of ch. 1. Thus the interval $|\lambda-\lambda_0| \leqslant M(1+2\varepsilon)$ contains a point of $\sigma_e(L)$. Since this is true for every $\varepsilon>0$, the interval $|\lambda-\lambda_0| \leqslant M$ must either contain a point of $\sigma_e(L)$ or a point which is the limit of points of $\sigma_e(L)$. By Theorem 7.11 of ch. 1 this point must be in $\sigma_e(L)$.

Next let λ_0 be any point in $\sigma(P_0)$. Then for each $\varepsilon>0$ there is a $\xi \in E^n$ such that $|P(\xi)-\lambda_0| < \varepsilon$ (Corollary 3.3 of ch. 4). By what we have just proved, the interval $|\lambda-P(\xi)| \leqslant M$ contains a point of $\sigma_e(L)$. Thus the interval $|\lambda-\lambda_0| < M+\varepsilon$ contains a point of $\sigma_e(L)$. Since this is true for each $\varepsilon>0$, the interval $|\lambda-\lambda_0| \leqslant M$ either contains a point of $\sigma_e(L)$ or contains a point which is the limit of points of $\sigma_e(L)$. This point must also be in $\sigma_e(L)$ by Theorem 7.11 of ch. 1. This completes the proof.

In applications the following consequence of Theorem 3.1 is useful.

Corollary 3.2. *Suppose there is a real number λ_1 and a sequence $\{T_k\}$ of non-intersecting spheres with radii $r_k \to \infty$ such that $q \in L^2(T_k)$ and*

$$\frac{1}{|T_k|} \int_{T_k} [q(x)-\lambda_1]^2 \,\mathrm{d}x \to M^2 \text{ as } k \to \infty . \tag{3.4}$$

Then for each λ_0 such that $\lambda_0-\lambda_1 \in \sigma(P_0)$ and each self-adjoint extension L of $P(\mathrm{D})+q$ on $C_0^\infty \cap D(q)$ the interval $[\lambda_0-M, \lambda_0+M]$ contains a point of $\sigma_e(L)$.

Proof. Set $\tilde{P}(\xi)=P(\xi)+\lambda_1$, $\tilde{q}(x)=q(x)-\lambda_1$. Then $\tilde{q}$ satisfies the hypotheses of Theorem 3.1. Since L is an extension of $\tilde{P}(\mathrm{D})+\tilde{q}$, we see that for each $\lambda_0 \in \sigma(\tilde{P}_0)$ the interval $[\lambda_0-M, \lambda_0+M]$ contains a point of $\sigma_e(L)$. We now merely note that λ_0 is in $\sigma(\tilde{P}_0)$ if and only if $\lambda_0-\lambda_1$ is in $\sigma(P_0)$. This completes the proof.

Corollary 3.3. *Assume that there is a sequence $\{T_k\}$ of non-intersecting spheres with radii tending to infinity such that*

$$\limsup_{k\to\infty} \sup_{T_k} q(x) = M \tag{3.5}$$

$$\liminf_{k\to\infty} \inf_{T_k} q(x) = m . \tag{3.6}$$

Set $\lambda_1=\frac{1}{2}(M+m)$, $\lambda_2=\frac{1}{2}(M-m)$. If $\lambda_0-\lambda_1 \in \sigma(P_0)$, then for each self-adjoint extension L of $P(\mathrm{D})+q$ on $C_0^\infty \cap D(q)$ the interval $[\lambda_0-\lambda_2, \lambda_0+\lambda_2]$ contains a point of $\sigma_e(L)$.

Proof. Let $\varepsilon > 0$ be given. For k sufficiently large

$$-\lambda_2 - \varepsilon = m - \varepsilon - \lambda_1 \leqslant q(x) - \lambda_1 \leqslant M + \varepsilon - \lambda_1 = \lambda_2 + \varepsilon, \qquad x \in T_k .$$

Therefore

$$|q(x) - \lambda_1| \leqslant \lambda_2 + \varepsilon , \qquad x \in T_k$$

for k sufficiently large. This means that

$$\frac{1}{|T_k|} \int_{T_k} [q(x) - \lambda_1]^2 \mathrm{d}x \leqslant (\lambda_2 + \varepsilon)^2$$

for such k. Since ε was arbitrary, we have

$$\limsup_{k \to \infty} \frac{1}{|T_k|} \int_{T_k} [q(x) - \lambda_1]^2 \mathrm{d}x \leqslant \lambda_2^2 ,$$

which implies via Corollary 3.2 that for $\lambda_0 - \lambda_1 \in \sigma(P_0)$ the interval $[\lambda_0 - \lambda_2, \lambda_0 + \lambda_2]$ contains a point of $\sigma_e(L)$. This completes the proof.

Corollary 3.4. *If $q(x)$ satisfies* (3.5) *and* (3.6) *for a sequence $\{T_k\}$ of non-intersecting spheres with radii tending to infinity, then for each $\lambda_0 \geqslant 0$ and each self-adjoint extension L of $-\Delta + q$ on $C_0^\infty \cap D(q)$ the interval $[\lambda_0 - \lambda_2, \lambda_0 + \lambda_2]$ contains a point of $\sigma_e(L)$.*

Proof. We note that $\sigma(-\Delta) = [0, \infty)$ (see §5 of ch. 4) and apply Corollary 3.3.

§4. Essentially self-adjoint operators

Until now we have been concerned with the existence of self-adjoint extensions having certain properties. We now turn our attention to the question as to when there is only one self-adjoint extension. As we saw in ch. 1, §7, this is equivalent to asking when the closure of an operator is self-adjoint, i.e., when the operator is essentially self-adjoint. Corollaries 6.2 and 7.5 of ch. 5 give sufficient conditions. In this section we shall use another set of conditions for elliptic operators which cover many applications.

Theorem 4.1. *Consider the operator*

$$L(x, \mathrm{D}) = P(\mathrm{D}) + \sum_{j=1}^{r} q_j(x) Q_j(\mathrm{D}) , \tag{4.1}$$

where $P(\mathrm{D})$ is an elliptic operator of order m with real coefficients, the $Q_j(\mathrm{D})$

are operators of order $m_j < m$ with real coefficients and the $q_j(x)$ are real valued functions such that $q_j \in N_{2m-2m_j}$ and if $2(m-m_j) \leqslant n$, then

$$N_{2m-2m_j,\delta}(q_j) \to 0 \quad as \quad \delta \to 0, \qquad 1 \leqslant j \leqslant r. \tag{4.2}$$

Then $L(x, \mathrm{D})$ on C_0^∞ is essentially self-adjoint in L^2.

Proof. Since $D(P_0) = H^{m,2}$ (Lemma 4.3 of ch. 6), it suffices to show that $H^{m,2} \subset D(q_j Q_{j0})$ and that

$$\Sigma \|q_j Q_{j0} v\| \leqslant a \|P_0 v\| + b \|v\|, \qquad v \in H^{m,2}, \tag{4.3}$$

with $a < 1$ (Lemma 7.8 of ch. 1). The former follows from Theorem 4.4 of ch. 6, while the latter will be proved if we can show that for each j and $\varepsilon > 0$ there is a constant K such that

$$\|q_j Q_j(\mathrm{D}) v\| \leqslant \varepsilon \|P(\mathrm{D}) v\| + K \|v\|, \qquad v \in S. \tag{4.4}$$

By Lemmas 7.5 and 7.4 of ch. 7 for each $\varepsilon > 0$ there is a constant C such that

$$\begin{aligned} \|q_j Q_j(\mathrm{D}) v\| &\leqslant \varepsilon \|Q_j(\mathrm{D}) v\|_{m-m_j,2} + C \|Q_j(\mathrm{D}) v\| \\ &\leqslant \varepsilon C' \|v\|_{m,2} + KC'' \|v\|_{m_j,2} \\ &\leqslant \varepsilon (C'+1) \|v\|_{m,2} + C''' \|v\| . \end{aligned}$$

Thus by Theorem 3.1 of ch. 3 for each $\varepsilon > 0$ there is a constant K such that (4.4) holds. This completes the proof.

For second order operators we shall give more elaborate criteria in §2 of the next chapter.

§5. Finite negative spectrum

Let $P(\mathrm{D})$ be a constant coefficient operator such that

$$P(\xi) \geqslant 0, \qquad \xi \in E^n. \tag{5.1}$$

Then P_{02} is self-adjoint and its spectrum is contained in $[0, \infty)$. In this section we shall give a condition on a function $q(x)$ so that $P(\mathrm{D}) + q$ should have a self-adjoint extension with negative spectrum consisting of at most a finite number of eigenvalues.

Theorem 5.1. *Let $P(\mathrm{D})$ be an operator satisfying* (5.1), *and put*

$$K_\varepsilon(x) = \bar{F}\{1/[P(\xi)+\varepsilon]\}, \qquad \varepsilon > 0. \tag{5.2}$$

Let $q(x)$ be a real valued function, and define $\sigma(x)$ by (3.1). Set

$$B_\varepsilon = \int\int \sigma(x)^2 \sigma(y)^2 |K_\varepsilon(x-y)|^2 \, dx \, dy, \qquad \varepsilon > 0. \tag{5.3}$$

If $B_\varepsilon < \infty$ for some $\varepsilon > 0$, then $P(D)+q$ on $C_0^\infty \cap D(q)$ has a self-adjoint extension H such that $\sigma(H) \cap (-\infty, -\varepsilon]$ consists of at most B_ε eigenvalues. If

$$B = \limsup_{\varepsilon \to 0} B_\varepsilon < \infty, \tag{5.4}$$

then $\sigma(H) \cap (-\infty, 0)$ consists of at most B eigenvalues.

Corollary 5.2. *If*

$$\tilde{B}_\varepsilon = \varepsilon^{2n-4} \int\int \sigma(x)^2 \sigma(y)^2 G_2(\varepsilon|x-y|)^2 \, dx \, dy < \infty,$$

then the operator $-\Delta + q$ on $C_0^\infty \cap D(q)$ has a self-adjoint extension having at most $\tilde{B}$ eigenvalues in the interval $(-\infty, \varepsilon]$. If

$$\frac{16\pi^n}{\Gamma(\frac{1}{2}(n-2))^2} \tilde{B} = \int\int \sigma(x)^2 \sigma(y)^2 |x-y|^{4-2n} \, dx \, dy < \infty, \tag{5.6}$$

then it has a self-adjoint extension with at most $\tilde{B}$ eigenvalues in $(-\infty, 0)$.

Proof. In this case $P(\xi) = |\xi|^2$. Thus

$$K_\varepsilon(x) = \bar{F}[1/(|\xi|^2 + \varepsilon)] = \varepsilon^{n-2} G_2(\varepsilon x). \tag{5.7}$$

Note that

$$\limsup_{\varepsilon \to 0} \tilde{B}_\varepsilon \leqslant \tilde{B}. \tag{5.8}$$

The proof of Theorem 5.1 will be based on the following theorems.

Theorem 5.3. *Let A be a self-adjoint operator such that $\sigma(A) \subset (0, \infty)$. Suppose there is a closable operator B and a compact operator K such that $D(A) \cap D(B*B)$ is dense and*

$$BA^{-1}B^* x = Kx, \quad x \in D(BA^{-1}B^*). \tag{5.9}$$

Let N be the number of eigenvalues of K (counting multiplicities) in the interval $[1, \|K\|]$. Then $A - B^*B$ has a self-adjoint extension H such that $\sigma(H) \cap (-\infty, 0]$ consists of at most N eigenvalues (counting multiplicities).

Theorem 5.4. *Let $K(x, y)$ be a real valued function in $L^2(E^{2n})$ and let the operator C be defined by*

$$Cu(x) = \int K(x, y)u(y)\, dy. \tag{5.10}$$

Then C is a compact operator. If $\alpha_1, \alpha_2 \ldots$ are its eigenvalues, then

$$\sum_{k=1}^{\infty} \alpha_k^2 \leqslant \|K\|^2. \tag{5.11}$$

Before proving Theorems 5.3 and 5.4, let us show how they imply Theorem 5.1.

Proof of Theorem 5.1. Set $A = P_0 + \varepsilon$, $B = \sigma$ and

$$K(x, y) = \sigma(x)\sigma(y)K_\varepsilon(x-y).$$

Then the operator C defined by (5.10) agrees with $\sigma A^{-1}\sigma$ on $D(\sigma A^{-1}\sigma)$. In fact if $v = \sigma A^{-1}\sigma u$, then

$$v = \sigma\, \bar{F}\{F(\sigma u)/[P(\xi)+\varepsilon]\} = \sigma[K_\varepsilon * (\sigma u)] = Cu.$$

By Theorem 5.4, C is a compact operator and has at most B_ε eigenvalues in the interval $[1, \infty)$. Thus by Theorem 5.3, $P(\mathrm{D}) + \varepsilon - \sigma^2$ on $C_0^\infty \cap D(\sigma^2)$ has a self-adjoint extension S such that $\sigma(S) \cap (-\infty, 0]$ consists of at most B_ε eigenvalues. Thus $H = S - \varepsilon$ is a self-adjoint extension of $P_0 - \sigma^2$ such that $\sigma(H) \cap (-\infty, -\varepsilon]$ consists of at most B_ε eigenvalues. Next note that

$$([P(\mathrm{D})+q]\varphi, \varphi) \geqslant (H\varphi, \varphi), \quad \varphi \in C_0^\infty \cap D(q).$$

Since H is bounded from below, it is regularly accretive (Theorem 7.7 of ch. 1). Thus there is a Hilbert space W continuously embedded in L^2 such that

$$\|\varphi\|_W^2/C \leqslant (H\varphi, \varphi) + N\|\varphi\|^2 \leqslant C\|\varphi\|_W^2, \qquad \varphi \in D(H).$$

Set

$$\|\varphi\|_{\tilde{W}}^2 = \|\varphi\|_W^2 + \|\rho\varphi\|^2, \qquad \varphi \in C_0^\infty \cap D(q),$$

and let $\tilde{W}$ denote the completion of $C_0^\infty \cap D(q)$ with respect to this norm. Then we have

$$\|\varphi\|_{\tilde{W}}^2/C \leqslant ([P(\mathrm{D})+q]\varphi, \varphi) + N\|\varphi\|^2 \leqslant C\|\varphi\|_{\tilde{W}}^2, \qquad \varphi \in C_0^\infty \cap D(q).$$

It now follows by Theorem 6.6 of ch. 1 that $P(\mathrm{D}) + q$ on $C_0^\infty \cap D(q)$ has a regularly accretive extension T. This extension is self-adjoint by Theorem 7.6 of ch. 1. Moreover $D(T) \subset D(H)$ and

$$(Hu, u) \leqslant (Tu, u), \qquad u \in D(T).$$

Now suppose T had $m > B_\varepsilon$ eigenvalues in $(-\infty, -\varepsilon]$. Then the corresponding eigenfunctions would span an m dimensional subspace $V \subset D(T)$ such that $(Tv, v) < \varepsilon \|v\|^2$ for $v \in V$. This would mean that H had m eigenvalues in $(-\infty, -\varepsilon]$, contrary to what was just proved. This contradiction proves the theorem.

In proving Theorem 5.3 we shall make use of

Lemma 5.5. *Let E be a linear operator on a Hilbert space X and suppose there is a compact operator K on X such that*

$$\|Eu\|^2 = (Ku, u), \qquad u \in D(E).$$

Then there is a compact operator K_1 on X such that

$$Eu = K_1 u, \qquad u \in D(E).$$

Proof. Let u be an element in $\overline{D(E)}$. Then there is a sequence $\{u_k\}$ of elements in $D(E)$ such that $u_k \to u$. Thus

$$\|E(u_j - u_k)\|^2 \leqslant \|K(u_j - u_k)\| \, \|u_j - u_k\| \to 0.$$

Hence $Eu_k \to w$. Define $K_1 u = w$. The operator K_1 is bounded from $D(E)$ to X and

$$\|K_1 u\|^2 = (Ku, u), \qquad u \in \overline{D(E)}.$$

If $\{u_k\}$ is a sequence of elements in $\overline{D(E)}$ such that $\|u_k\| \leqslant C$, then there is a subsequence $\{v_m\}$ of $\{u_k\}$ such that $\{Kv_m\}$ converges. Thus

$$\|K_1(v_m - v_n)\|^2 = (K[v_m - v_n], v_m - v_n) \to 0.$$

Hence K_1 is compact from $\overline{D(E)}$ to X. Define K_1 to vanish on $D(E)^\perp$. Clearly K_1 is compact on X. The proof is complete.

We can now give the

Proof of Theorem 5.3. Set

$$a(u, v) = (Au, v) - (Bu, Bv), \qquad u, v \in D(A) \cap D(B)$$

$$(u, v)_W = (Au, v) + (Bu, Bv), \qquad u, v \in D(A) \cap D(B).$$

Set $E = A^{-\frac{1}{2}} B^*$. By (5.9)

$$\|Eu\|^2 = (BA^{-1}B^* u, u) = (Ku, u), \qquad u \in D(E).$$

Thus by Lemma 5.5 there is a compact operator K_1 such that

$$A^{-\frac{1}{2}}B^*u = K_1 u, \qquad u \in D(A^{-\frac{1}{2}}B^*).$$

Thus

$$BA^{-\frac{1}{2}}u = K_1^* u, \qquad u \in D(BA^{-\frac{1}{2}}). \tag{5.12}$$

This implies that there is a constant C such that

$$2\|Bu\|^2 \leqslant (Au, u) + 2C\|u\|^2, \qquad u \in D(A) \cap D(B). \tag{5.13}$$

In fact, if (5.13) did not hold, there would be a sequence $\{u_k\}$ of elements in $D(A) \cap D(B)$ such that

$$\|A^{\frac{1}{2}}u_k\| = 1, \quad 2\|Bu_k\|^2 > 1 + k\|u_k\|^2. \tag{5.14}$$

By (5.12) there would be a subsequence $\{v_m\}$ of $\{u_k\}$ such that $Bv_m = BA^{-\frac{1}{2}}A^{\frac{1}{2}}v_m$ converges. Thus $Bv_m \to w$ and $v_m \to 0$. Since B is closable, $w=0$. But $2\|w\|^2 \geqslant 1$ by (5.14). Thus (5.13) holds. Once (5.13) is known, we have

$$a(u) = (Au, u) - \|Bu\|^2 \geqslant \tfrac{1}{2}(Au, u) - C\|u\|^2$$
$$\|u\|_W^2 = (Au, u) + \|Bu\|^2 \leqslant \tfrac{3}{2}(Au, u) + C\|u\|^2.$$

This shows that $A - B^*B$ on $D(A) \cap D(B^*B)$ has a regularly accretive extension H (Theorem 6.6 of ch. 1). Furthermore this extension is self-adjoint (Theorem 7.6 of ch. 1).

Now for $u \in D(A) \cap D(B^*B)$ we have

$$\begin{aligned} 0 &\leqslant ([A - B^*B]u, A^{-1}[A - B^*B]u) \\ &= (Au, A^{-1}[A - B^*B]u) - (B^*Bu, [I - A^{-1}B^*B]u) \\ &= (u, [A - B^*B]u) - (Bu, [I - BA^{-1}B^*]Bu). \end{aligned}$$

Thus

$$(Bu, [I-K]Bu) \leqslant (u, Hu), \qquad u \in D(H). \tag{5.15}$$

If $\lambda_1, \ldots, \lambda_n$ are eigenvalues of H in $(-\infty, 0]$ and $u_1, \ldots, u_n$ are corresponding orthonormal eigenelements, then the Bu_k are linearly independent. For if

$$u = \Sigma\, \alpha_k u_k,$$

then

$$Hu = \Sigma\, \alpha_k \lambda_k u_k,$$

and

$$(u, Hu) = \Sigma\, \lambda_k \alpha_k^2 \leqslant 0.$$

Thus if $Bu=0$, we have $(u, Au)=0$. This implies $u=0$. Thus the Bu_k span an n-dimensional subspace V such that

$$(v, [I-K]v) \leqslant 0\,, \qquad v \in V\,,$$

or

$$(Kv, v) \leqslant \|v\|^2\,, \qquad v \in V\,.$$

This means that K has at least n eigenvalues in $[1, \|K\|]$ (Theorem 7.19 of ch. 1). This proves the theorem.

Proof of Theorem 5.4. Let $\{u_k\}$ be an orthonormal sequence in $L^2(E^n)$. There is a complete orthonormal sequence $\{v_m\}$ in $L^2(E^n)$. Hence

$$\begin{aligned}\Sigma\|Cu_k\|^2 &= \Sigma|(Cu_k, v_m)|^2 \\ &= \Sigma\left|\int K(x, y)u_k(y)\overline{v_m(x)}\,\mathrm{d}x\,\mathrm{d}y\right|^2 \\ &\leqslant \int|K(x, y)|^2\,\mathrm{d}x\,\mathrm{d}y = \|K\|^2\,,\end{aligned}$$

since the sequence $\{u_k(y)\overline{v_m(x)}\}$ is orthonormal in $L^2(E^{2n})$. This proves (5.11), since $\alpha_k = \|Cu_k\|$ if α_k is an eigenvalue and u_k is a corresponding eigenfunction. Set

$$C_k u = \sum_1^k (u, v_m)\,Cv_m\,, \qquad u \in L^2(E^n)\,.$$

Since

$$u = \sum_1^\infty (u, v_m)\,v_m\,,$$

we have

$$Cu = \sum_1^\infty (u, v_m)\,Cv_m\,.$$

Thus

$$\begin{aligned}\|Cu - C_k u\|^2 &\leqslant \left[\sum_{k+1}^\infty |(u, v_m)|\,\|Cv_m\|\right]^2 \\ &\leqslant \left(\sum_{k+1}^\infty |(u, v_m)|^2\right)\left(\sum_{k+1}^\infty \|Cv_m\|^2\right) \leqslant \|u\|^2 \sum_{k+1}^\infty \|Cv_m\|^2\,.\end{aligned}$$

Thus

$$\|C - C_k\| \to 0 \text{ as } k \to \infty\,.$$

Now each operator C_k is of finite rank and hence compact (Lemma 2.7 of ch. 1). Thus C is the limit in norm of compact operators. By Theorem 2.16 of ch. 1, C is compact as well. This completes the proof.

CHAPTER 9

SECOND ORDER OPERATORS

§1. Introduction

In this chapter we shall study second order operators of the form

$$L(x, \mathrm{D}) = \sum_{j,k=1}^{n} a_{jk}[\mathrm{D}_j+b_j(x)][\mathrm{D}_k+b_k(x)]+q(x), \tag{1.1}$$

where (a_{jk}) is a real symmetric matrix and the $b_k(x)$ and $q(x)$ are real valued functions locally in L^2. Clearly every symmetric linear second order partial differential operator can be put in this form. We have chosen to concentrate on such operators because many problems in quantum mechanics are concerned with them (see ch. 11).

Our basic assumptions are that (1.1) is elliptic and bounded from below. In this section we prove some basic inequalities. Then in §2 we show that $L(x, \mathrm{D})$ on C_0^∞ is essentially self-adjoint in L^2 under very mild regularity assumptions on the coefficients. The hypotheses of §§1, 2 are assumed throughout the chapter. In §3 we make some further observations and prove some inequalities to be employed later. In §4 we compare operators of the form (1.1) and describe the relationships between their spectra under various assumptions. In particular we give conditions for the essential spectrum of $L(x, \mathrm{D})$ to be the interval $[0, \infty)$ or to be empty. In §5 we show how information about the spectrum can be obtained from considering a quadratic form related to (1.1). This quadratic form is studied in detail in §6. It is shown how the essential spectrum is essentially controlled by the behavior of the coefficients at infinity. In §7 we show how spectra can be "added" if the variables can be "separated" in a suitable way. In §8 we carry out the details for the operator (1.1) when the coefficients allow such separation. In §9 we give the method of "clusters" which has important applications. Applications of the results of this chapter are given in ch. 11.

Our first assumption is that $L(x, \mathrm{D})$ is elliptic, i.e., that

$$\Sigma\, a_{jk}\xi_j\xi_k > 0 \tag{1.2}$$

whenever $\xi=(\xi_1, \ldots, \xi_n)$ is a real non-vanishing vector. It is easy to see that this implies that there is a constant $c_0>0$ such that

$$\Sigma\, a_{jk}\zeta_j\bar{\zeta}_k \geqslant c_0\, \Sigma\, |\zeta_k|^2 \tag{1.3}$$

for any complex vector $\zeta=(\zeta_1, \ldots, \zeta_n)$. In fact if $\zeta=\xi+\mathrm{i}\eta$, where ξ and η are real vectors, we have

$$\begin{aligned}\Sigma\, a_{jk}(\xi_j+\mathrm{i}\eta_j)(\xi_k-\mathrm{i}\eta_k) &= \Sigma\, a_{jk}\,\xi_j\xi_k+\Sigma\, a_{jk}\,\eta_j\eta_k \\ &\quad +\mathrm{i}\,\Sigma\, a_{jk}\,\eta_j\xi_k-\mathrm{i}\,\Sigma\, a_{jk}\,\xi_j\eta_k\,.\end{aligned}$$

Since $a_{jk}=a_{kj}$, this gives

$$\Sigma\, a_{jk}\zeta_j\bar{\zeta}_k = \Sigma\, a_{jk}\,\xi_j\xi_k+\Sigma\, a_{jk}\,\eta_j\eta_k\,. \tag{1.4}$$

Thus the left-hand side of (1.3) is positive on the sphere $|\zeta|=1$. Since it is continuous, it has a positive minimum c_0 there. This gives (1.3).

Our second assumption involves $q(x)$. As before we set

$$\rho(x)^2 = \max\,[q(x), 0], \quad \sigma(x)^2 = \rho(x)^2-q(x)\,. \tag{1.5}$$

We assume that $\sigma\in N_2$. Moreover by Theorem 7.3 of ch. 7 there is a constant K_2 depending only on r such that

$$\|\sigma\varphi\|^2 \leqslant K_2 N_{2,\delta}(\sigma)\|\varphi\|_{1,2}^2+C_\delta N_2(\sigma)\|\varphi\|^2\,, \qquad \varphi\in C_0^\infty\,, \tag{1.6}$$

where the constant C_δ depends only on n and δ. We assume

$$\liminf_{\delta\to 0} N_{2,\delta}(\sigma)< c_0/K_2\,, \tag{1.7}$$

where c_0 is the constant in (1.3).

Theorem 1.1. *There are constants* $c_1<c_0$ *and* C *such that*

$$\|\sigma\varphi\|^2 \leqslant c_1\,\Sigma\,\|(\mathrm{D}_j+b_j)\varphi\|^2+C\,\|\varphi\|\,, \qquad \varphi\in C_0^\infty\,. \tag{1.8}$$

We shall base the proof of Theorem 1.1 on a very simple but useful lemma.

Lemma 1.2. *Let* τ *be a function such that*

$$\|\tau\varphi\|^2 \leqslant C_1\,\Sigma\,\|\mathrm{D}_k\varphi\|^2+C_2\|\varphi\|^2, \qquad \varphi\in C_0^\infty\,. \tag{1.9}$$

Then

$$\|\tau\varphi\|^2 \leqslant C_1\,\Sigma\,\|(\mathrm{D}_k+b_k)\varphi\|^2+C_2\|\varphi\|^2, \qquad \varphi\in C_0^\infty\,. \tag{1.10}$$

Proof. Let $\varphi\in C_0^\infty$ be given and let ψ be a function in C_0^∞ such that $0\leqslant\psi\leqslant 1$ and $\psi=1$ in the support of φ. For $\varepsilon>0$ set

$$v=(|\varphi|^2+\varepsilon^2)^{\frac{1}{2}}, \qquad w=\psi v .$$

Then $w\in C_0^\infty$. Since $v^2=\varphi\bar{\varphi}+\varepsilon^2$, we have

$$\begin{aligned} 2v\mathrm{D}_k v &= \mathrm{D}_k(v^2) = \bar{\varphi}\mathrm{D}_k\varphi+\varphi\mathrm{D}_k\bar{\varphi} \\ &= \bar{\varphi}(\mathrm{D}_k+b_k)\varphi-\varphi\overline{(\mathrm{D}_k+b_k)\varphi} . \end{aligned} \tag{1.11}$$

Hence

$$|\mathrm{D}_k v| \leqslant |(\mathrm{D}_k+b_k)\varphi| . \tag{1.12}$$

Since $\psi=1$ in the support of φ and $v=\varepsilon$ when $\varphi=0$, we see that $v=\varepsilon$ when $\mathrm{D}_k\psi\neq 0$. Thus

$$\mathrm{D}_k w = v\mathrm{D}_k\psi+\psi\mathrm{D}_k v = \varepsilon\mathrm{D}_j\psi+\psi\mathrm{D}_k v . \tag{1.13}$$

Consequently we have

$$|\mathrm{D}_k w| \leqslant \varepsilon|\mathrm{D}_k\psi|+|(\mathrm{D}_k+b_k)\varphi| . \tag{1.14}$$

Since $w\in C_0^\infty$, we have by (1.9)

$$\begin{aligned} \|\tau w\|^2 &\leqslant C_1\Sigma\|\mathrm{D}_k w\|^2+C_2\|w\|^2 \\ &\leqslant C_1\Sigma\|(\mathrm{D}_k+b_k)\varphi\|^2+\varepsilon^2 C_1\Sigma\|\mathrm{D}_k\psi\|^2 \\ &\quad +2\varepsilon C_1\Sigma\|(\mathrm{D}_k\psi)(\mathrm{D}_k+b_k)\varphi\|+C_2\|w\|^2 . \end{aligned}$$

We now let $\varepsilon\to 0$. Then $w\to|\varphi|$. This gives (1.10) in the limit and completes the proof.

It is now a simple matter to give the proof of Theorem 1.1. By (1.7) there is a constant $c_1<c_0$ and a $\delta>0$ such that

$$N_{2,\delta}(\sigma)=c_1/K_2 .$$

Hence by (1.6)

$$\|\sigma\varphi\|^2\leqslant \|\varphi\|_{1,2}^2+C_\delta\|\varphi\|^2, \qquad \varphi\in C_0^\infty . \tag{1.15}$$

Now

$$\begin{aligned} \|\varphi\|_{1,2}^2 &= \int(1+|\xi|^2)|F\varphi|^2\,\mathrm{d}\xi = \int(\Sigma|\xi_k F\varphi|^2+|F\varphi|^2)\,\mathrm{d}\xi \\ &= \Sigma\|\mathrm{D}_k\varphi\|^2+\|\varphi\|^2 . \end{aligned} \tag{1.16}$$

Thus (1.15) becomes

$$\|\sigma\varphi\|^2\leqslant c_1\Sigma\|\mathrm{D}_k\varphi\|^2+C\|\varphi\|^2, \qquad \varphi\in C_0^\infty .$$

If we now apply Lemma 1.2, we obtain (1.8). The proof is complete.

Set

$$P(x, \mathrm{D}) = \sum_{j,k=1}^{n} a_{jk}[\mathrm{D}_j + b_j(x)][\mathrm{D}_k + b_k(x)]. \tag{1.17}$$

Then by (1.3) we have

$$(P(x, \mathrm{D})\varphi, \varphi) \geqslant c_0 \Sigma \|(\mathrm{D}_k + b_k)\varphi\|^2. \tag{1.18}$$

As a consequence of this and Theorem 1.1 we have

Corollary 1.3. *There is a constant* K_0 *such that*

$$\left(1 - \frac{c_1}{c_0}\right)(P(x, \mathrm{D})\varphi, \varphi) + \|\rho\varphi\|^2 \leqslant (L(x, \mathrm{D})\varphi, \varphi) + K_0\|\varphi\|^2, \quad \varphi \in C_0^\infty. \tag{1.19}$$

Proof. By definition

$$(P(x, \mathrm{D})\varphi, \varphi) = (L(x, \mathrm{D})\varphi, \varphi) - \|\rho\varphi\|^2 + \|\sigma\varphi\|^2,$$

and by (1.8) and (1.18)

$$\|\sigma\varphi\|^2 \leqslant \frac{c_1}{c_2}(P(x, \mathrm{D})\varphi, \varphi) + C\|\varphi\|^2.$$

Combining the last two expressions we get (1.19).

Corollary 1.4.

$$\|\rho\varphi\|^2 + (c_0 - c_1)\Sigma\|(\mathrm{D}_k + b_k)\varphi\|^2 \leqslant (L(x, \mathrm{D})\varphi, \varphi) + K_0\|\varphi\|_2, \quad \varphi \in C_0^\infty. \tag{1.20}$$

Proof. Combine (1.17) and (1.18).

Remark. In many applications $q(x)$ satisfies

$$\liminf_{|x| \to \infty} q(x) > -\infty. \tag{1.21}$$

If $\sigma \in N_\beta^{\mathrm{loc}}$ for some $\beta < 2$, this implies

$$N_{2,\delta}(\sigma) \to 0 \text{ as } \delta \to 0. \tag{1.22}$$

For (1.21) implies that $\sigma(x)$ is bounded for $|x|$ large, and consequently $\sigma \in N_\beta$. This implies (1.22) by the remark following Lemma 7.4 of ch. 7.

§2. Essential self-adjointness

In this section we give additional hypotheses on $q(x)$ and the $b_k(x)$ so that the operator $L(x, \mathrm{D})$ given by (1.1) is essentially self-adjoint. In stating our hypotheses it is convenient to let N_α^{loc} denote the set of those functions h such that $\varphi h \in N_\alpha$ for each $\varphi \in C_0^\infty$. Set

$$b(x) = \Sigma a_{jk} b_j(x) b_k(x), \qquad e(x) = \Sigma a_{jk} \mathrm{D}_j b_k(x). \tag{2.1}$$

We assume that there is an α satisfying

$$0 < \alpha < 4 \tag{2.2}$$

such that

(a) $q \in N_\alpha^{\mathrm{loc}}$
(b) $b \in N_\alpha^{\mathrm{loc}}$
(c) $e \in N_\alpha^{\mathrm{loc}}$.

In particular it follows from these assumptions that the coefficients of $L(x, \mathrm{D})$ are locally in L^2. Hence $L(x, \mathrm{D})$ is defined on C_0^∞. We have

Theorem 2.1. *Under the hypotheses of* §1 *and* (a)–(c), *the operator* $L(x, \mathrm{D})$ *on* C_0^∞ *is essentially self adjoint in* L^2.

In proving Theorem 2.1 we shall make use of the following two lemmas.

Lemma 2.2. *If* $u \in L^2$ *and*

$$|(u, L(x, \mathrm{D})\varphi)| \leqslant C\|\varphi\|, \qquad \varphi \in C_0^\infty,$$

then $\psi u \in H^{2,2}$ *for each* $\psi \in C_0^\infty$.

Lemma 2.3. *Let* P_0 *and* L_0 *denote the closures in* L^2 *of* $P(x, \mathrm{D})$ *and* $L(x, \mathrm{D})$, *respectively* (*see* (1.1) *and* (1.17)). *If* $u \in H^{2,2}$, *then* $\psi u \in D(P_0)$ *and* $D(L_0)$ *for each* $\psi \in C_0^\infty$ *and*

$$P_0(\psi u) = P(x, \mathrm{D})(\psi u)\ ;\ L_0(\psi u) = L(x, \mathrm{D})(\psi u).$$

Let us postpone the proofs of these lemmas until the end of this section. Using them we give the proof of Theorem 2.1. Let K_0 be the constant in (1.19) and let A denote the operator $L(x, \mathrm{D}) + K_1$ on C_0^∞, where $K_1 = K_0 + 1$. It suffices to show that A is essentially self-adjoint. By Corollary 1.3

$$\|\varphi\| \leqslant \|A\varphi\|, \qquad \varphi\in C_0^\infty . \tag{2.4}$$

Hence by Lemma 7.5 of ch. 1 it will follow that A is essentially self-adjoint if we can show that its range is dense in L^2. To this end let $u\in L^2$ be any function such that

$$(u, A\varphi)=0, \qquad \varphi\in C_0^\infty . \tag{2.5}$$

If $\bar{A}$ denotes the closure of $\bar{A}$ in L^2, then (2.5) implies

$$(u, \bar{A}v)=0, \qquad v\in D(\bar{A}) . \tag{2.6}$$

Another implication of (2.5) is

$$|(u, L(x,\mathrm{D})\varphi)| \leqslant K_1\|u\|\,\|\varphi\|, \qquad \varphi\in C_0^\infty , \tag{2.7}$$

which in turn implies by Lemma 2.2 that $\varphi u\in H^{2,2}$ for each $\varphi\in C_0^\infty$. Moreover for each $\psi\in C_0^\infty$ there is a $\varphi\in C_0^\infty$ which equals one on the support of ψ. Hence $\psi u=\psi\varphi u$, and since φu is in $H^{2,2}$, we see that the function ψu is in $D(L_0)$ with $L_0(\psi u)=L(x,\mathrm{D})(\psi u)$ (Lemma 2.3). Thus by (2.6)

$$(u, L(x,\mathrm{D})(\psi^2 u))=-K_1\|\psi u\|^2 \tag{2.8}$$

for all real valued functions $\psi\in C_0^\infty$. Moreover by (1.19)

$$C_2(P(x,\mathrm{D})\psi u, \psi u)+\|\psi u\|^2 \leqslant (L(x,\mathrm{D})(\psi u), \psi u)+K_1\|\psi u\|^2 , \tag{2.9}$$

where $C_2=(c_0-c_1)/c_0$. Now

$$\begin{aligned}
&\int\Sigma\, a_{jk}(\mathrm{D}_j+b_j)(\psi^2u)\overline{(\mathrm{D}_k+b_k)u}\,\mathrm{d}x\\
&\quad=\int\Sigma\, a_{jk}[\psi(\mathrm{D}_j+b_j)+\psi_j](\psi u)\overline{(\mathrm{D}_k+b_k)u}\,\mathrm{d}x\\
&\quad=\int\Sigma\, a_{jk}(\mathrm{D}_j+b_j)(\psi u)\overline{[(\mathrm{D}_k+b_k)(\psi u)-\psi_k u]}\,\mathrm{d}x\\
&\qquad+\int\Sigma\, a_{jk}\psi_j u\overline{[(\mathrm{D}_k+b_k)(\psi u)-\psi_k u]}\,\mathrm{d}x ,
\end{aligned} \tag{2.10}$$

where $\psi_j=\mathrm{D}_j\psi$, $1\leqslant j\leqslant n$. Since (a_{jk}) is a symmetric positive definite matrix, we have

$$\begin{aligned}
&2\,|\Sigma\, a_{jk}(\;\;_j+b_j)(\psi u)\psi_k|\\
&\quad\leqslant 2[\Sigma\, a_{jk}(\mathrm{D}_j+b_j)(\psi u)\overline{(\mathrm{D}_k+b_k)(\psi u)}]^{\frac12}[\Sigma\, a_{jk}\psi_j\psi_k]^{\frac12}\\
&\quad\leqslant \varepsilon\Sigma\, a_{jk}(\mathrm{D}_j+b_j)(\psi u)\overline{(\mathrm{D}_k+b_k)(\psi u)}+\frac{1}{\varepsilon}\Sigma\, a_{jk}\psi_j\psi_k
\end{aligned}$$

for any $\varepsilon > 0$. Thus by (2.10)

$$(1-\varepsilon)\int \Sigma\, a_{jk}(D_j+b_j)(\psi u)\overline{(D_k+b_k)(\psi u)}\,dx$$
$$\leqslant \operatorname{Re}\int \Sigma\, a_{jk}(D_j+b_j)(\psi^2 u)\overline{(D_k+b_k)u}\,dx + \left(1+\frac{1}{\varepsilon}\right)\int \Sigma\, a_{jk}\psi_j\psi_k|u|^2\,dx\,.$$

Then

$$(1-\varepsilon)(P(x,D)(\psi u),\psi u) \leqslant \operatorname{Re}\,(P(x,D)(\psi^2 u),u) + \left(1+\frac{1}{\varepsilon}\right)\int \Sigma\, a_{jk}\psi_j\psi_k|u|^2\,dx\,. \quad (2.11)$$

Now by (2.8)

$$(P(x,D)(\psi^2 u),u) = (L(x,D)(\psi^2 u),u) - \|\rho\psi u\|^2 + \|\sigma\psi u\|^2$$
$$= -K_1\|\psi u\|^2 - \|\rho\psi u\|^2 + \|\sigma\psi u\|^2\,.$$

Thus (2.11) becomes

$$(L(x,D)(\psi u),\psi u) + K_1\|\psi u\|^2 \leqslant \varepsilon(P(x,D)(\psi u),\psi u) + \left(1+\frac{1}{\varepsilon}\right)\int \Sigma\, a_{jk}\psi_j\psi_k|u|^2\,dx\,. \quad (2.12)$$

If we now take $\varepsilon = C_2$ and apply (2.9), we get

$$\|\psi u\|^2 \leqslant C_3\int \Sigma\, a_{jk}\psi_j\psi_k|u|^2\,dx\,, \quad (2.13)$$

where $C_3 = (C_2+1)/C_2$. Now let $\psi(x)$ be a function in C_0^∞ such that

$$0 \leqslant \psi(x) \leqslant 1\,, \qquad x\in E^n$$
$$\psi(x) = 1\,, \qquad |x| < 1$$
$$\psi(x) = 0\,, \qquad |x| > 2\,,$$

and set

$$\psi_R(x) = \psi(x/R)\,, \qquad R > 0\,.$$

If we substitute ψ_R for ψ in (2.13), we obtain

$$\int_{|x|<R} |u(x)|^2\,dx \leqslant \frac{C_3 M^2 N}{R^2}\int_{|x|>|R} |u(x)|^2\,dx\,, \quad (2.14)$$

where $M = \max|D_k\psi|$ and $N = \Sigma|a_{jk}|$. Letting $R\to\infty$, we see that $\|u\| = 0$. This shows that every u satisfying (2.5) must vanish. Consequently $R(A)$ is dense in L^2. This completes the proof.

Remark. An examination of the proof of Theorem 7.3 of ch. 7 shows that we may take

$$K_2 = \limsup_{|x|\to 0} G_2(x)/\omega_2(x). \tag{2.15}$$

It remains to prove Lemmas 2.2 and 2.3. In proving the former we shall use

Lemma 2.4. *Let s be a number satisfying* $\frac{1}{4}\alpha < s-1 < 1$, *and let* ψ *be a real valued function in* C_0^∞. *Set*

$$\begin{aligned} Q(x, \mathrm{D})\varphi &= \psi \Sigma a_{jk} b_j \mathrm{D}_k(\psi\varphi) + \psi \Sigma a_{jk} \mathrm{D}_j(b_k \psi\varphi) + (b+q)\psi^2 \varphi \\ &= 2\psi \Sigma a_{jk} b_j \mathrm{D}_k(\psi\varphi) + (b+e+q)\psi^2 \varphi. \end{aligned} \tag{2.16}$$

Then there is a constant C such that

$$\|Q(x, \mathrm{D})\varphi\|_{-t,2} \leqslant C\|\varphi\|_{s-t,2}, \qquad \varphi \in C_0^\infty, \tag{2.17}$$

for each real t satisfying $0 \leqslant t \leqslant s$.

Proof. It suffices to prove

$$\|Q(x, \mathrm{D})\varphi\| \leqslant C\|\varphi\|_{s,2}, \qquad \varphi \in C_0^\infty, \tag{2.18}$$

$$\|Q(x, \mathrm{D})\varphi\|_{-s,2} \leqslant C\|\varphi\|, \qquad \varphi \in C_0^\infty. \tag{2.19}$$

For then $Q(x, \mathrm{D})$ can be extended to a bounded operator from $H^{s,2}$ to $L^2 = H^{0,2}$ and to a bounded operator from L^2 to $H^{-s,2}$. It will then follow from Corollary 4.15 of ch. 2 that $Q(x, \mathrm{D})$ is a bounded operator from $H^{(1-\theta)s,2}$ to $H^{-\theta s,2}$ for each θ satisfying $0 \leqslant \theta \leqslant 1$. If we take $\theta = t/s$, we obtain (2.17).

Moreover (2.18) implies (2.19). For by the symmetry of $Q(x, \mathrm{D})$ we have

$$\begin{aligned} |(Q(x, \mathrm{D})\varphi, v)| &= |(\varphi, Q(x, \mathrm{D})v)| \leqslant \|\varphi\| \, \|Q(x, \mathrm{D})v\| \\ &\leqslant C\|\varphi\| \, \|v\|_{s,2}. \end{aligned}$$

This gives (2.19) in view of Theorem 4.12 of ch. 2.

Now (2.18) will follow from

$$\|\psi^2 b\varphi\| + \|\psi^2 e\varphi\| + \|\psi^2 q\psi\| + \Sigma\|\psi b_j \mathrm{D}_k(\psi\varphi)\| \leqslant C\|\varphi\|_{s,2}, \qquad \varphi \in C_0^\infty. \tag{2.20}$$

To prove (2.20) note that by hypotheses (a)–(c), $\psi^2 b$, $\psi^2 e$ and $\psi^2 q$ are in $N_\alpha \subset N_{2s}$. Since $2s > 2 + \frac{1}{2}\alpha > \alpha$, $N_\alpha \subset N_{2s}$. Thus the inequality for the first three terms of (2.20) follow from Theorem 7.3 of ch. 7. Moreover, $b_k^2 \leqslant b/c_0$ for each by (1.3). Thus each ψb_k is in N_β for each $\beta > \frac{1}{2}\alpha$. Since $2s - 2 > \frac{1}{2}\alpha$, we have $\psi b_k \in N_{2s-2}$ for each k. Thus by Theorem 7.3 of ch. 7

$$\|\psi b_j \mathrm{D}_k(\psi\varphi)\| \leqslant C\|\mathrm{D}_k(\psi\varphi)\|_{s-1,2} \leqslant C\|\psi\varphi\|_{s,2} \leqslant C'\|\varphi\|_{s,2}$$

by Theorem 4.6 and Lemma 4.8 of ch. 2. This gives (2.20) and completes the proof of the lemma.

We can now give the

Proof of Lemma 2.2. Let Ω be a bounded domain in E^n and set

$$P(\mathrm{D}) = \sum_{j,k=1}^{n} a_{jk}\mathrm{D}_j\mathrm{D}_k\,. \tag{2.21}$$

If ψ is a real valued function in C_0^∞ which equals 1 on Ω, then by (2.16)

$$L(x,\mathrm{D})\varphi = P(\mathrm{D})\varphi + Q(x,\mathrm{D})\varphi\,, \qquad \varphi \in C_0^\infty(\Omega)\,. \tag{2.22}$$

Thus by (2.3)

$$|(u, P(\mathrm{D})\varphi)| \leqslant |(u, Q(x,\mathrm{D})\varphi)| + C\|\varphi\|\,, \qquad \varphi \in C_0^\infty(\Omega)\,. \tag{2.23}$$

Let s be any number satisfying $\frac{1}{4}\alpha < s-1 < 1$. By Lemma 2.4

$$|(u, P(\mathrm{D})\varphi)| \leqslant C\|\varphi\|_{s,2}\,, \qquad \varphi \in C_0^\infty(\Omega)\,.$$

In view of the fact that $P(\mathrm{D})$ is elliptic, it now follows from Theorem 3.2 of ch. 3 that $\varphi u \in H^{\delta,2}$ for each $\varphi \in C_0^\infty(\Omega)$, where $\delta = 2-s > 0$. Since Ω was arbitrary, it follows that $\varphi u \in H^{\delta,2}$ for each $\varphi \in C_0^\infty$.

Again let Ω be a bounded domain and let ψ be a real valued function in C_0^∞ which equals one on Ω. Since $\psi u \in H^{\delta,2}$ and $\psi u = u$ on Ω, we have by (2.23)

$$\begin{aligned} |(u, P(\mathrm{D})\varphi)| &\leqslant \|\psi u\|_{\delta,2}\|Q(x,\mathrm{D})\varphi\|_{-\delta,2} + C\|\varphi\| \\ &\leqslant C(\|\psi u\|_{\delta,2}+1)\|\varphi\|_{s-\delta,2}\,, \qquad \varphi \in C_0^\infty(\Omega)\,, \end{aligned}$$

where we have applied Lemma 2.4 with $t=\delta$. Again we apply Theorem 3.2 of ch. 3 to conclude that φu is in $H^{2\delta,2}$ for each $\varphi \in C_0^\infty(\Omega)$, and since Ω was arbitrary, for each $\varphi \in C_0^\infty$. If we now reapply (2.23) and Lemma 2.4 we get

$$\begin{aligned} |(u, P(\mathrm{D})\varphi)| &\leqslant \|\psi u\|_{2\delta,2}\|Q(x,\mathrm{D})\varphi\|_{-2\delta,2} + C\|\varphi\| \\ &\leqslant C(\|\psi u\|_{2\delta,2}+1)\|\varphi\|_{s-2\delta,2}\,, \qquad \varphi \in C_0^\infty(\Omega)\,, \end{aligned}$$

which allows us to conclude via Theorem 3.2 of ch. 3 that $\varphi u \in H^{3\delta,2}$ for each $\varphi \in C_0^\infty$. Continuing in this way we reach $\varphi u \in H^{2,2}$ for $\varphi \in C_0^\infty$. This completes the proof of the lemma.

Proof of Lemma 2.3. Since $u \in H^{2,2}$, there is a sequence $\{\varphi_k\}$ of functions in C_0^∞ such that

$$\|\varphi_k - u\|_{2,2} \to 0 \text{ as } k \to \infty .$$

By Theorem 4.6 of ch. 2,

$$\|\psi(\varphi_k - u)\|_{2,2} \to 0 \text{ as } k \to \infty .$$

Since the sequence $\{\psi\varphi_k\}$ vanishes outside the support of ψ, we see by (2.18) and (2.22) that $L(x, \mathrm{D})(\psi\varphi_k)$ converges in L^2 to $L(x, \mathrm{D})(\psi u)$. This shows that ψu is in $D(L_0)$ and that $L_0(\psi u) = L(x, \mathrm{D})(\psi u)$. Similarly $\psi u \in D(P_0)$ and $P_0(\psi u) = P(x, \mathrm{D})(\psi u)$.

As a consequence of Theorem 2.1 we have

Corollary 2.5. *Let L denote the closure in L^2 of $L(x, \mathrm{D})$ on C_0^∞ and let K_0 be the constant in* (1.19). *Then $\sigma(L)$ is contained in the half-line $\lambda + K_0 \geqslant 0$.*

Proof. By (1.18)

$$-(\lambda + K_0)\|\varphi\| \leqslant \|(L-\lambda)\varphi\| , \qquad \varphi \in C_0^\infty .$$

For $\lambda + K_0 < 0$ this shows that $L - \lambda$ is one–to–one. Since L is self-adjoint, $L - \lambda$ is also onto. Hence $\lambda \in \rho(L)$.

The following consequence of Theorem 2.1 is useful in applications.

Theorem 2.6 *Assume that q, b and e are in N_α^{loc} for some $\alpha < 4$ and that* (1.2) *and* (1.21) *hold. Then $L(x, \mathrm{D})$ on C_0^∞ is essentially self-adjoint in L^2.*

Proof. Note that $q \in N_\alpha^{\mathrm{loc}}$ implies that $\sigma \in N_\beta^{\mathrm{loc}}$ for some $\beta < 2$ (Lemma 5.4 of ch. 6). Thus (1.22) holds by the remark at the end of §1. We now apply Theorem 2.1.

§3. Some observations

In this section we discuss further properties of the operator (1.1). Throughout we shall assume the hypotheses of §§1, 2, viz., that (1.3) and (1.7) hold as well as assumptions (a)–(c) of §2. In particular we shall make use of all of the results proved in §§1, 2.

Lemma 3.1. *For each bounded domain $\Omega \subset E^n$ there is a constant C such that*

$$\|\varphi\|_{2,2} \leqslant C(\|L(x, \mathrm{D})\varphi\| + \|\varphi\|) , \qquad \varphi \in C_0^\infty(\Omega) . \tag{3.1}$$

Proof. Since $P(\mathrm{D})$ is a second order elliptic operator, we have by Theorem 3.1 of ch. 3

$$\|\varphi\|_{2,2} \leqslant C(\|P(\mathrm{D})\varphi\| + \|\varphi\|), \qquad \varphi \in C_0^\infty . \tag{3.2}$$

Let ψ be a function in C_0^∞ which is identically one on Ω. Then

$$L(x, \mathrm{D})\varphi = P(\mathrm{D})\varphi + Q(x, \mathrm{D})\varphi, \qquad \varphi \in C_0^\infty(\Omega),$$

where $Q(x, \mathrm{D})$ is given by (2.16). Thus by Lemma 2.4

$$\|P(\mathrm{D})\varphi\| \leqslant \|L(x, \mathrm{D})\varphi\| + C\|\varphi\|_{s,2}, \qquad \varphi \in C_0^\infty(\Omega),$$

where s is any number satisfying $\frac{1}{4}\alpha < s-1 < 1$ and C' depends on s. Pick any such value of s. Then by Lemma 7.4 of ch. 7 for any $\varepsilon > 0$ there is a constant C_ε such that

$$\|\varphi\|_{s,2} \leqslant \varepsilon\|\varphi\|_{2,2} + C_\varepsilon\|\varphi\|, \qquad \varphi \in C_0^\infty .$$

Thus

$$\|\varphi\|_{2,2} \leqslant C\|L(x, D)\varphi\| + CC'C_\varepsilon\|\varphi\| + \varepsilon CC'\|\varphi\|_{2,2} .$$

We take ε so small that $\varepsilon CC' < 1$. This implies (3.1).

Lemma 3.2. *For each $\psi \in C_0^\infty$ there is a constant C such that*

$$\|\psi\varphi\|_{2,2} \leqslant C(\|L(x, \mathrm{D})\varphi\| + \|\varphi\|), \qquad \varphi \in C_0^\infty . \tag{3.3}$$

Proof. Let $\psi \in C_0^\infty$ be given and let ζ be a real valued function in C_0^∞ which equals one in the support of ψ. Set

$$|||u|||_j = \sum_{|\mu| \leqslant j} \|\zeta^{|\mu|}\mathrm{D}^\mu u\|, \qquad u \in H^{j,2}, \qquad j = 0, 1, \dots . \tag{3.4}$$

Then one has

$$\|\zeta^j u\|_{j,2} \leqslant C|||u|||_j, \qquad j = 0, 1, \dots \tag{3.5}$$

$$|||u|||_j \leqslant \|\zeta^j u\|_{j,2} + |||u|||_{j-1}, \qquad j = 1, 2, \dots \tag{3.6}$$

$$\sum_{j=1}^n \|\zeta\mathrm{D}_j u\|^2 \leqslant C|||u|||_2\|u\| . \tag{3.7}$$

Inequalities (3.5) and (3.6) are immediate consequences of the definition (3.4) and Theorem 4.6 of ch. 2. To prove (3.7) note that

$$\begin{aligned}\|\zeta\mathrm{D}_j u\|^2 &= (\mathrm{D}_j(\zeta^2\mathrm{D}_j u), u)\\ &\leqslant (\|\zeta^2\mathrm{D}_j^2 u\| + \|2\zeta(\mathrm{D}_j\zeta)\mathrm{D}_j u\|)\|u\| \leqslant C|||u|||_2\|u\| .\end{aligned}$$

Now by Lemma 3.1 there is a constant C such that

$$\|\zeta^2\varphi\|_{2,2} \leqslant C(\|L(x, \mathrm{D})(\zeta^2\varphi)\| + \|\zeta^2\varphi\|), \qquad \varphi \in C_0^\infty. \tag{3.8}$$

But a simple calculation gives

$$L(x, \mathrm{D})(\zeta^2\varphi) = \zeta^2 L(x, \mathrm{D})\varphi + \psi\zeta \Sigma a_{jk}(\mathrm{D}_j\zeta)(\mathrm{D}_k + b_k)\varphi$$
$$+ 2\varphi \Sigma a_{jk}[(\mathrm{D}_j\zeta)\mathrm{D}_k\zeta + \zeta\mathrm{D}_j\mathrm{D}_k\zeta]. \tag{3.9}$$

Thus

$$\|L(x, \mathrm{D})(\zeta^2\varphi)\| \leqslant \|\zeta^2 L(x, \mathrm{D})\varphi\| + C(\Sigma\|\zeta\mathrm{D}_k\varphi\|$$
$$+ \Sigma\|(\mathrm{D}_j\zeta)b_k\zeta\varphi\| + \|\varphi\|). \tag{3.10}$$

As we noted in the proof of Lemma 2.4, $(\mathrm{D}_j\zeta)b_k$ is in N_2 for each j, k. Hence by Theorem 7.3 of ch. 7

$$\Sigma\|(\mathrm{D}_j\zeta)b_k\zeta\varphi\| \leqslant C\|\zeta\varphi\|_{1,2}.$$

Hence by (3.5) and (3.10)

$$\|L(x, \mathrm{D})(\zeta^2\varphi)\| \leqslant C(\|L(x, \mathrm{D})\varphi\| + |||\varphi|||_1). \tag{3.11}$$

If we now apply (3.6) and (3.11) to (3.8), we get

$$|||\varphi|||_2 \leqslant C_1(\|L(x, \mathrm{D})\varphi\| + |||\varphi|||_1). \tag{3.12}$$

But by (3.7) there is a constant C_2 such that

$$|||\varphi|||_1 \leqslant (1/2C_1)|||\varphi|||_2 + C_2\|\varphi\|.$$

Combining this with (3.12) we obtain

$$|||\varphi|||_2 \leqslant C_3(\|L(x, \mathrm{D})\varphi\| + \|\varphi\|). \tag{3.13}$$

We now note that $\psi\varphi = \psi\zeta^2\varphi$ for all $\varphi \in C_0^\infty$. Consequently by (3.5)

$$\|\psi\varphi\|_{2,2} = \|\psi\zeta^2\varphi\|_{2,2} \leqslant C_4\|\zeta^2\varphi\|_{2,2} \leqslant C_5|||\varphi|||_2.$$

Applying this to (3.13) we obtain (3.3). The proof is complete.

Next set

$$\|\varphi\|_W^2 = \Sigma\|(\mathrm{D}_k + b_k)\varphi\|^2 + \|\rho\varphi\|^2 + \|\varphi\|^2, \qquad \varphi \in C_0^\infty, \tag{3.14}$$

and let W represent the completion of C_0^∞ with respect to this norm. Clearly W is a Hilbert space which is a subspace of L^2, and

$$\|u\| \leqslant \|u\|_W, \qquad u \in W. \tag{3.15}$$

We also have

Lemma 3.3 *There is a constant* C *such that*

$$\|\varphi\|_W^2/C \leqslant [(L(x, \mathrm{D})\varphi, \varphi)+K_1\|\varphi\|^2] \leqslant C\|\varphi\|_W^2\,, \tag{3.16}$$

where $K_1 = K_0 + 1$ (*see* (1.19)).

Proof. This is immediate from Corollary 1.4.

Lemma 3.4. *For each* $\psi \in C_0^\infty$ *there is a constant* C *such that*

$$\|\psi\varphi\|_W \leqslant C\|\varphi\|_{1,2}\,, \qquad \varphi \in C_0^\infty \tag{3.17}$$

$$\|\psi\varphi\|_{1,2} \leqslant C\|\varphi\|_W\,, \qquad \varphi \in C_0^\infty\,. \tag{3.18}$$

In order to prove Lemma 3.4 we need the simple

Lemma 3.5. $\rho \in N_2^{\mathrm{loc}}$.

Proof of Lemma 3.5. By hypothesis (a) of §2 and Lemma 3.4 of ch. 6, $|q|^{\frac{1}{2}}$ is in N_β^{loc} for each $\beta > \frac{1}{2}\alpha$. In particular it is in N_2^{loc}. Since $\rho^2 \leqslant |q| + \sigma^2$,

$$N_{2,x}(\rho) \leqslant N_{2,x}(|q|^{\frac{1}{2}}) + N_{2,x}(\sigma)\,.$$

The result now follows from the fact that $\sigma \in N_2$.

Proof of Lemma 3.4. We have

$$\begin{aligned}\|(\mathrm{D}_k+b_k)(\psi\varphi)\| &\leqslant \|\mathrm{D}_k(\psi\varphi)\| + \|\psi b_k\varphi\| \\ &\leqslant \|\psi\varphi\|_{1,2} + C\|\varphi\|_{1,2} \leqslant C'\|\varphi\|_{1,2}\,,\end{aligned}$$

since ψb_k is in N_2. Moreover

$$\|\rho\psi\varphi\| \leqslant C\|\varphi\|_{1,2} \tag{3.19}$$

since $\psi\rho$ is in N_2 (Lemma 3.5). This proves (3.17). To prove (3.18) note that by (1.16)

$$\begin{aligned}\|\psi\varphi\|_{1,2}^2 &= \Sigma\|\mathrm{D}_k(\psi\varphi)\|^2 + \|\psi\varphi\|^2 \\ &\leqslant C(\Sigma\|(\mathrm{D}_k+b_k)(\psi\varphi)\|^2 + \Sigma\|\psi b_k\varphi\|^2 + \|\psi\varphi\|^2) \\ &\leqslant C'(\Sigma\|\psi(\mathrm{D}_k+b_k)\varphi\|^2 + \Sigma\|\psi b_k\varphi\|^2 + \|\varphi\|^2)\,.\end{aligned} \tag{3.20}$$

Let ζ be a function in C_0^∞ which equals one on the support of ψ. Then ζb_k is in N_β for each $\beta > \frac{1}{2}\alpha$. Thus there is an $s < 1$ and a constant C_6 such that

$$\Sigma\|\zeta b_k\varphi\| \leqslant C_6\|\varphi\|_{s,2}\,, \qquad \varphi \in C_0^\infty\,.$$

By Lemma 7.4 of ch. 7 there is a constant C_7 such that

$$\Sigma\|\zeta b_k\psi\varphi\|^2 \leqslant (1/2C')\|\varphi\|_{1,2}^2 + C_7\|\varphi\|^2\,.$$

Hence (3.20) yields

$$\|\psi\varphi\|_{1,2}^2 \leqslant C'(\Sigma\|\psi(\mathrm{D}_k+b_k)\varphi\|^2+(C_7+1)\|\varphi\|^2)+\tfrac{1}{2}\|\psi\varphi\|_{1,2}^2 .$$

This implies (3.18).

Lemma 3.6. *There is a constant C such that*

$$\|\tau\varphi\|^2 \leqslant CN_2(\tau)(\Sigma\|(\mathrm{D}_k+b_k)\varphi\|^2+\|\varphi\|^2), \qquad \tau\in N_2,\ \varphi\in C_0^\infty. \tag{3.21}$$

Proof. By Theorem 7.3 of ch. 7 there is a constant C such that

$$\|\tau\varphi\|^2 \leqslant CN_2(\tau)\|\varphi\|_{1,2}^2, \qquad \tau\in N_2,\ \varphi\in C_0^\infty .$$

In view of (1.16) this is equivalent to

$$\|\tau\varphi\|^2 \leqslant CN_2(\tau)(\Sigma\|\mathrm{D}_k\varphi\|^2+\|\varphi\|^2) .$$

By Lemma 1.2 this implies (3.21).

Theorem 3.7. *Let τ be a function in N_4^{loc} such that $N_{2,x}(\tau)$ is finite for $|x|$ large. If*

$$\limsup_{|x|\to\infty} N_{2,x}(\tau) < \infty , \tag{3.22}$$

then there is a constant C such that

$$\|\tau\varphi\| \leqslant C(\|L(x,\mathrm{D})\varphi\|+\|\varphi\|), \qquad \varphi\in C_0^\infty . \tag{3.23}$$

Proof. Let $\psi\in C_0^\infty$ be a function in C_0^∞ such that $0\leqslant\psi\leqslant 1$ and

$$\psi(x)=1 , \qquad |x|<1 . \tag{3.24}$$

For $R>0$ set $\psi_R(x)=\psi(x/R)$. By hypothesis, $\tau(1-\psi_R)$ is in N_2 for R sufficiently large. Thus by Lemmas 3.6 and 3.3

$$\begin{aligned}\|\tau(1-\psi_R)\varphi\|^2 \leqslant C\|\varphi\|_W^2 &\leqslant C'[(L(x,\mathrm{D})\varphi,\varphi)+K_0\|\varphi\|^2] \\ &\leqslant C''\|\varphi\|(\|L(x,\mathrm{D})\varphi\|+\|\varphi\|) .\end{aligned}$$

Thus

$$\|\tau(1-\psi_R)\varphi\| \leqslant C''(\|L(x,\mathrm{D})\varphi\|+\|\varphi\|) . \tag{3.25}$$

On the other hand $\psi_R\tau$ is in N_4 so that by Theorem 7.3 of ch. 7

$$\|\psi_R\tau\varphi\| \leqslant C\|\varphi\|_{2,2}, \qquad \varphi\in C_0^\infty .$$

Let ζ be a function in C_0^∞ which equals one in the support of ψ_R. Then by Lemma 3.2

$$\|\psi_R \tau\zeta\varphi\| \leqslant C\|\zeta\varphi\|_{2,2} \leqslant C'''(\|L(x, \mathrm{D})\varphi\| + \|\varphi\|)\,. \tag{3.26}$$

Since $\psi_R\zeta = \psi_R$, we obtain (3.23) by combining (3.25) and (3.26). This completes the proof.

Theorem 3.8. *If* $\tau \in N_\beta^{\mathrm{loc}}$ *for* $\beta < 2$ *and*

$$N_{2,x}(\tau) \to 0 \;\; as \;\; |x| \to \infty\,, \tag{3.27}$$

then τ *is a compact operator from* W *to* L^2.

Proof. Let ψ_R be defined as in the preceding proof and let C be the constant in (3.21). Let $\{u_k\}$ be a bounded sequence in W and let $\varepsilon > 0$ be given. By (3.27) we can take R so large that

$$N_2[(1-\psi_R)\tau] < \varepsilon/4CK\,,$$

where $\|u_k\|_W \leqslant K$. Thus

$$\|(1-\psi_R)\tau u\| \leqslant \|u\|_W/4K\,, \qquad u \in W\,. \tag{3.28}$$

Moreover if ζ is a function in C_0^∞ which equals one in the support of ψ_R, then there is an $s < 1$ such that

$$\|\psi_R \tau u\| \leqslant C'\|\zeta^2 u\|_{s,2}\,, \qquad u \in W \tag{3.29}$$

(Theorem 7.3 of ch. 7). Now by Lemma 3.4, $\{\zeta u_k\}$ is a bounded sequence in $H^{1,2}$. Hence there is a subsequence $\{v_j\}$ of $\{u_k\}$ such that $\zeta^2 v_j$ converges in $H^{s,2}$ (Theorem 4.4 of ch. 2). In particular

$$\|\zeta^2(v_j - v_n)\|_{s,2} < \varepsilon/2C'$$

for j, n sufficiently large. Thus by (3.29)

$$\|\psi_R \tau(v_j - v_n)\| < \tfrac{1}{2}\varepsilon$$

for j, n large. Moreover by (3.28)

$$\|(1-\psi_R)\tau(v_j - v_n)\| < \tfrac{1}{2}\varepsilon$$

for all j, n. Thus

$$\|\tau(v_j - v_n)\| < \varepsilon \quad \text{for } j, n \text{ large.}$$

Hence $\{\tau v_j\}$ converges.

Lemma 3.9. *If* $\psi \in C_0^\infty$ *and* $\gamma > 1$, *then* ψ *is a compact operator from* $H^{\gamma,2}$ *to* W.

Proof. Let ζ be a function in C_0^∞ which equals one in the support of ψ and let $\{u_k\}$ be a bounded sequence of functions in $H^{\gamma,2}$. Then by Theorem 4.4 of ch. 2 there is a subsequence $\{v_j\}$ of $\{u_k\}$ such that $\{\zeta v_j\}$ converges in $H^{1,2}$. Moreover by Lemma 3.4

$$\|\psi(v_j - v_n)\|_W \leqslant C\|\zeta(v_j - v_n)\|_{1,2} \to 0 \text{ as } j, k \to \infty .$$

Thus $\{\psi v_j\}$ converges in W. This completes the proof.

Next let L denote the closure in L^2 of $L(x, D)$ on C_0^∞. We have

Lemma 3.10. *If $\lambda \in \rho(L)$, $\psi \in C_0^\infty$ and $\tau \in N_\alpha^{\mathrm{loc}}$, then $\psi\tau(L-\lambda)^{-1}$ is a compact operator in L^2.*

Proof. Let ζ be a function in C_0^∞ which equals one in the support of ψ. Then $\zeta(L-\lambda)^{-1}$ is a bounded operator from L^2 to $H^{2,2}$ (Lemma 3.2). Since $\psi\tau$ is in N_α and has compact support, it is a compact operator from $H^{2,2}$ to L^2 (Theorem 4.1 of ch. 6). Since $\psi\tau(L-\lambda)^{-1} = \psi\tau\zeta(L-\lambda)^{-1}$, it is a compact operator in L^2.

Lemma 3.11. *If $\lambda \in \rho(L)$, $\psi \in C_0^\infty$ and*

$$|\tau(x)| \leqslant C\rho(x), \qquad x \in E^n, \tag{3.30}$$

then $\psi\tau(L-\lambda)^{-1}$ is a compact operator in L^2.

Proof. Let $\zeta \in C_0^\infty$ be as in the proof of Lemma 3.10. Then $\zeta(L-\lambda)^{-1}$ is a bounded operator from L^2 to $H^{2,2}$ (Lemma 3.2). By Lemma 3.9, ψ is a compact operator from $H^{2,2}$ to W. Finally τ is a bounded operator from W to L^2 in view of (3.14) and (3.30).

Lemma 3.12. *$D(L) \subset W$ and*

$$\|u\|_W^2 \leqslant [(Lu, u) + K_0\|u\|^2] . \tag{3.31}$$

Proof. This follows immediately from Lemma 3.3.

Corollary 3.13. *If $\lambda \in \rho(L)$,*

$$\|u\|_W \leqslant C\|(L-\lambda)u\| , \qquad u \in D(L) .$$

Lemma 3.14. *There is a constant $K_3 > 1$ such that*

$$|(L(x, D)\varphi, \psi)| \leqslant K_3\|\varphi\|_W\|\psi\|_W . \tag{3.32}$$

Proof. Apply Lemma 3.3 and Lemma 6.10 of ch. 1.

Lemma 3.15. *If*

$$N_{2,x}(\sigma) \to 0 \quad as \quad |x| \to \infty ,$$

then for each $\varepsilon > 0$ *there is an* R *so large that*

$$(L(x, \mathrm{D})\varphi, \varphi) + \varepsilon\|\varphi\|^2 \geqslant 0 \tag{3.34}$$

for all $\varphi \in C_0^\infty$ *which vanish for* $|x| < R$.

Proof. Let $\psi(x)$ be a function in C_0^∞ which equals 1 for $|x| < \frac{1}{2}$ and vanishes for $|x| > 1$. Set $\psi_R(x) = \psi(x/R)$. By Lemma 3.6 and (1.18)

$$\|\sigma\varphi\|^2 \leqslant CN_2(\sigma)[(P(x, \mathrm{D})\varphi, \varphi) + \|\varphi\|^2] . \tag{3.35}$$

Let $\varepsilon > 0$ be given. We may assume $\varepsilon < 1$. Take R so large that $CN_2[\sigma(1-\psi_R)] < \varepsilon$; this can be done by (3.33). If $\varphi \in C_0^\infty$ vanishes for $|x| < R$, then $\varphi\psi_R = 0$. Thus

$$\begin{aligned}(L(x, \mathrm{D})\varphi, \varphi) &\geqslant (P(x, \mathrm{D})\varphi, \varphi) - \|\sigma(1-\psi_R)\varphi\|^2 \\ &\geqslant (1-\varepsilon)(P(x, \mathrm{D})\varphi, \varphi) - \varepsilon\|\varphi\|^2\end{aligned}$$

by (3.35). This gives (3.34), and the proof is complete.

Theorem 3.16. *Let* $\{Z(R)\}$ *be a family of open subsets of* E^n *depending on a parameter* R, $R_0 \leqslant R < \infty$. *Suppose that* τ *is a real valued function in* N_2 *with*

$$\sup_{x \in Z(R)} N_{2,x}(\tau) \to 0 \text{ as } R \to \infty . \tag{3.36}$$

If

$$(L(x, \mathrm{D})\varphi, \varphi) \geqslant \lambda_0\|\varphi\|^2 , \qquad \varphi \in C_0^\infty , \tag{3.37}$$

then

$$\liminf_{R \to \infty} \inf_{\varphi \in C_0^\infty(Z(R))} \frac{([L(x, \mathrm{D}) - \tau^2]\varphi, \varphi)}{\|\varphi\|^2} \geqslant \lambda_0 . \tag{3.38}$$

Proof. By Corollary 1.4 and Lemma 3.6 there are positive constants C_1, C_2 such that

$$\|h\varphi\|^2 \leqslant C_1 N_2(h)[(L(x, \mathrm{D})\varphi, \varphi) + C_2\|\varphi\|^2] , \qquad h \in N_2, \varphi \in C_0^\infty . \tag{3.39}$$

Set

$$\begin{aligned}\tau_R(x) &= \tau(x) , \qquad & x \in Z(R) \\ &= 0 , \qquad & x \notin Z(R) .\end{aligned}$$

Then τ_R is in N_2 and $N_2(\tau_R) \to 0$ as $R \to \infty$. Let ε satisfying $0 < \varepsilon < 1$ be given. Then there is an R so large that $C_1 N_2(\tau_R) < \varepsilon$. Thus by (3.39)

$$\|\tau_R\varphi\|^2 \leqslant \varepsilon[(L(x, \mathrm{D})\varphi, \varphi) + C_2\|\varphi\|^2] , \qquad \varphi \in C_0^\infty .$$

Consequently

$$([L(x, \mathrm{D})-\tau_R^2]\varphi, \varphi) \geqslant (1-\varepsilon)(L(x, \mathrm{D})\varphi, \varphi)-\varepsilon C_2\|\varphi\|^2$$
$$\geqslant [(1-\varepsilon)\lambda_0-\varepsilon C_2]\|\varphi\|^2, \qquad \varphi \in C_0^\infty .$$

Since $\tau_R = \tau$ in $Z(R)$, this gives

$$([L(x, \mathrm{D})-\tau^2]\varphi, \varphi) \geqslant [(1-\varepsilon)\lambda_0-\varepsilon C_2]\|\varphi\|^2, \qquad \varphi \in C_0^\infty(Z(R)) .$$

Since this is true for each value of R sufficiently large, we see that the left-hand side of (3.38) is $\geqslant (1-\varepsilon)\lambda_0 - \varepsilon C_2$. Since this is true for every value of ε, the proof is complete.

Another useful result is the following.

Theorem 3.17. *Suppose that the functions b_k in $L(x, \mathrm{D})$ are constants, and let $P(\mathrm{D})$ be defined by* (2.21). *Let T denote the closure of the operator $P(\mathrm{D})+q$ on C_0^∞. Then $\sigma(T)=\sigma(L)$.*

Proof. Set

$$h(x) = \exp\left\{ \mathrm{i} \sum_1^n b_k x_k \right\}. \tag{3.40}$$

Then for $\varphi \in C_0^\infty$

$$\mathrm{D}_k(h\varphi) = h(\mathrm{D}_k + b_k)\varphi .$$

Consequently

$$P(\mathrm{D})(h\varphi) = \Sigma\, a_{jk}\mathrm{D}_j\mathrm{D}_k(h\varphi) = \Sigma\, a_{jk}\mathrm{D}_j[h(\mathrm{D}_k+b_k)\varphi]$$
$$= h\,\Sigma\, a_{jk}(\mathrm{D}_j+b_j)(\mathrm{D}_k+b_k)\varphi .$$

This gives

$$(T-\lambda)(h\varphi) = h(L-\lambda)\varphi, \qquad \varphi \in C_0^\infty,\ \lambda \text{ complex.} \tag{3.41}$$

Now suppose $u \in D(L)$. Then there is a sequence $\{\varphi_k\}$ of functions in C_0^∞ such that $\varphi_k \to u$ and $L\varphi_k \to Lu$ (Theorem 2.1). Therefore $h\varphi_k \to hu$ and $T(h\varphi_k) = hL\varphi_k \to hLu$. This shows that $hu \in D(T)$ and $T(hu) = hLu$. Conversely, let v be a function in $D(T)$. Then there is a sequence $\{\psi_k\}$ of functions in C_0^∞ such that $\psi_k \to v$ and $T\psi_k \to Tv$. Consequently $h^{-1}\psi_k \to h^{-1}v$ and $L(h^{-1}\psi_k) = h^{-1}T\psi_k \to h^{-1}Tv$. This implies that $h^{-1}v \in D(L)$ and $L(h^{-1}v) = h^{-1}Tv$.

Next suppose $\lambda \in \rho(T)$. Then for each $f \in L^2$ there is a $v \in D(T)$ such that $(T-\lambda)v = hf$ and

$$\|v\| \leqslant C\|hf\| = C\|f\|, \tag{3.42}$$

where the constant C does not depend on v or f. Set $u=h^{-1}v$. Then $u\in D(L)$ by our previous remarks, and

$$(L-\lambda)u=h^{-1}(T-\lambda)v=f$$

by (3.41). Moreover by (3.42)

$$\|u\|=\|v\|\leqslant C\|f\|.$$

This shows that $\lambda\in\rho(L)$. Conversely, if $\lambda\in\rho(L)$, then for each $f\in L^2$ there is a $u\in D(L)$ such that $(L-\lambda)u=h^{-1}f$ and

$$\|u\|\leqslant C\|h^{-1}f\|=C\|f\|,$$

where the constant does not depend on u or f. Set $v=hu$. Then $v\in D(T)$ and

$$(T-\lambda)v=h(L-\lambda)u=f.$$

Moreover

$$\|v\|=\|u\|\leqslant C\|f\|.$$

Thus $\lambda\in\rho(L)$. This completes the proof of the theorem.

§4. Comparison of operators

Let $L(x, \mathrm{D})$ denote the operator (1.1) satisfying all of the hypotheses of §§1, 2. Let $\tilde{L}$ be another such operator given by

$$\tilde{L}(x,\mathrm{D})=\sum_{j,k=1}^{n}a_{jk}[\mathrm{D}_j+\tilde{b}_j(x)][\mathrm{D}_k+\tilde{b}_k(x)]+\tilde{q}(x), \tag{4.1}$$

and assume that $\tilde{q}$ and the $\tilde{b}_j$ are real valued functions also satisfying the hypotheses of §§1, 2. (The coefficients a_{jk} are to be the same for both operators.) By Theorem 2.1 both $L(x, \mathrm{D})$ and $\tilde{L}(x, \mathrm{D})$ are essentially self-adjoint on C_0^∞. Let L and $\tilde{L}$ denote their (unique) self-adjoint extensions. In this section we shall discuss conditions which will guarantee that

$$\sigma_{\mathrm{e}}(\tilde{L})=\sigma_{\mathrm{e}}(L). \tag{4.2}$$

Our main result is

Theorem 4.1. *Set*

$$c_j(x)=b_j(x)-\tilde{b}_j(x), \qquad 1\leqslant j\leqslant n, \tag{4.3}$$

and assume that $q(x)-\tilde{q}(x)=\tau(x)\,\tilde{\tau}(x)$, *where* τ *and* $\tilde{\tau}$ *are real valued functions such that*

$$\tau=\tau_1+\tau_2,\quad \tau_1\in N_\alpha^{\mathrm{loc}},\qquad |\tau_2(x)|\leqslant C\rho(x),\quad x\in E^n \tag{4.4}$$

$$\tilde{\tau}\in N_4^{\mathrm{loc}} \text{ with } N_{2,x}(\tilde{\tau}) \text{ finite for } |x| \text{ large} \tag{4.5}$$

$$\limsup_{|x|\to\infty} N_{2,x}(\tau_1)<\infty \tag{4.6}$$

$$N_{2,x}(\tilde{\tau})\to 0 \text{ as } |x|\to\infty$$

$$N_{2,x}(c_j)\to 0 \text{ as } |x|\to\infty,\qquad 1\leqslant j\leqslant n\,. \tag{4.8}$$

Then (4.2) *holds.*

Before we prove Theorem 4.1 let us derive some useful consequences. Set

$$L_1(x,\mathrm{D})=P(x,\mathrm{D})+\rho^2=L(x,\mathrm{D})+\sigma^2\,,$$

$$L_2(x,\mathrm{D})=P(x,\mathrm{D})\,,$$

$$L_3(\mathrm{D})=P(\mathrm{D})\,,$$

and let $L_1\,L_2$, L_3 denote the respective closures in L^2 of these operators on C_0^∞. Clearly $L_3=P_0$.

Theorem 4.2. *If*

$$N_{2,x}(\sigma)\to 0 \text{ as } |x|\to\infty\,, \tag{4.9}$$

then

$$\sigma_{\mathrm{e}}(L)=\sigma_{\mathrm{e}}(L_1)\,. \tag{4.10}$$

Proof. Take $\tilde{L}=L_1$ in Theorem 4.1. Then $c_j=0$ for each j and $\tau=\tilde{\tau}=\tau_1=-\sigma$, $\tau_2=0$. Note that (4.9) becomes (4.7). Since $\sigma\in N_2$, all of the hypotheses of Theorem 4.1 are satisfied.

Theorem 4.3. *If*

$$N_{2,x}(\rho)\to 0 \text{ as } |x|\to\infty\,, \tag{4.11}$$

then

$$\sigma_{\mathrm{e}}(L_1)=\sigma_{\mathrm{e}}(L_2)\,. \tag{4.12}$$

Proof. Take $L=L_1$, $\tilde{L}=L_2$ in Theorem 4.1. Then the $c_j=0$ and $\tau=\tilde{\tau}=\tau_2=\rho$, $\tau_1=0$. Note that (4.11) becomes (4.7). Since $\rho\in N_2^{\mathrm{loc}}$ (Lemma 3.5), all of the hypotheses are satisfied.

Theorem 4.4. *If*

$$N_{2,x}(b_j) \to 0 \quad as \ |x| \to \infty, \qquad 1 \leqslant j \leqslant n, \tag{4.13}$$

then

$$\sigma_e(L_2) = \sigma_e(L_3). \tag{4.14}$$

Proof. Take $L = L_2$, $\tilde{L} = L_3$ in Theorem 4.1. Then $c_j = b_j$ and $\tau = \tilde{\tau} = \tau_1 = \tau_2 = 0$. Then (4.13) becomes (4.8). All of the hypotheses are satisfied.

Lemma 4.5.

$$\sigma(P_0) = [0, \infty). \tag{4.15}$$

Proof. By Corollary 3.3 of ch. 4, $\sigma(P_0)$ consists of all values taken on by $P(\xi)$ for $\xi \in E^n$. By (1.2) these values are non-negative. On the other hand let λ be any non-negative number. If $\xi \neq 0$ is a real vector, then $\lambda_0 = P(\xi)$ is not zero by (1.2). For t real we have $P(t\xi) = t^2 \lambda_0$. Let t be such that $t^2 = \lambda/\lambda_0$. Hence $P(t\xi) = \lambda$. This shows that $\lambda \in \sigma(P_0)$. Thus $\sigma(P_0)$ consists of all non-negative numbers.

Corollary 4.9. *If* (4.9), (4.11) *and* (4.13) *hold, then*

$$\sigma_e(L) = [0, \infty). \tag{4.16}$$

Proof. Combine Theorems 4.2–4.4 and Lemma 4.5.

Corollary 4.7. *If* (4.9) *holds, then*

$$\sigma_e(L) \subset [0 \ \infty). \tag{4.17}$$

Proof. Since

$$(L_1(x, D)\varphi, \varphi) \geqslant c_0 \Sigma \|(D_k + b_k)\varphi\|^2 + \|\rho\varphi\|^2 \geqslant 0,$$

we see that $\sigma(L_1) \subset [0, \infty)$. Apply Theorem 4.2.

Theorem 4.8. *If*

$$\lambda_0 \leqslant \liminf_{|x| \to \infty} q(x), \tag{4.18}$$

then

$$\sigma_e(L) \subset [\lambda_0, \infty). \tag{4.19}$$

Proof. Set $\tilde{q}(x) = q(x) - \lambda_0$, $\tilde{L}(x, D) = P(x, D) + \tilde{q}$. Let $\tilde{\rho}$, $\tilde{\sigma}$ be the corresponding functions defined by (1.5). Then

$$\tilde{\sigma}^2 \leqslant |\lambda_0| + \sigma^2 \tag{4.20}$$

and

$$\tilde{\sigma}(x) \to 0 \text{ as } |x| \to \infty . \tag{4.21}$$

Thus $\tilde{\sigma} \in N_2$, and (1.7) and (4.9) hold. We can now apply Corollary 4.7 to conclude that $\sigma_e(\tilde{L}) \subset [0, \infty)$. Since $\tilde{L} = L - \lambda_0$, this gives (4.19) and the proof is complete.

Corollary 4.9. *If $q(x) \to \infty$ as $|x| \to \infty$, then $\sigma(L)$ is discrete, i.e., $\sigma(L)$ consists of at most a denumerable set of eigenvalues of finite multiplicity having no limit point other than possibly $+\infty$.*

Proof. Let λ_0 be any positive number. Then by Theorem 4.8, $\sigma_e(L) \subset [\lambda_0, \infty)$. Thus the part of $\sigma(L)$ in the interval $(-\infty, \lambda_0 - 1)$ consists of at most a finite number of eigenvalues of finite multiplicity (Corollary 7.12 of ch. 1). Since λ_0 was arbitrary, the result follows.

We should note that the strength of Theorem 4.1 lies in the fact that nothing is assumed concerning the behavior of the $b_k(x)$ at infinity. However, if we are willing to make assumptions of this sort we can improve Corollary 4.6.

Theorem 4.10. *Assume that $b^{\frac{1}{2}} \in N_4$ and that*

$$N_{2,\delta}(\sigma) + N_{2,\delta}(b_k) \to 0 \text{ as } \delta \to 0, \qquad 1 \leqslant k \leqslant n , \tag{4.22}$$

$$N_{4,x}(b^{\frac{1}{2}}) + N_{4,x}(\rho) \to 0 \text{ as } |x| \to \infty . \tag{4.23}$$

If (4.9) and (4.13) hold, then $\sigma_e(L) = [0, \infty)$.

Proof. We appeal to Theorem 2.4 of ch. 8. We write $L(x, \mathrm{D})$ in the form

$$L(x, \mathrm{D})\varphi = P(\mathrm{D})\varphi + \Sigma\, a_{jk} \mathrm{D}_j(b_k \varphi) + \Sigma\, a_{jk} b_j \mathrm{D}_k \varphi + b\varphi + q\varphi . \tag{4.24}$$

In the notation of (1.10) of ch. 8 we can take $N = n + 1$,

$$\begin{aligned} P_j(\mathrm{D}) &= \mathrm{D}_j , && 1 \leqslant j \leqslant n \\ &= 1 , && j = N , \end{aligned}$$

$$q_{jN} = \sum_k a_{jk} b_k , \; q_{Nk} = \sum_j a_{jk} b_j , \qquad 1 \leqslant j \leqslant n$$

$$q_{NN} = b + q, \; \rho_N^2 = b + \rho^2, \; \sigma_N = \sigma ,$$

with all others vanishing. We take $h_{1jN} = q_{jN}$, $h_{2jN} = 1$, $1 \leqslant j \leqslant n$. All of the hypotheses of Theorems 1.7 and 2.4 of ch. 8 are satisfied. Thus our conclusion is proved.

Proof of Theorem 4.1. By Corollary 2.5 there is a constant λ so large that both $(L+\lambda)^{-1}$ and $(\tilde{L}+\lambda)^{-1}$ exist. Let R be the range of $L(x, \mathrm{D})+\lambda$ on C_0^∞. Note that R is dense in L^2. For if $u\in L^2$ is orthogonal to R, then

$$(u, [L+\lambda]v)=0\,, \qquad v\in D(L)\,,$$

since L is the closure of $L(x, \mathrm{D})$ on C_0^∞. Since L is self-adjoint, $u\in D(L)$ and $(L+\lambda)u=0$. Consequently $u=0$. Now on R we have

$$(\tilde{L}+\lambda)^{-1}-(L+\lambda)^{-1}=(L+\lambda)^{-1}(L-\tilde{L})(\tilde{L}+\lambda)^{-1}\,. \tag{4.25}$$

Since

$$L(x, \mathrm{D})-\tilde{L}(x, \mathrm{D})=\Sigma\, a_{jk}c_j(\mathrm{D}_k+\tilde{b}_k)+\Sigma\, a_{jk}(\mathrm{D}_j+b_j)c_k+\tau\tilde{\tau}\,, \tag{4.26}$$

we have

$$\begin{aligned}(\tilde{L}+\lambda)^{-1}-(L+\lambda)^{-1}&=\Sigma\, a_{jk}[(L+\lambda)^{-1}c_j](\mathrm{D}_k+\tilde{b}_k)(\tilde{L}+\lambda)^{-1}\\&\quad+\Sigma\, a_{jk}[(L+\lambda)^{-1}(\mathrm{D}_j+b_j)]\,c_k(\tilde{L}+\lambda)^{-1}\\&\quad+(L+\lambda)^{-1}\tau\tilde{\tau}(\tilde{L}+\lambda)^{-1}\end{aligned} \tag{4.27}$$

on R. We shall show that the right-hand side of (4.27) is a compact operator on L^2. Since the left-hand side is a bounded operator defined everywhere on L^2, it will follow that it is compact on L^2. The Theorem will follow from Theorem 4.7 of ch. 1.

Let $\psi(x)$ be a function in C_0^∞ which equals one for $|x|<1$, and set $\psi_R(x)=\psi(x/R)$. Now by Lemma 3.10 each of the operators $\psi_R c_j(L+\lambda)^{-1}$, $\psi_R c_k(\tilde{L}+\lambda)^{-1}$ and $\psi_R\tau(L+\lambda)^{-1}$ is a compact operator on L^2. By taking adjoints we see that the same is true of $(L+\lambda)^{-1}c_j\psi_R$, $(L+\lambda)^{-1}\tau\psi_R$ (Theorem 2.17 of ch. 1). Moreover, the operators $(\mathrm{D}_j+b_j)(L+\lambda)^{-1}$, $(\mathrm{D}_k+\tilde{b}_k)(\tilde{L}+\lambda)^{-1}$ and $\tau_2(L+\lambda)^{-1}$ are bounded on L^2 (see (3.14) and Corollary 3.13). Again by adjoints $(L+\lambda)^{-1}(\mathrm{D}_j+b_j)$ are bounded. Similarly $\tilde{\tau}(\tilde{L}+\lambda)^{-1}$, $\tau_1(L+\lambda)^{-1}$ and $(L+\lambda)^{-1}\tau$ are bounded (Theorem 3.7). Thus our theorem will be proved if

$$\|(L+\lambda)^{-1}c_j(1-\psi_R)\|\to 0 \text{ as } R\to\infty \tag{4.28}$$

$$\|(1-\psi_R)c_k(\tilde{L}+\lambda)^{-1}\|\to 0 \text{ as } R\to\infty \tag{4.29}$$

$$\|(1-\psi_R)\tilde{\tau}(\tilde{L}+\lambda)^{-1}\|\to 0 \text{ as } R\to\infty \tag{4.30}$$

(see Theorem 2.16 of ch. 1). By (4.7) and (4.8)

$$N_2[(1-\psi_R)\tilde{\tau}]+\Sigma\, N_2[(1-\psi_R)c_k]\to 0 \text{ as } R\to\infty\,.$$

If we now apply Lemma 3.6 and Corollary 3.13 we obtain (4.29) and (4.30).

Taking adjoints we get (4.28) (see Theorem 2.18 of ch. 1). This completes the proof.

Theorem 4.11. *Assume that there is a sequence* $\{T_i\}$ *of spheres in* E^n *with radii tending to infinity such that*

$$\int_{T_i} b(x)\mathrm{d}x/|T_i| \to 0 \quad as \quad i \to \infty \tag{4.31}$$

and

$$\int_{T_i} [b(x)+q(x)+e(x)]^2 \mathrm{d}x/|T_i| \to 0 \quad as \quad i \to \infty\,, \tag{4.32}$$

where $|T_i|$ *is the volume of* T_i. *Then*

$$\sigma_e(L) \supset [0, \infty)\,. \tag{4.33}$$

Proof. We write $L(x, \mathrm{D})$ in the form

$$L(x, \mathrm{D}) = P(\mathrm{D}) + \Sigma\, b_j Q_j(\mathrm{D}) + b + q + e\,, \tag{4.34}$$

where

$$Q_j(\xi) = 2 \sum_k a_{jk} \xi_k\,, \qquad 1 \leqslant j \leqslant n\,. \tag{4.35}$$

We now apply Theorem 6.7 of ch. 4. Note that (4.31) implies

$$\int_{T_i} \Sigma\, b_k(x)^2 \mathrm{d}x/|T_i| \to 0 \quad \text{as } i \to \infty\,.$$

In applying Theorem 4.2 one may find the following criterion useful.

Lemma 4.12. *For each* $\delta > 0$ *set*

$$\begin{aligned} \sigma_\delta(x) &= \sigma(x) \quad for \quad \sigma(x) > \delta \\ &= 0 \qquad for \quad \sigma(x) \leqslant \delta\,. \end{aligned}$$

Suppose than $n > 2$ *and that for each* $\delta > 0$

$$\int_R^{R+1} \sigma_\delta(rz)^2 \,\mathrm{d}r \to 0 \quad as \quad R \to \infty,\ |z| = 1,\ uniformly\ in\ z. \tag{4.36}$$

Then (4.9) *holds.*

Proof. Let $\varepsilon > 0$ be given. Take M so great that

$$\int_R^{R+1} \sigma_\varepsilon(rz)^2 \,\mathrm{d}r < \varepsilon\,, \qquad |z| = 1,\ R > M\,. \tag{4.37}$$

Let y be any vector in E^n such that $|y| > M+1$, and set $y_0 = y/|y|$. Then

$$\int_{|x-y|<1} \sigma_\varepsilon(x)^2 |x-y|^{2-n} \mathrm{d}x$$

$$\leqslant \int_{\substack{|z|=1 \\ |z-y_0|<1/|y|}} \mathrm{d}z \int_{|y|-1}^{|y|+1} \sigma_\varepsilon(rz)^2 |rz-y|^{2-n} r^{n-1} \mathrm{d}r$$

$$= \int_{\substack{|z|=1 \\ |z-y_0|<1/R}} |z-y_0|^{2-n} \mathrm{d}z \int_{R-1}^{R+1} \sigma_\varepsilon(rz)^2 r \,\mathrm{d}r$$

$$\leqslant 2\varepsilon(R+1)C \int_0^{1/R} \mathrm{d}\lambda \leqslant 4\varepsilon C,$$

where C is independent of ε and we put $R = |y|$. Thus

$$\int_{|x-y|<1} \sigma(x)^2 |x-y|^{2-n} \mathrm{d}x \leqslant 4\varepsilon C + \varepsilon^2 \int_{|z|<1} |z|^{2-n} \mathrm{d}z$$

for $|y|$ sufficiently large. Since this is true for every $\varepsilon > 0$, the lemma is proved.

Corollary 4.13. *Assume that for each* $\delta > 0$ *the integral*

$$\int_0^\infty \sigma_\delta(rz)^2 \,\mathrm{d}r$$

exists and is uniformly bounded for $|z| = 1$. *Then* (4.9) *holds.*

Proof. The assumptions clearly imply (4.36).

§5. Estimating the essential spectrum

In this section we shall introduce a method of finding the minimum point of $\sigma_e(L)$. In later sections the method will be refined to produce sharper results under more specific hypotheses.

Let $L(x, \mathrm{D})$ be given by (1.1), and assume that the hypotheses of §§1, 2 are satisfied. Let $p_i(x), \ldots, p_r(x)$ be functions on E^n each of which is continuous and non-negative and has first order derivatives in $L^\infty(E^n)$. Assume that each function $p_j(x)$ is infinitely differentiable at points where it does not vanish. Set

$$M = \max_{j,k} \|\mathrm{D}_j p_k\|_\infty, \qquad a = 2(1 + 16nM^2). \tag{5.1}$$

Set $t=2^r$, and for each $R>0$ and $K>0$ let $Z_{1,K}(R), \ldots, Z_{t,K}(R)$ denote the sets of E^n each defined by r inequalities, the kth inequality stating either

$$p_k(x) > R \tag{5.2}$$

or

$$p_k(x) < R+K\,, \tag{5.3}$$

with all possible combination taken. We divide the sets $Z_{i,K}(R)$ into two categories, the first consisting of those which are unbounded for R and K sufficiently large and the second consisting of the rest. Note that the second category may be empty, but not the first. By reordering the $Z_{i,K}(R)$ if necessary we may assume that there is an integer k satisfying $1 \leqslant k \leqslant t$ such that $Z_{i,K}(R)$ is in the first category for $1 \leqslant i \leqslant k$ and in the second for $k < i \leqslant t$. Set

$$\nu_i = \limsup_{\substack{R\to\infty \\ K\to\infty}} \inf_{\varphi \in C_0(Z_{i,K}(R))} \frac{(L(x, \mathrm{D})\varphi, \varphi)}{\|\varphi\|^2}\,, \qquad 1 \leqslant i \leqslant k\,, \tag{5.4}$$

$$\nu = \min_{1 \leqslant i \leqslant k} \nu_i\,. \tag{5.5}$$

As before, let L denote the closure in L^2 of $L(x, \mathrm{D})$ on C_0^∞. We have

Theorem 5.1. *Under the above hypotheses*

$$\sigma_e(L) \subset [\nu, \infty)\,. \tag{5.6}$$

Our proof of Theorem 5.1 will be based upon some observations concerning the quadratic form

$$J(\varphi, \Omega) = \int_\Omega (\Sigma |(\mathrm{D}_k + b_k(x))\varphi(x)|^2 + (\rho(x)^2+1)|\varphi(x)|^2)\,\mathrm{d}x\,, \tag{5.7}$$

where Ω is an arbitrary subset of E^n. When $\Omega = E^n$ we shall write $J(\varphi)$. Note that

$$J(\varphi) = \|\varphi\|_W^2 \tag{5.8}$$

(see (3.14)). We shall use the following two results.

Lemma 5.2. *Let* $\{u_k\}$ *be a sequence of functions in* $D(L)$ *such that*

$$\|u_k\| + \|Lu_k\| \leqslant C\,, \qquad k = 1, 2, \ldots\,. \tag{5.9}$$

Then for any bounded set Ω *we can find a subsequence* $\{v_j\}$ *of* $\{u_k\}$ *and a*

$v \in H^{1,2}$ *such that*

$$J(v_j - v, \Omega) \to 0 \quad as \quad k \to \infty. \tag{5.10}$$

Theorem 5.3. *For each $\varepsilon > 0$ there is a bounded set $\Omega \subset E^n$ such that*

$$\begin{aligned}(v-\varepsilon)\|\varphi\|^2 \leqslant (L(x, \mathrm{D})\varphi, \varphi) + \varepsilon(1 + |v - \varepsilon|)J(\varphi) \\ + a^r(t-k)(|v-\varepsilon| + K_3)J(\varphi, \Omega), \qquad \varphi \in C_0^\infty,\end{aligned} \tag{5.11}$$

where K_3 is the constant in (3.32).

The proof of Lemma 5.2 will be given at the end of this section; that of Theorem 5.3 will be given in §6. We now show how they can be used to give the

Proof of Theorem 5.1. Let λ be any point of $\sigma_e(L)$. Since L is self-adjoint (Theorem 2.1), we see by Theorem 7.14 of ch. 1 that there is a sequence $\{\varphi_k\}$ of functions in C_0^∞ such that

$$\|\varphi_k\| = 1, \; \varphi_k \rightharpoonup 0, \quad [L(x, \mathrm{D}) - \lambda]\varphi_k \to 0. \tag{5.12}$$

In particular there is a constant C_8 such that

$$\|\varphi_k\| + \|L(x, \mathrm{D})\varphi_k\| \leqslant C_8. \tag{5.13}$$

Let ε satisfying $0 < \varepsilon \leqslant 1$ be given. By Theorem 5.3 there is a bounded set Ω and a constant C such that (5.11) holds. By Lemma 5.2 there is a subsequence of $\{\varphi_k\}$ (also denoted by $\{\varphi_k\}$) and a $v \in H^{1,2}$ such that

$$J(\varphi_k - v, \Omega) \to 0 \quad \text{as} \quad k \to \infty. \tag{5.14}$$

Thus by (5.7)

$$\int_\Omega |\varphi_k(x) - v(x)|^2 \mathrm{d}x \to 0 \quad \text{as} \quad k \to \infty.$$

This implies that

$$(\varphi_k - v, u)_\Omega \to 0 \quad \text{as} \quad k \to \infty, \qquad u \in L^2(\Omega).$$

But $\varphi_k \rightharpoonup 0$. Thus $(v, u)_\Omega = 0$ for all $u \in L^2(\Omega)$ and consequently $v = 0$ on Ω. In view of (5.14) this gives

$$J(\varphi_k, \Omega) \to 0 \quad \text{as} \quad k \to \infty. \tag{5.15}$$

If we now apply (5.11) to the sequence $\{\varphi_k\}$, we obtain

$$\nu-\varepsilon \leqslant (L(x, \mathrm{D})\varphi_k, \varphi_k)+CJ(\varphi_k, \Omega)+\varepsilon(1+|\nu-\varepsilon|)J(\varphi_k)\,. \tag{5.16}$$

Moreover by (3.16), (5.8) and (5.13) there is a constant C_9 such that

$$J(\varphi_k) \leqslant C_9\,. \tag{5.17}$$

Letting $k\to\infty$ in (5.16) and applying (5.12), (5.15) and (5.17), we have

$$\nu-\varepsilon \leqslant \lambda+\varepsilon(1+|\nu-\varepsilon|)C_9\,. \tag{5.18}$$

Since ε was arbitrary, we see that $\nu \leqslant \lambda$. Since λ was any point of $\sigma_e(L)$, we see that $\sigma_e(L)$ is contained in the interval $[\nu, \infty)$. This completes the proof.

As an illustration, let us consider the case $n=1$ and take $p(x)=|x|$. Then the sets $Z_{1,K}(R)$ and $Z_{2,K}(R)$ are the regions $|x|>R$ and $|x|<R+K$, respectively. Thus

$$\nu = \limsup_{R\to\infty} \inf_{\substack{\varphi\in C_0^\infty \\ \varphi=0 \text{ for } |x|<R}} \frac{(L(x, \mathrm{D})\varphi, \varphi)}{\|\varphi\|^2}\,. \tag{5.19}$$

Suppose for instance that (4.18) holds. Then for each $\varepsilon>0$ there is an $N>0$ such that

$$q(x)>\lambda_0-\varepsilon \quad \text{for} \quad |x|>N\,.$$

Consequently

$$(L(x, \mathrm{D})\varphi, \varphi) \geqslant (q\varphi, \varphi) \geqslant (\lambda_0-\varepsilon)\|\varphi\|^2\,, \quad \varphi\in C_0^\infty(Z_{1,K}(R))\,, \quad R>N\,.$$

This shows that $\nu \geqslant \lambda_0$. Thus we have another proof of Theorem 4.7.

§6. The quadratic form $J(\varphi)$

In this section we shall study the quadratic form given by (5.7). Its properties will be investigated, and a proof of Theorem 5.3 will be given. As before the hypotheses of §§1, 2 are assumed throughout.

First we note that by (5.8) and Lemma 3.14

$$|(L(x, \mathrm{D})\varphi, \psi)|^2 \leqslant K_3^2 J(\varphi)J(\psi)\,, \qquad \varphi, \psi\in C_0^\infty\,. \tag{6.1}$$

Secondly we note that if h is a bounded function on Ω and has first derivatives also bounded there, then

$$J(h\varphi, \Omega) \leqslant 2\left(\max_\Omega |h|^2+\max_\Omega \sum_{j=1}^n |\mathrm{D}_j h|^2\right) J(\varphi, \Omega)\,. \tag{6.2}$$

We shall find the following theorem very useful.

Theorem 6.1. *Let $p(x)$ be a non-negative, continuous function on E^n having first derivatives in $L^\infty(E^n)$ and such that $p(x)$ is infinitely differentiable at points where $p(x) \neq 0$. Set*

$$M = \max_j \|D_j p\|_\infty . \tag{6.3}$$

Let $\varepsilon > 0$ be given, and let K be an integer greater than $2K_3(1+64nM^2)/\varepsilon$. Then for each $\varphi \in C_0^\infty$ and each $R > 0$ one can find functions ζ, ψ in C_0^∞ such that

$$\varphi = \zeta + \psi \tag{6.4}$$

$$\zeta(x) = 0 \quad \textit{for } p(x) < R \tag{6.5}$$

$$\psi(x) = 0 \quad \textit{for } p(x) > R+K \tag{6.6}$$

$$|\zeta(x)| \leqslant |\varphi(x)|, \quad |\psi(x)| \leqslant |\varphi(x)| \tag{6.7}$$

$$J(\zeta, \Omega) \leqslant 2(1+16nM^2) J(\varphi, \Omega), \qquad \Omega \subset E^n \tag{6.8}$$

$$J(\psi, \Omega) \leqslant 2(1+16nM^2) J(\varphi, \Omega), \qquad \Omega \subset E^n, \tag{6.9}$$

$$|(L(x, D)\zeta, \psi)| < \varepsilon J(\varphi) \tag{6.10}$$

$$|(\zeta, \psi)| < \varepsilon J(\varphi). \tag{6.11}$$

Proof. Let $w_1(t)$, $w_2(t)$ be functions in $C^\infty(E^1)$ with the following properties:

$$w_1(t) = 1 \quad \text{for} \quad t < \tfrac{1}{3} \tag{6.12}$$

$$\qquad = 0 \quad \text{for} \quad t > \tfrac{2}{3} \tag{6.13}$$

$$w_2(t) = 1 \quad \text{for} \quad \tfrac{1}{3} < t < \tfrac{2}{3} \tag{6.14}$$

$$\qquad = 0 \quad \text{for} \quad t < 0 \text{ and for } t > 1 \tag{6.15}$$

$$0 \leqslant w_j(t) \leqslant 1, \quad |w_j'(t)| \leqslant 4, \qquad j = 1, 2. \tag{6.16}$$

Clearly such functions can be constructed (see Theorem 1.2 of ch. 2). For each integer k such that $1 \leqslant k \leqslant K$ let Σ_k denote the set of those $x \in E^n$ such that $R+k-1 < p(x) < R+k$. The sets Σ_k are open in virtue of the continuity of $p(x)$ (Theorem 1.15 of ch. 2). There exists at least one integer k satisfying $1 \leqslant k \leqslant K$ such that

$$J(\varphi, \Sigma_k) \leqslant J(\varphi)/K, \tag{6.17}$$

for otherwise we would have $J(\varphi, \cup \Sigma_k) > J(\varphi)$. Choose one such value of k and hold it fixed. Set

$$g_j(x) = w_j(R+k-p(x)), \qquad j=1,2. \tag{6.18}$$

Since the $g_j(x)$ vanish in the neighborhood of points x where $p(x)=0$, they are infinitely differentiable on E^n. By (6.3) and (6.16)

$$0 \leqslant g_j(x) \leqslant 1, \quad |D_k g_j| \leqslant 4M, \qquad j=1,2, \ 1 \leqslant k \leqslant n. \tag{6.19}$$

Note that

$$g_1(x) = 1 \quad \text{for} \quad R+k-\tfrac{1}{3} < p(x) \tag{6.20}$$

$$= 0 \quad \text{for} \quad p(x) < R+k-\tfrac{2}{3}. \tag{6.21}$$

$$g_2(x) = 1 \quad \text{for} \quad R+k-\tfrac{2}{3} < p(x) < R+k-\tfrac{1}{3} \tag{6.22}$$

$$= 0 \quad \text{for} \quad p(x) > R+k \text{ and for } p(x) < R+k-1. \tag{6.23}$$

Moreover by (6.2) and (6.19)

$$J(g_j\varphi, \Omega) \leqslant 2(1+16nM^2)J(\varphi, \Omega), \qquad j=1,2. \tag{6.24}$$

Since

$$|D_k(g_1 g_2)| = |g_1 D_k g_2 + g_2 D_k g_1| \leqslant 8M, \tag{6.25}$$

we have

$$J(g_1 g_2 \varphi, \Omega) \leqslant 2(1+64nM^2)J(\varphi, \Omega),$$

and similarly

$$J(g_2(1-g_1)\varphi, \Omega) \leqslant 2(1+64nM^2)J(\varphi, \Omega). \tag{6.27}$$

Set $\zeta = g_1\varphi$, $\psi = \varphi - \zeta$. Then (6.4)–(6.9) follow from (6.19)–(6.24). It remains to verify (6.10) and (6.11). To this end note that $\bar{\psi}L(x, D)\varphi$ vanishes for $p(x) < R+k-\frac{2}{3}$ and for $p(x) > R+k-\frac{1}{3}$ (see (6.20) and (6.21)). Thus the points x where it does not vanish satisfy $R+k-\frac{2}{3} < p(x) < R+k-\frac{1}{3}$. However, for such points $g_2(x)$ is identically one (see (6.22)). Thus

$$\begin{aligned} |(L(x, D)\zeta, \psi)|^2 &= |(L(x, D)(g_2\zeta), g_2\psi)|^2 \\ &\leqslant K_3^2 J(g_2\zeta) J(g_2\psi) \\ &= K_3^2 J(g_2\zeta, \Sigma_k) J(g_2\psi, \Sigma_k) \end{aligned}$$

since g_2 vanishes outside Σ_k (see (6.23)). Thus by (6.17), (6.26) and (6.27)

$$|(L(x, D)\zeta, \psi)|^2 \leqslant 4K_3^2(1+64nM^2)^2 J(\varphi, \Sigma_k)^2 < \varepsilon^2 J(\varphi)^2 .$$

This gives (6.10). Inequality (6.11) is obtained in a similar way. In fact we have

$$\begin{aligned}|(\zeta, \psi)|^2 = |(g_2\zeta, g_2\psi)|^2 &\leqslant J(g_2\zeta, \Sigma_k) J(g_2\psi, \Sigma_k) \\ &\leqslant 4(1+64nM^2)^2 J(\varphi, \Sigma_k)^2 < \varepsilon^2 J(\varphi)^2 ,\end{aligned}$$

since $K_3 \geqslant 1$. This completes the proof.

The next important step towards the proof of Theorem 5.3 is

Theorem 6.2. *Let $p_1(x), \ldots, p_r(x)$ be functions on E^n which satisfy the hypotheses of Theorem* 5.1. *Set*

$$M = \max_{j,k} \|D_j p_k\|_\infty , \tag{6.28}$$

$$a = 2(1+16nM^2) , \qquad b = 2(1+64nM^2) . \tag{6.29}$$

Let $\varepsilon > 0$ be given, and let K be any integer greater than $2K_3 b(2^r a^r - 1)/\varepsilon(2a-1)$. Set $t = 2^r$ and for each $R > 0$ let $Z_{1,K}(R), \ldots, Z_{t,K}(R)$ denote the sets of E^n each defined by r inequalities, the j-th inequality stating either

$$p_j(x) > R \tag{6.30}$$

or

$$p_j(x) < R + K , \tag{6.31}$$

with all possible combinations taken. Then for each $\varphi \in C_0^\infty$ and each $R > 0$ there are functions $\varphi_1, \ldots, \varphi_t$ such that $\varphi_i \in C_0^\infty(Z_{i,K}(R))$ and

$$\varphi = \sum_{i=1}^{t} \varphi_i \tag{6.32}$$

$$|\varphi_i(x)| \leqslant |\varphi(x)| , \qquad x \in E^n , \ 1 \leqslant i \leqslant t , \tag{6.33}$$

$$J(\varphi_i, \Omega) \leqslant a^r J(\varphi, \Omega), \qquad \Omega \subset E^n , \ 1 \leqslant i \leqslant t , \tag{6.34}$$

$$\left| \sum_1^t (L(x, D)\varphi_i, \varphi_i) - (L(x, D)\varphi, \varphi) \right| < \varepsilon J(\varphi) \tag{6.35}$$

$$\left| \sum_1^t \|\varphi_i\|^2 - \|\varphi\|^2 \right| < \varepsilon J(\varphi) . \tag{6.36}$$

Proof. We proceed by induction on r. Set

$$C_k = 2K_3 b(2^k a^k - 1)/(2a-1) , \qquad k = 1, 2, \ldots , \tag{6.37}$$

and note that

$$C_{k+1} = C_k + 2^k a^k C_1\,, \qquad k=1, 2, \ldots\,. \tag{6.38}$$

The theorem states that if $K > C_r/\varepsilon$, then there are functions $\varphi_i \in C_0^\infty(Z_{i,K}(R))$ such that (6.32)–(6.36) hold. We first verify this for $r=1$. Let $\varepsilon > 0$ be given, and let K be an integer $> C_1/\varepsilon$. Set $\varepsilon_1 = \frac{1}{2}\varepsilon$. Then $K > K_3 b/\varepsilon_1$. Applying Theorem 6.1 we obtain $\varphi_1, \varphi_2 \in C_0^\infty$ such that $\varphi = \varphi_1 + \varphi_2$ and

$$\varphi_1(x) = 0 \quad \text{for} \quad p_1(x) < R \tag{6.39}$$

$$\varphi_2(x) = 0 \quad \text{for} \quad p_1(x) > R+K \tag{6.40}$$

$$|\varphi_i(x)| \leqslant |\varphi(x)|\,, \qquad i=1, 2\,, \tag{6.41}$$

$$J(\varphi_i, \Omega) \leqslant aJ(\varphi, \Omega),\ \Omega \subset E^n\,, \qquad i=1, 2\,, \tag{6.42}$$

$$|(L(x, \mathrm{D})\varphi_1, \varphi_2)| < \varepsilon_1 J(\varphi) \tag{6.43}$$

$$|(\varphi_1, \varphi_2)| < \varepsilon_1 J(\varphi)\,. \tag{6.44}$$

Consequently we have

$$\begin{aligned} &|(L(x, \mathrm{D})\varphi, \varphi) - (L(x, \mathrm{D})\varphi_1, \varphi_1) - (L(x, \mathrm{D})\varphi_2, \varphi_2)| \\ &\qquad \leqslant 2|(L(x, \mathrm{D})\varphi_1, \varphi_2)| < 2\varepsilon_1 J(\varphi) = \varepsilon J(\varphi) \end{aligned}$$

and

$$|\|\varphi\|^2 - \|\varphi_1\|^2 - \|\varphi_2\|^2| \leqslant 2|(\varphi_1, \varphi_2)| < \varepsilon J(\varphi)\,.$$

This proves the theorem for $r=1$.

Next assume that the theorem is proved for $r=k$. We shall prove it for $r=k+1$. Let $\varepsilon > 0$ be given, and assume $K > C_{k+1}/\varepsilon$. Set $\varepsilon_2 = \varepsilon C_k/C_{k+1}$. Then $K > C_k/\varepsilon_2$.

Let $Z_{1,K}^{(k)}(R), \ldots, Z_{2^k,K}^{(k)}(R)$ be the sets corresponding to $p_1(x), \ldots, p_k(x)$. Then by our induction hypotheses there are functions $\varphi_i \in C_0^\infty(Z_{i,K}^{(k)}(R))$ such that

$$\varphi = \Sigma\, \varphi_i \tag{6.45}$$

$$|\varphi_i(x)| \leqslant |\varphi(x)|\,, \quad J(\varphi_i, \Omega) \leqslant a^k J(\varphi, \Omega)\,, \tag{6.46}$$

$$|\Sigma(L(x, \mathrm{D})\varphi_i, \varphi_i) - (L(x, \mathrm{D})\varphi, \varphi)| < \varepsilon_2 J(\varphi) \tag{6.47}$$

$$|\Sigma\|\varphi_i\|^2 - \|\varphi\|^2| < \varepsilon_2 J(\varphi)\,. \tag{6.48}$$

Next set $\varepsilon_3 = \varepsilon C_1/C_{k+1}$. Thus $K > C_1/\varepsilon_3$. Since the theorem has been proved for $r=1$, for each i we can find functions $\varphi_{i1}, \varphi_{i2}$ in C_0^∞ such that

$$\varphi_i = \varphi_{i1} + \varphi_{i2}\,, \qquad 1 \leqslant i \leqslant 2^k \tag{6.49}$$

$$\varphi_{i1}(x) = 0 \quad \text{for} \quad p_{k+1}(x) < R\,, \tag{6.50}$$

$$\varphi_{i2}(x) = 0 \quad \text{for} \quad p_{k+1}(x) > R + K\,, \tag{6.51}$$

$$|\varphi_{ij}| \leqslant |\varphi_{i1}|\,, \quad J(\varphi_{ij}, \Omega) \leqslant aJ(\varphi_i, \Omega)\,, \tag{6.52}$$

$$|(L(x, \mathrm{D})\varphi_{i1}, \varphi_{i1}) + (L(x, \mathrm{D})\varphi_{i2}, \varphi_{i2}) - (L(x, \mathrm{D})\varphi_i, \varphi_i)| < \varepsilon_3 J(\varphi_i)\,, \tag{6.53}$$

$$|\,\|\varphi_{i1}\|^2 + \|\varphi_{i2}\|^2 - \|\varphi_i\|^2| < \varepsilon_3 J(\varphi_i)\,. \tag{6.54}$$

In particular, we have by (6.46) and (6.52)

$$|\varphi_{ij}| \leqslant |\varphi|\,, \quad J(\varphi_{ij}, \Omega) \leqslant a^{k+1} J(\varphi, \Omega)\,,$$

and by (6.47) and (6.53)

$$\begin{aligned} &\left| \sum_{i=1}^{2^k} \sum_{j=1}^{2} (L(x, \mathrm{D})\varphi_{ij}, \varphi_{ij}) - (L(x, \mathrm{D})\varphi, \varphi) \right| \\ &\quad \leqslant \sum_{i=1}^{2^k} \left| \sum_{j=1}^{2} (L(x, \mathrm{D})\varphi_{ij}, \varphi_{ij}) - (L(x, \mathrm{D})\varphi_i, \varphi_i) \right| \\ &\qquad + \left| \sum_{i=1}^{2^k} (L(x, \mathrm{D})\varphi_i, \varphi_i) - (L(x, \mathrm{D})\varphi, \varphi) \right| \\ &\quad < 2^k \varepsilon_3 \max_i J(\varphi_i) + \varepsilon_2 J(\varphi) \leqslant (2^k a^k \varepsilon_3 + \varepsilon_2) J(\varphi) = \varepsilon J(\varphi)\,, \end{aligned}$$

since

$$2^k a^k \varepsilon_3 + \varepsilon_2 = \varepsilon(2^k a^k C_1 + C_k)/C_{k+1} = \varepsilon$$

by (6.38). Similarly we have by (6.48) and (6.54)

$$\begin{aligned} \left| \sum_{i=1}^{2^k} \sum_{j=1}^{2} \|\varphi_{ij}\|^2 - \|\varphi\|^2 \right| &\leqslant \sum_{i=1}^{2^k} \left| \sum_{j=1}^{2} \|\varphi_{ij}\|^2 - \|\varphi_i\|^2 \right| + \left| \sum_{i=1}^{2^k} \|\varphi_i\|^2 - \|\varphi\|^2 \right| \\ &< \varepsilon J(\varphi)\,. \end{aligned}$$

It therefore follows that the theorem is true for $r = k + 1$. The proof is complete.

We can now give the

Proof of Theorem 5.3. Let $\varepsilon > 0$ be given. Then there exist R, K so large that

$$(L(x, \mathrm{D})\varphi, \varphi) \geqslant (\nu_i - \varepsilon)\|\varphi\|^2, \quad \varphi \in C_0^\infty(Z_{i,K}(R)),\ 1 \leqslant i \leqslant k. \tag{6.55}$$

We may take $K > C_r/\varepsilon$ (see 6.37). Set

$$\Omega = \bigcup_{k+1}^{t} Z_{i,K}(R). \tag{6.56}$$

Ω is a bounded set. By Theorem 6.2 there are functions $\varphi_i \in C_0^\infty(Z_{i,K}(R))$ such that (6.32)–(6.36) hold. By (6.34), (6.36), (6.55) and (6.56)

$$\begin{aligned}\sum_1^k (L(x, \mathrm{D})\varphi_i, \varphi_i) &> \sum_1^k (\nu_i - \varepsilon)\|\varphi_i\|^2 \geqslant (\nu - \varepsilon)\sum_1^k \|\varphi_i\|^2 \\ &= (\nu - \varepsilon)\sum_1^t \|\varphi_i\|^2 - (\nu - \varepsilon)\sum_{k+1}^t \|\varphi_i\|^2 \\ &\geqslant (\nu - \varepsilon)\|\varphi\|^2 - \varepsilon|\nu - \varepsilon|\, J(\varphi) - |\nu - \varepsilon| \sum_{k+1}^t J(\varphi_i, Z_{i,K}(R)) \\ &\geqslant (\nu - \varepsilon)\|\varphi\|^2 - \varepsilon|\nu - \varepsilon|\, J(\varphi) - a^r|\nu - \varepsilon|\,(t-k)\, J(\varphi, \Omega).\end{aligned}$$

Moreover by (6.1), (6.34) and (6.56)

$$\begin{aligned}\sum_{k+1}^t |(L(x, \mathrm{D})\varphi_i, \varphi_i)| &\leqslant K_3 \sum_{k+1}^t J(\varphi_i, Z_{i,K}(R)) \\ &\leqslant a^r K_3 (t-k)\, J(\varphi, \Omega).\end{aligned}$$

Combining the last two inequalities we get

$$\begin{aligned}\sum_1^t (L(x, \mathrm{D})\varphi_i, \varphi_i) \geqslant (\nu - \varepsilon)\|\varphi\|^2 - \varepsilon|\nu - \varepsilon|\psi(\varphi) \\ - a^r(t-k)(|\nu - \varepsilon| + K_3)\, J(\varphi, \Omega).\end{aligned}$$

If we now apply inequality (6.35), we obtain (5.11). The proof is complete.

§7. Adding of spectra

Let n_1, n_2 be non-negative integers, and set $n = n_1 + n_2$. Let $x^{(i)} = (x_1^{(i)}, \ldots, x_{n_i}^{(i)})$ denote the coordinates in $E_i = E^{n_i}$ and consider E^n as $E^{n_1} \times E^{n_2}$ (see ch. 1, §1). Thus each point of E is represented as $x = (x^{(1)}, x^{(2)}) = (x_1^{(1)}, \ldots, x_{n_1}^{(1)}, x_1^{(2)}, \ldots, x_{n_2}^{(2)})$.

Let S_i be a self-adjoint operator on $L^2(E_i)$, $i=1, 2$, and let S be a self-adjoint operator on $L^2(E^n)$ such that $u(x^{(1)})\,v(x^{(2)}) \in D(S)$ whenever $u(x^{(1)}) \in D(S_1)$ and $v(x^{(2)}) \in D(S_2)$ with

$$S(uv) = vS_1 u + uS_2 v\,. \tag{7.1}$$

One would expect that there is a relationship between $\sigma(S)$, $\sigma(S_1)$ and $\sigma(S_2)$. This is indeed the case. In fact we have

Lemma 7.1 *If $\lambda_1 \in \sigma(S_1)$ and $\lambda_2 \in \sigma(S_2)$, then $\lambda_1+\lambda_2 \in \sigma(S)$.*

Proof. There is a sequence $\{u_k(x^{(1)})\}$ of functions in $D(S_1)$ such that

$$\|u_k\| = 1, \qquad (S_1-\lambda_1)u_k \to 0$$

(Lemma 7.10 of ch. 1). Similarly there is a sequence $\{v_k(x^{(2)})\}$ of functions in $D(S_2)$ such that

$$\|v_k\| = 1, \qquad (S_2-\lambda_2)v_k \to 0\,.$$

Thus $u_k v_k \in D(S)$ and $\|u_k v_k\| = 1$. Moreover

$$(S-\lambda_1-\lambda_2)u_k v_k = v_k(S_1-\lambda_1)u_k + u_k(S_2-\lambda_2)v_k \to 0\,.$$

This shows that $\lambda_1+\lambda_2 \in \sigma(S)$ and the proof is complete.

We can give a converse of Lemma 7.1. Let M be the set of linear combinations of functions of the form uv, $u \in D(S_1)$, $v \in D(S_2)$. We have

Theorem 7.2. *Assume in addition that S is the closure of its restriction to M. Then $\lambda \in \sigma(S)$ if and only if $\lambda=\lambda_1+\lambda_2$ with $\lambda_i \in \sigma(S_i)$, $i=1, 2$. In symbols*

$$\sigma(S) = \sigma(S_1)+\sigma(S_2)\,. \tag{7.2}$$

Proof. By Lemma 7.1 all we need show is that $\lambda \in \rho(S)$ if there is a $\delta > 0$ such that

$$|\lambda-\lambda_1-\lambda_2| > \delta$$

for all $\lambda_1 \in \sigma(S_1)$ and $\lambda_2 \in \sigma(S_2)$. Let $\{E_i(\lambda_i)\}$ be the spectral family associated with S_i, $i=1, 2$ (see ch. 1, §7). Then for $u_i \in D(S_i)$

$$\begin{aligned} S(u_1 u_2) &= u_2 S_1 u_1 + u_1 S_2 u_2 \\ &= u_2 \int_{-\infty}^{\infty} \lambda_1 \,\mathrm{d}E_1(\lambda_1)u_1 + u_1 \int_{-\infty}^{\infty} \lambda_2 \,\mathrm{d}E_2(\lambda_2)u_2 \\ &= \int_{-\infty}^{\infty}\int_{-\infty}^{\infty} (\lambda_1+\lambda_2)\,\mathrm{d}E_1(\lambda_1)u_1\,\mathrm{d}E_2(\lambda_2)u_2\,. \end{aligned}$$

Hence for $w \in M$ we have

$$Sw = \int_{-\infty}^{\infty} \int_{-\infty}^{\infty} (\lambda_1 + \lambda_2) \mathrm{d}E_1(\lambda_1) \mathrm{d}E_2(\lambda_2) w\,, \tag{7.3}$$

and consequently

$$\|(S-\lambda)w\|^2 = \int_{-\infty}^{\infty} \int_{-\infty}^{\infty} (\lambda_1 + \lambda_2 - \lambda)^2 \mathrm{d}\|E_1(\lambda_1)\, E_2(\lambda_2) w\|^2$$

$$\geqslant \delta^2 \int_{-\infty}^{\infty} \int_{-\infty}^{\infty} \mathrm{d}\|E_1(\lambda_1) E_2(\lambda_2) w\|^2 = \delta^2 \|w\|^2\,.$$

Thus

$$\|w\| \leqslant \|(S-\lambda)w\|/\delta\,, \qquad w \in M\,. \tag{7.4}$$

Now if f is any function in $D(S)$, there is a sequence $\{w_k\}$ of functions in M such that $w_k \to f$ and $Sw_k \to Sf$ in $L^2(E^n)$. Consequently (7.4) implies

$$\|f\| \leqslant \|(S-\lambda)f\|/\delta\,, \qquad f \in D(S)\,. \tag{7.5}$$

This in turn implies that $\lambda \in \rho(S)$ (Lemma 7.10 of ch. 1), and the proof is complete.

As an application we have

Theorem 7.3. *Suppose that the functions b_k in $L(x, \mathrm{D})$ are constants and that the hypotheses of* §§1, 2 *are satisfied. Assume that there is a vector $\xi \neq 0$ in E^n such that*

$$q(x + t\xi) = q(x)\,, \qquad x \in E^n,\ t \text{ real}\,. \tag{7.6}$$

Then

$$\sigma(L) = [\varkappa, \infty)\,, \tag{7.7}$$

where

$$\varkappa = \min_{\lambda \in \sigma(L)} \lambda\,. \tag{7.8}$$

In proving the theorem we shall make use of the following simple but useful lemma. When dependence on n is to be noted we write $N_\alpha(E^n)$ and $N_\alpha(q, E^n)$ for N_α and $N_\alpha(q)$, respectively.

Lemma 7.4. *Let m be a positive integer $< n$, and let α be a positive real number. Then a function depending only on $x_1, \ldots, x_m$ is in $N_\alpha(E^n)$ if and only if it is in $N_\alpha(E^m)$. Similarly, it is in $N_\alpha^{\mathrm{loc}}(E^n)$ if and only if it is in $N_\alpha^{\mathrm{loc}}(E^m)$.*

We save the proof of the lemma until the end of the section. We use it now in giving the

Proof of Theorem 7.3. Assume first that

$$L(x, \mathrm{D}) = \Sigma\, \mathrm{D}_k^2 + q(x) \tag{7.9}$$

and

$$\xi = (|\xi|, 0, \ldots, 0)\,. \tag{7.10}$$

Then if we take $t = -x_1/|\xi|$ in (7.6), we have

$$q(x) = q(x + t\xi) = q(0, x_2, \ldots, x_n)\,.$$

Thus $q(x)$ does not depend on x_1. By Lemma 7.4, q is in $N_\alpha^{\mathrm{loc}}(E^{n-1})$ and $\sigma \in N_2(E^{n-1})$, where E^{n-1} denotes the space determined by the coordinates $x_2, \ldots, x_n$. Hence the operator

$$\sum_{k=2}^{n} \mathrm{D}_k^2 + q(x) \tag{7.11}$$

on $C_0^\infty(E^{n-1})$ is essentially self-adjoint in $L^2(E^{n-1})$ (Theorem 2.1). Moreover, the operator D_1^2 on $C_0^\infty(E^1)$ is essentially self-adjoint on $L^2(E^1)$, where E^1 denotes the space determined by x_1. Let S_1 denote the closure of D_1^2 on $C_0^\infty(E^1)$ in $L^2(E^1)$ and let S_2 denote the closure of the operator (7.11) on $C_0^\infty(E^{n-1})$ in $L^2(E^{n-1})$. By Lemma 4.5

$$\sigma(S_1) = [0, \infty)\,. \tag{7.12}$$

Set

$$\nu = \min_{\lambda \in \sigma(S_2)} \lambda\,. \tag{7.13}$$

Then

$$(S_2\varphi, \varphi) \geqslant \nu \|\varphi\|^2\,, \qquad \varphi \in C_0^\infty(E^{n-1})\,,$$

which implies

$$(S_2\varphi, \varphi) \geqslant \nu \|\varphi\|^2\,, \qquad \varphi \in C_0^\infty(E^n)\,. \tag{7.14}$$

By (7.12) we have in view of Lemma 7.17 of ch. 1

$$(S_1\varphi, \varphi) \geqslant 0\,, \qquad \varphi \in C_0^\infty\,. \tag{7.15}$$

Adding (7.14) and (7.15) we get

$$(L(x, \mathrm{D})\varphi, \varphi) \geqslant \nu \|\varphi\|^2\,, \qquad \varphi \in C_0^\infty\,.$$

Since $L(x, \mathrm{D})$ on C_0^∞ is essentially self-adjoint, we see that

$$(Lu,u) \geqslant \nu \|u\|^2 , \qquad u \in D(L) , \tag{7.16}$$

and consequently

$$\sigma(L) \subset [\nu, \infty) .$$

On the other hand, suppose $\lambda \geqslant \nu$. Then $\lambda - \nu$ is in $\sigma(S_1)$ by (7.12) and $\nu \in \sigma(S_2)$ by (7.13). Thus $\lambda \in \sigma(L)$ by Lemma 7.1. This shows that $\nu = \varkappa$ and (7.7) holds for the special case considered.

To give the proof for the general case, note that we may assume that the b_j all vanish (Theorem 3.17). Thus $L(x, \mathrm{D})$ is of the form

$$L(x, \mathrm{D}) = P(\mathrm{D}) + q(x) . \tag{7.17}$$

Let A be an affine transformation of E^n onto itself which transforms the operator $P(\mathrm{D})$ into the Laplace operator

$$-\Delta = \Sigma \mathrm{D}_k^2 , \tag{7.18}$$

and let R be a rotation of E^n onto E^n which takes the vector $A\xi$ into the vector $(|A\xi|, 0, \ldots, 0)$. Note that R takes the Laplacian into itself. Set

$$\tilde{q}(RAx) = q(x) . \tag{7.19}$$

Then

$$\tilde{q}(RAx + tRA\xi) = \tilde{q}[RA(x + t\xi)] = q(x + t\xi) = q(x) = \tilde{q}(RAx) . \tag{7.20}$$

Hence the function $\tilde{q}$ satisfies (7.6) for the vector $RA\xi$. Since the transformation RA takes (7.17) into the operator (7.9) and the spectrum of the operator is invariant under the transformation, the result follows.

We now consider a more general situation. For each vector $\xi \in E^n$ let the operator $T(\xi)$ be defined by

$$T(\xi) f(x) = f(x - \xi) , \qquad x \in E^n . \tag{7.21}$$

If $f \in L^2$, it is obvious that

$$\|T(\xi) f\| = \|f\| , \qquad \xi \in E^n .$$

In particular $T(\xi)$ is a bounded operator on L^2 for each $\xi \in E^n$. We have

Theorem 7.5. *Let S_i be a self-adjoint operator on $L^2(E_i)$ which is the closure of its restriction to $D(S_i) \cap C_0^\infty(E_i)$, $i = 1, 2$. Let S be a self-adjoint operator on $L^2(E^n)$ such that $u(x^{(1)}) v(x^{(2)}) \in D(S)$ for $u \in D(S_1) \cap C_0^\infty(E_1)$ and $v \in D(S_2) \cap C_0^\infty(E_2)$ with*

$$S(uv) = vS_1 u + uS_2 v + \sum_j q_j(x) Q_j(\mathrm{D}) u R_j(\mathrm{D}) v\,, \tag{7.22}$$

where the q_j are functions locally in $L^2(E^n)$, the $Q_j(\mathrm{D})$ are constant coefficient operators in E_1 and the $R_j(\mathrm{D})$ are constant coefficient operators in E_2. Assume that there is a sequence $\{\xi_{(k)}\}$ of vectors in E_2 such that

$$\int_{|x-y|<1} |q_j(x^{(1)}, x^{(2)} + \xi_{(k)})|^2 \,\mathrm{d}x \to 0 \text{ as } k \to \infty \tag{7.23}$$

for each $y \in E^n$ and each j. Suppose that

$$T(\xi_{(k)}) S_2 \varphi = S_2 T(\xi_{(k)}) \varphi\,, \qquad \varphi \in C_0^\infty(E^n), \qquad k = 1, 2, \ldots\,. \tag{7.24}$$

Then $\lambda_1 \in \sigma(S_1)$ and $\lambda_2 \in \sigma(S_2)$ imply $\lambda_1 + \lambda_2 \in \sigma(S)$.

Proof. We first note that for $\varphi(x^{(1)}) \in C_0^\infty(E_1)$ and $\psi(x^{(2)}) \in C_0^\infty(E_2)$ one has

$$\begin{aligned} &\|q_j T(\xi_{(k)}) Q_j(\mathrm{D}) \varphi R_j(\mathrm{D}) \psi\|^2 \\ &\quad = \int_\Omega |q_j(x^{(1)}, x^{(2)} + \xi_{(k)}) Q_j(\mathrm{D}) \varphi(x^{(1)}) R_j(\mathrm{D}) \psi(x^{(2)})|^2 \,\mathrm{d}x \\ &\quad \leqslant \|Q_j(\mathrm{D})\varphi\|_\infty^2 \, \|R_j(\mathrm{D})\psi\|_\infty^2 \int_\Omega |q_j(x^{(1)}, x^{(2)} + \xi_{(k)})|^2 \,\mathrm{d}x\,, \end{aligned}$$

where Ω is a bounded domain. Since Ω can be covered by a finite number of spheres of the form $|x-y| < 1$, we see by (7.23) that

$$q_j T(\xi_{(k)}) Q_j(\mathrm{D}) \varphi R_j(\mathrm{D}) \psi \to 0 \text{ as } k \to \infty \tag{7.25}$$

for each j.

Now suppose $\lambda_i \in \sigma(S_i)$, $i = 1, 2$. Then by Lemma 7.13 of ch. 1 there is a sequence $\{u_m(x^{(1)})\}$ of functions in $D(S_1) \cap C_0^\infty(E_1)$ such that

$$\|u_m\| = 1, \quad (S_1 - \lambda) u_m \to 0\,, \tag{7.26}$$

and a sequence $\{v_m(x^{(2)})\}$ of functions in $D(S_2) \cap C_0^\infty(E_2)$ such that

$$\|v_m\| = 1\,, \quad (S_2 - \lambda_2) v_m \to 0\,. \tag{7.27}$$

Then $\|u_m v_m\| = 1$ and

$$\begin{aligned} &(S - \lambda_1 - \lambda_2) T(\xi_{(k)}) u_m v_m \\ &\qquad = T(\xi_{(k)}) v_m (S_1 - \lambda_1) u_m + u_m T(\xi_{(k)}) (S_2 - \lambda_2) v_m \\ &\qquad\quad + \sum_j q_j T(\xi_{(k)}) Q_j(\mathrm{D}) u_m R_j(\mathrm{D}) v_m \end{aligned} \tag{7.28}$$

by (7.24). Thus

$$\|(S-\lambda_1-\lambda_2)T(\xi_{(k)})u_m v_m\| \leqslant \|(S_1-\lambda_1)u_m\| + \|(S_2-\lambda_2)v_m\| \\ + \sum_j \|q_j T(\xi_{(k)})Q_j(\mathrm{D})u_m R_j(\mathrm{D})v_m\| . \tag{7.29}$$

Let $\varepsilon > 0$ be given. Take m so large that the first two terms on the right-hand side are each $< \frac{1}{3}\varepsilon$. Then take k so large that the third is $< \frac{1}{3}\varepsilon$. All this can be done by (7.25), (7.26) and (7.27). Thus for each $\varepsilon > 0$ there is a $w \in D(S)$ such that $\|w\| = 1$ and $\|(S-\lambda_1-\lambda_2)w\| < \varepsilon$. This shows that $\lambda_1+\lambda_2 \in \sigma(S)$, and the proof is complete.

We now turn to the proof of Lemma 7.4. In giving it we shall make use of

Lemma 7.6. *For* $0 \leqslant \rho \leqslant 1$, α *real and* $s \geqslant 0$ *set*

$$H(\rho) = \int_0^1 (\rho^2+t^2)^{\frac{1}{2}\alpha} t^s \mathrm{d}t . \tag{7.30}$$

Then there is a constant C *depending only on* α *and* s *such that*

$$H(\rho)/C \leqslant \rho^{\alpha+s+1} \leqslant CH(\rho) , \qquad \alpha+s < -1 \tag{7.31}$$

$$H(\rho)/C \leqslant (1-\log \rho) \leqslant CH(\rho) , \qquad \alpha+s = -1 \tag{7.32}$$

$$H(\rho)/C \leqslant 1 \leqslant CH(\rho) , \qquad \alpha+s > -1 . \tag{7.33}$$

Proof. For $\alpha+s < -1$ we have

$$H(\rho) = \rho^{\alpha+s+1} \int_0^{1/\rho} (1+y^2)^{\frac{1}{2}\alpha} y^s \mathrm{d}y .$$

Hence

$$\rho^{\alpha+s+1} \int_0^1 (1+y^2)^{\frac{1}{2}\alpha} y^s \mathrm{d}y \leqslant H(\rho) \leqslant \rho^{\alpha+s+1} \int_0^\infty (1+y^2)^{\frac{1}{2}\alpha} y^2 \mathrm{d}y .$$

Since $\alpha+s < -1$, the integral is convergent. This gives (7.31).

If $\alpha+s = -1$, then

$$H(\rho) \leqslant \int_0^1 (\rho^2+t^2)^{-\frac{1}{2}} \mathrm{d}t = \log\left[t+(\rho^2+t^2)^{\frac{1}{2}}\right]\Big|_0^1 \leqslant \log(3/\rho) .$$

On the other hand by L'Hospital's rule

$$\lim_{\rho\to 0}\frac{-\log\rho}{\int_0^1(\rho^2+t^2)^{\frac12\alpha}t^s\,\mathrm{d}t}=\lim_{\rho\to 0}\frac{-\rho^{-1}}{\alpha\rho\int_0^1(\rho^2+t^2)^{\frac12\alpha-1}t^s\,\mathrm{d}t}$$

$$=\lim_{\rho\to 0}\frac{-1}{\alpha\int_0^{1/\rho}(1+y^2)^{\frac12\alpha-1}y^s\,\mathrm{d}y}=-\left[\alpha\int_0^{\infty}(1+y^2)^{\frac12\alpha-1}y^s\,\mathrm{d}y\right]^{-1}$$

This proves (7.32). For $-1<\alpha+s\leqslant s$ we have

$$\int_0^1(1+t^2)^{\frac12\alpha}t^s\,\mathrm{d}t\leqslant H(\rho)\leqslant\int_0^1 t^{\alpha+s}\,\mathrm{d}t\,,$$

and for $\alpha>0$ we have

$$\int_0^1 t^{\alpha+s}\,\mathrm{d}t\leqslant H(\rho)\leqslant\int_0^1(1+t^2)^{\frac12\alpha}t^s\,\mathrm{d}t\,.$$

This gives (7.33), and the proof is complete.

For functions f, g we let $f\sim g$ mean that there is a constant C such that

$$\frac{|f|}{|g|}+\frac{|g|}{|f|}\leqslant C\,. \tag{7.34}$$

Lemma 7.7. *For any $m<n$ and $\alpha>0$*

$$\int_{\sum_{m+1}^{n}x_i^2<1}\omega_\alpha^{(n)}(x)\,\mathrm{d}x_{m+1}\ldots\mathrm{d}x_n\sim\omega_\alpha^{(m)}\left[\left(\sum_1^m x_i^2\right)^{\frac12}\right] \tag{7.35}$$

(*see* (7.8) *of ch.* 7).

Proof. Converting to spherical coordinates we have by Lemma 7.6

$$\int_{\sum_{m+1}^{n}x_i^2<1}\omega_\alpha^{(n)}(x)\,\mathrm{d}x_{m+1}\ldots\mathrm{d}x_n=C\int_0^1\omega_\alpha^{(m)}\left[\left(\sum_1^m x_i^2+t^2\right)^{\frac12}\right]t^{n-m-1}\,\mathrm{d}t$$

$$\sim\omega_\alpha^{(m)}\left[\left(\sum_1^m x_i^2\right)^{\frac12}\right]$$

for $\alpha<n$. For $\alpha\geqslant n$ we have

$$\int_{\sum_{m+1}^{n}x_i^2<1}\omega_\alpha^{(n)}(x)\,\mathrm{d}x_{m+1}\ldots\mathrm{d}x_n\sim C\sim\omega_\alpha^{(m)}\left[\left(\sum_1^m x_i^2\right)^{\frac12}\right].$$

This proves (7.35).

Proof of Lemma 7.4. Let $h(x_1, \ldots, x_m)$ be a function depending only on $x_1, \ldots, x_m$. Then by Lemma 7.7

$$\begin{aligned}
&\int_{|x|<1} |h(x_1-y_1, \ldots, x_m-y_m)|^2 \omega_\alpha^{(n)}(x)\,\mathrm{d}x \\
&\quad \leqslant \int_{\sum_{m+1}^{n} x_i^2<1} \int_{\sum_1^m x_i^2<1} |h(x_1-y_1, \ldots, x_m-y_m)|^2 \omega_\alpha^{(n)}(x)\,\mathrm{d}x \\
&\quad \sim \int_{\sum_{m+1}^{n} x_i^2<1} |h(x_1-y_1, \ldots, x_m-y_m)|^2 \,\omega_\alpha^{(m)}\left[\left(\sum_1^m x_i^2\right)^{\frac{1}{2}}\right] \mathrm{d}x_1 \ldots \mathrm{d}x_m \\
&\quad = N_\alpha(h, E^m)\,.
\end{aligned}$$

Conversely

$$\begin{aligned}
N_\alpha(h, E^m) &\sim \int_{\sum_1^m x_i^2<1} \int_{\sum_1^m x_i^2<1} |h(x_1-y_1, \ldots, x_m-y_m)|^2 \omega_\alpha^{(n)}(x)\,\mathrm{d}x \\
&\leqslant CN_2(h, E^n)\,,
\end{aligned}$$

since the volume $\Sigma_1^m x_i^2 < 1$, $\Sigma_{m+1}^n x_i^2 < 1$ can be covered by a finite number of spheres of radius one. These inequalities prove the first assertion in the lemma.

Next suppose $h \in N_\alpha^{\text{loc}}(E^m)$ and let $\varphi(x)$ be any function in $C_0^\infty(E^n)$. Let R be so large that $\varphi(x)$ vanishes for $|x| > R$. Let ψ be a function in $C_0^\infty(E^m)$ which is identically one for $\Sigma_1^m x_i^2 < R^2$. Then $\varphi h = \varphi\psi h$. But ψh is in $N_\alpha(E^m) \subset N_\alpha(E^n)$ by what has already been proved. Hence $\varphi h \in N_\alpha(E^n)$. Since this is true for each $\varphi \in C_0^\infty$, we see that $h \in N_\alpha^{\text{loc}}(E^n)$. Conversely, suppose $h \in N_\alpha^{\text{loc}}(E^n)$. Let ψ be a function in $C_0^\infty(E^m)$ and let v be a function in $C_0^\infty(E^{n-m})$ which is identically one for $\Sigma_{m+1}^n x_i^2 < 1$. Then ψv is in $C_0^\infty(E^n)$, and consequently $\psi v h$ is in $N_\alpha(E^n)$. By Lemma 7.7

$$\begin{aligned}
&\int_{\sum_1^m x_i^2<1} |\psi(x_1-y_1, \ldots, x_m-y_m) h(x_1-y_1, \ldots, x_m-y_m)|^2 \\
&\qquad\qquad \times \omega_\alpha^{(m)} \left[\left(\sum_1^m x_i^2\right)^{\frac{1}{2}}\right] \mathrm{d}x_1 \ldots \mathrm{d}x_m \\
&\sim \int_{\sum_1^m x_i^2<1} \int_{\sum_{m+1}^n x_i^2<1} |\psi(x_1-y_1, \ldots, x_m-y_m) h(x_1-y_1, \ldots, x_m-y_m)|^2 \\
&\qquad\qquad \times |v(x_{m+1}, \ldots, x_n)|^2 \omega_\alpha^{(n)}(x)\,\mathrm{d}x \\
&\sim N_\alpha(\psi v h, E^n)\,,
\end{aligned}$$

since $v \equiv 1$ for $\Sigma_{m+1}^n x_i^2 < 1$. Thus ψh is in $N_\alpha(E^m)$. Since this is true for each $\psi \in C_0^\infty(E^m)$, the proof is complete.

§8. Separation of coordinates

We now consider a special case of the operator (1.1) in which the coordinates are grouped in "brunches". Let $n_1, \ldots, n_N$ be positive integers such that $n_1+\ldots+n_N=n$. Let E_i denote the Euclidean space E^{n_i} with coordinates $x^{(i)}=(x_1^{(i)}, \ldots, x_{n_i}^{(i)})$, $1 \leqslant i \leqslant N$. We consider E^n as the cartesian product $E_1 \times \ldots \times E_N$ (see §1 of ch. 1). Thus every $x \in E^n$ can be represented in the form $x=(x^{(1)}, \ldots, x^{(N)})$, where $x^{(i)} \in E_i$. The cartesian product of all the E_j for $j \neq i$ will be denoted by E^{n-n_i}.

Let $D_j^{(i)}=-(-1)^{\frac{1}{2}}\partial/\partial x_j^{(i)}$ and

$$D^{(i)} = (D_1^{(i)}, \ldots, D_{n_i}^{(i)}).$$

In each E_i we consider an operator of the form

$$L_i(x^{(i)}, D^{(i)}) = \sum_{j,k=1}^{n_i} a_{jk}^{(i)}[D_j^{(i)}+b_j^{(i)}(x^{(i)})][D_k^{(i)}+b_k^{(i)}(x^{(i)})]+q_i(x^{(i)}). \quad (8.1)$$

We assume that each operator $L_i(x^{(i)}, D^{(i)})$ satisfies in E_i the hypotheses of §§1, 2. For each q_i we define the functions ρ_i, σ_i by (1.5).

In addition we assume that there are real valued functions $V(x)$, $V_1(x)$, …, $V_n(x)$, $U_1(x)$, …, $U_n(x)$ such that

(a) $V(x)=V_i(x)+U_i(x)$, $\quad 1 \leqslant i \leqslant N$

(b) $V_i, U_i \in N_\alpha^{\text{loc}}$, $\quad 1 \leqslant i \leqslant N$

(c) $V_i(x)$ does not depend on $x^{(i)}$.

(d) Set

$$Q_i(x)^2 = -\min[V_i(x), 0], \quad S_i(x)^2 = -\min[U_i(x), 0]. \quad (8.2)$$

Then the Q_i and S_i are in N_2 with

$$\Sigma N_{2,\delta}(Q_i)+\Sigma N_{2,\delta}(S_i) \to 0 \text{ as } \delta \to 0. \quad (8.3)$$

We set

$$T_i(x,D) = \sum_{j \neq i} L_j(x^{(j)})+V_i(x), \quad 1 \leqslant i \leqslant N \quad (8.4)$$

$$L(x, D) = \sum_1^N L_i(x^{(i)}, D^{(i)})+V(x). \quad (8.5)$$

Lemma 8.1. *We have*

(A) *The closure T_i of $T_i(x, D)$ on $C_0^\infty(E^{n-n_i})$ is self-adjoint for $1 \leqslant i \leqslant N$;*

(B) *The closure* L *of* $L(x, \mathrm{D})$ *on* $C_0^\infty(E^n)$ *is self-adjoint.*

Proof. From the fact that the $b_j^{(i)}$ and q_i satisfy the hypotheses of §§1, 2 in E_i we can conclude via Lemma 7.4 that they satisfy them in E^{n-n_j}, $j \neq i$. The ellipticity of $T_i(x, \mathrm{D})$ in E^{n-n_i} follows from the ellipticity of each $L_j(x^{(j)}, \mathrm{D}^{(j)})$ in E_j. Another application of Lemma 7.4 to (b) and (d) shows that $V_i(x) \in N_\alpha^{\mathrm{loc}}(E^{n-n_i})$ and $Q_i(x) \in N_2(E^{n-n_i})$ with (8.3) holding in E^{n-n_i}. These facts imply that $T_i(x, \mathrm{D})$ satisfies in E^{n-n_i} all of the hypotheses of §§1, 2. Consequently it is essentially self-adjoint in $C_0^\infty(E^{n-n_i})$ (Theorem 2.1). This establishes (A).

With respect to (B) note that Lemma 7.4 implies that the operator

$$\sum_1^N L_i(x^{(i)}, \mathrm{D}^{(i)})$$

satisfies the hypotheses of §§1, 2. By (b) and (d) the addition of $V(x)$ does not alter the situation. Thus the lemma is established.

The present section will be devoted to the study of the relationship between $\sigma(L)$ and the $\sigma(T_i)$. Our first result is

Theorem 8.2. *Assume that*

$$N_{2,x}(\sigma_i, E_i) \to 0 \text{ as } |x^{(i)}| \to \infty, \qquad 1 \leqslant i \leqslant N. \tag{8.6}$$

Assume also that for each i *and fixed* $x^{(i)}$, $j \neq i$ *the function* $S_i(x)$ *is in* $N_2(E_i)$ *with*

$$N_{2,x}(S_i, E_i) \to 0 \text{ as } |x| \to \infty, \tag{8.7}$$

uniformly in the $x^{(i)}$, $j \neq i$. *Set*

$$\varkappa = \min_{1 \leqslant i \leqslant N} \min_{\lambda \in \sigma(T_i)} \lambda. \tag{8.8}$$

Then

$$\sigma_e(L) \subset [\varkappa, \infty). \tag{8.9}$$

Proof. Let $\varepsilon > 0$ be given. By (8.6), (8.7) and Lemma 3.15 we can find an R so large that

$$([L_i(x^{(i)}, \mathrm{D}^{(i)}) - S_i^2]\varphi, \varphi) + \varepsilon\|\varphi\|^2 \geqslant 0 \tag{8.10}$$

for all $\varphi \in C_0^\infty(E_i)$ which vanish for $|x^{(i)}| < R$, $1 \leqslant i \leqslant N$. Let $W_i(R)$ denote the set $|x^{(i)}| > R$ in E^n. Then (8.10) implies

$$([L_i(x^{(i)}, \mathrm{D}^{(i)}) - S_i^2]\varphi, \varphi) + \varepsilon\|\varphi\|^2 \geqslant 0, \ \varphi \in C_0^\infty(W_i(R)), \ 1 \leqslant i \leqslant N. \tag{8.11}$$

Set $p_i(x)=|x^{(i)}|$, $1\leqslant i\leqslant N$ and define the sets $Z_{j,K}(R)$ as in the beginning of §5. Note that all of the hypotheses of Theorem 5.1 are satisfied. Also note that each of the sets $Z_{j,K}(R)$ in the first category is contained in one of the sets $W_i(R)$. Now by hypothesis

$$(T_i(x,\mathrm{D})\varphi,\varphi)\geqslant\varkappa\|\varphi\|^2,\qquad \varphi\in C_0^\infty(E^{n-n_i}),\ 1\leqslant i\leqslant N$$

(see Lemma 7.17 of ch. 1). This implies

$$(T_i(x,\mathrm{D})\varphi,\varphi)\geqslant\varkappa\|\varphi\|^2,\qquad \varphi\in C_0^\infty,\quad 1\leqslant i\leqslant N. \tag{8.12}$$

Adding (8.11) and (8.12) we get

$$(L(x,D)\varphi,\varphi)\geqslant(\varkappa-\varepsilon)\|\varphi\|^2,\qquad \varphi\in C_0^\infty(W_i(R)),\qquad 1\leqslant i\leqslant N.$$

This implies $v_i\geqslant\varkappa-\varepsilon$ for each i and consequently $v\geqslant\varkappa-\varepsilon$. Since this is true for any $\varepsilon>0$, we have $v\geqslant\varkappa$. The result now follows from Theorem 5.1.

We now turn our attention to the converse theorem. We give sufficient conditions that

$$\sigma_{\mathrm{e}}(L)\supset[\varkappa,\infty). \tag{8.13}$$

Theorem 8.3. *Assume that there are real constants* $c_j^{(i)}$, $1\leqslant j\leqslant n_i$, $1\leqslant i\leqslant N$ *and a sequence* $\{\xi_{(m)}\}$ *of vectors in* E^n *such that for each* $y\in E^n$

$$\int_{|x-y|<1}[b_j^{(i)}\,x^{(i)}+\xi_{(m)}^{(i)})-c_j^{(i)}]^2\,\mathrm{d}x\to0 \text{ as } m\to\infty,\qquad 1\leqslant j\leqslant n_i,\ 1\leqslant i\leqslant N; \tag{8.14}$$

$$\int_{|x-y|<1}\Big[\sum_{j,k}a_{jk}^{(i)}\mathrm{D}_j^{(i)}b_k^{(i)}(x^{(i)}+\xi_{(m)}^{(i)})\Big]^2\mathrm{d}x\to0 \text{ as } m\to\infty,\ 1\leqslant i\leqslant N; \tag{8.15}$$

$$\int_{|x-y|<1}[q_i(x^{(i)}+\xi_{(m)}^{(i)})]^2\,\mathrm{d}x\to0 \text{ as } m\to\infty,\qquad 1\leqslant i\leqslant N; \tag{8.16}$$

$$\int_{|x-y|<1}[V_i(x^{(1)},\ldots,x^{(i)}+\xi_{(m)}^{(i)},\ldots,x^{(N)})]^2\,\mathrm{d}x\to0 \text{ as } m\to\infty,\qquad 1\leqslant i\leqslant N. \tag{8.17}$$

Then (8.13) *holds.*

Proof. By Theorem 7.11 of ch. 1 it suffices to show that

$$\sigma(L)\supset[\varkappa,\infty). \tag{8.18}$$

Let i be such that $\varkappa\in\sigma(T_i)$. Set $S_1=T_i$ and let S_2 be the closure of the operator

$$R_i(\mathrm{D}^{(i)}) = \sum_{j,k=1}^{n_i} a_{jk}^{(i)}(\mathrm{D}_j^{(i)}+c_j^{(i)})(\mathrm{D}_k^{(i)}+c_k^{(i)})$$

on $C_0^\infty(E_i)$. By Theorem 3.5 and (4.8) we have

$$\sigma(S_2) = [0, \infty)\,. \tag{8.19}$$

Moreover one has clearly

$$T(\xi)S_2 = S_2 T(\xi)\,, \qquad \xi \in E^n\,. \tag{8.20}$$

If $u \in C_0^\infty(E^{n-n_i})$ and $v \in C_0^\infty(E_i)$, then

$$\begin{aligned} L(x, \mathrm{D})(uv) &= vS_1 u + uS_2 v + 2u\,\Sigma\, a_{jk}^{(i)} h_j^{(i)}(\mathrm{D}_k^{(i)}+c_k^{(i)})v \\ &\quad + (\Sigma\, a_{jk}^{(i)}\mathrm{D}_j^{(i)} b_k^{(i)} + \Sigma\, a_{jk}^{(i)} h_j^{(i)} h_k^{(i)} + q_i + V_i)uv\,, \end{aligned} \tag{8.21}$$

where $h_j^{(i)} = b_j^{(i)} - c_j^{(i)}$. We now note that the hypotheses of Theorem 7.5 are satisfied in view of (8.14)–(8.17). If $\lambda \geqslant \varkappa$, $\lambda - \varkappa \in \sigma(S_2)$ by (8.19). Since $\varkappa \in \sigma(S_1)$, $\lambda = \varkappa + (\lambda - \varkappa) \in \sigma(L)$ by Theorem 7.5. This shows that (8.18) holds and the proof is complete.

Corollary 8.4. *If the hypotheses of Theorems* 8.2 *and* 8.3 *are satisfied, then*

$$\sigma_e(L) = [\varkappa, \infty)\,. \tag{8.22}$$

§9. Clusters

We now give a variation of Theorem 8.2 which will hold even when (8.2) is not satisfied. However, in order to carry out this approach one must assume that all of the E_i are of the same dimension, i.e., that $n_1 = \ldots = n_N = n_0$. We make this assumption throughout the section. Let C be a subset of the integers $\{0, \ldots, N\}$. We let E_C denote the subspace of E^n determined by the coordinates $x^{(j)}$, $j > 0$, $j \in C$, and $\tilde{E}_C$ the subspace determined by the coordinates $x^{(k)}$, $k \notin C$. We can consider E^n as the cartesian product $E_C \times \tilde{E}_C$. We set $x^{(0)} = 0 \in E^{n_0}$.

Theorem 9.1. *Suppose $V(x)$ is a real valued function with the following properties. For each proper subset C of the integers $\{1, \ldots, N\}$*

(a) $V(x) = V_C(x) + U_C(x)$.

(b) *V_C and U_C are in N_α^{loc}.*

(c) *If*

$$Q_C(x)^2 = -\min\,(U_C(x), 0],\quad S_C(x)^2 = -\min\,[V_C(x), 0]\,, \tag{9.1}$$

then Q_C *and* S_C *are in* N_2 *with*

$$N_{2,\delta}(Q_C)+N_{2,\delta}(S_C)\to 0 \text{ as } \delta\to 0. \tag{9.2}$$

(d) $N_{2,x}(S_C)\to 0$ *as* $\Delta_C(x)\to\infty$, *where*

$$\Delta_C(x) = \min_{\substack{j\in C\\ k\notin C}} \min \{|x^{(k)}-x^{(j)}|\,,\,|x^{(k)}|\}. \tag{9.3}$$

Set

$$L_C(x, \mathrm{D}) = \sum_{j\in C} L_j(x^{(j)}, \mathrm{D}^{(j)})+\sum_{k\notin C} P_k(x^{(k)}, \mathrm{D}^{(k)})+V_C \tag{9.4}$$

and let $L(x, \mathrm{D})$ *be given by* (8.5). *Let* L_C *and* L *denote the closures in* L^2 *of* $L_C(x, \mathrm{D})$ *and* $L(x, \mathrm{D})$ *in* C_0^∞, *respectively. Then* L_C *and* L *are self-adjoint. Set*

$$\varkappa_C = \min_{\lambda\in\sigma(L_C)} \lambda \tag{9.5}$$

$$\varkappa_0 = \min_C \varkappa_C\,, \tag{9.6}$$

where the minimum is taken of all proper subsets C *of* $\{1, \ldots, N\}$. *If* (8.6) *holds, then*

$$\sigma_e(L)\subset[\varkappa_0, \infty)\,. \tag{9.7}$$

Proof. By (a), (b) and (c), $L_C(x, \mathrm{D})$ and $L(x, \mathrm{D})$ satisfy the assumption of §§1, 2. Thus they are essentially self-adjoint. Set

$$p_{jk}(x)=|x^{(k)}-x^{(j)}|\,, \qquad 0\leqslant j<k\leqslant N \tag{9.8}$$

(recall that $x^{(0)}=0\in E^{n_0}$ by definition).

Let $Z_{i,K}(R)$ be the sets defined in the beginning of §5 corresponding to the $p_{jk}(x)$. All of the hypotheses of Theorem 5.1 are satisfied. Let $Z_{i,K}(R)$ be any one of these sets. Let C be the set of integers $j\geqslant 1$ such that $|x^{(j)}|$ is bounded for $x\in Z_{i,K}(R)$. If $Z_{i,K}(R)$ is in the first category, C is a proper subset of $\{1, \ldots, N\}$ and $|x^{(k)}|>R$ for $k\notin C$. Furthermore for each $j\in C$ and $k\notin C$

$$p_{jk}(x)>R\,, \qquad x\in Z_{i,K}(R)\,. \tag{9.9}$$

For otherwise we would have for some $j\in C$ and $k\notin C$

$$|x^{(k)}|\leqslant|x^{(k)}-x^{(j)}|+|x^{(j)}|<R+K+|x^{(j)}|\,.$$

This would imply that $|x^{(k)}|$ is bounded, contrary to the fact that k is not in C. Thus

$$\Delta_C(x)>R\,, \qquad x\in Z_{i,K}(R)\,. \tag{9.10}$$

We now apply Theorem 3.16. By (9.5) and Lemma 7.17 of ch. 1

$$(L_C(x, \mathrm{D})\,\varphi, \varphi) \geqslant \varkappa_C \|\varphi\|^2\,, \qquad \varphi \in C_0^\infty\,,$$

and by (d) and (9.10)

$$\sup_{x \in Z_{i,K}(R)} N_{2,x}(S_C) \to 0 \text{ as } R \to \infty\,.$$

Since $\Delta_C(x) \to \infty$ implies $|x^{(k)}| \to \infty$ for $k \notin C$, we also have by (8.6)

$$\sup_{x \in Z_{i,K}(R)} N_{2,x}(\sigma_k) \to 0 \text{ as } R \to \infty\,, \qquad k \notin C\,.$$

Thus by Theorem 3.16

$$\liminf_{\substack{R \to \infty \\ K \mapsto \infty}} \inf_{\varphi \in C_0^\infty(Z_{i,K}(R))} \frac{(L(x, \mathrm{D})\varphi, \varphi)}{\|\varphi\|^2} \geqslant \varkappa_C\,.$$

This means that $\nu_i \geqslant \varkappa_C \geqslant \varkappa_0$. Since this is true for all $Z_{i,K}(R)$ in the first category, we have $\nu \geqslant \varkappa_0$. The result now follows from Theorem 5.1. This completes the proof.

A very important special case of Theorem 9.1 is

Theorem 9.2. *Suppose we are given real valued functions*

$$V_{ij}(z)\,, \qquad 1 \leqslant i \leqslant j \leqslant N$$

defined in E^{n_0}, *each being in* $N_\alpha^{\mathrm{loc}}(E^{n_0})$. *If*

$$\sigma_{ij}(z)^2 = -\min\left[V_{ij}(z), 0\right]\,, \qquad 1 \leqslant i < j \leqslant N\,, \tag{9.11}$$

we assume

$$N_{2,\delta}(\sigma_{ij}, E^{n_0}) \to 0 \;\; as \;\; \delta \to 0\,, \qquad 1 \leqslant i < j \leqslant N\,, \tag{9.12}$$

and

$$N_{2,z}(\sigma_{ij}, E^{n_0}) \to 0 \;\; as \;\; |z| \to \infty, \qquad 1 \leqslant i < j \leqslant N\,. \tag{9.13}$$

Set

$$L(x, \mathrm{D}) = \sum_i L_i(x^{(i)}, \mathrm{D}^{(i)}) + \sum_{i,j} V_{ij}(x^{(j)} - x^{(i)})\,. \tag{9.14}$$

For each proper subset C *of* $\{1, \ldots, N\}$ *put*

$$V_C(x) = \sum_{i,j \in C} V_{ij}(x^{(j)} - x^{(i)}) + \sum_{i,j \notin C} V_{ij}(x^{(j)} - x^{(i)})\,, \tag{9.15}$$

$$L_C(x, \mathrm{D}) = \sum_{i \in C} L_i(x^{(i)}, \mathrm{D}^{(i)}) + \sum_{k \notin C} P_k(x^{(k)}, \mathrm{D}^{(k)}) + V_C\,, \tag{9.16}$$

and let L_C and L denote the closures in L^2 of $L_C(x, D)$ and $L(x, D)$ in C_0^∞, respectively. Then L_C and L are self-adjoint. Let $\varkappa_0$ be defined by (9.6) *and assume that* (8.6) *holds. Then* (9.7) *holds.*

Proof. We first note that $V_{ij}(x^{(j)}-x^{(i)})$ is in $N_\alpha^{\text{loc}}(E^n)$. To see this make the transformation

$$\hat{x}^{(k)} = x^{(k)}, \qquad k \neq j$$
$$\hat{x}^{(j)} = x^{(j)} - x^{(i)}.$$

Since $V_{ij}(z)$ is in $N_\alpha^{\text{loc}}(E^{n_0})$, $V_{ij}(\hat{x}^{(j)})$ is in $N_\alpha^{\text{loc}}(E^n)$ in the new coordinates (Lemma 7.4). Since the space N_α^{loc} is invariant under the coordinate transformation, the assertion is proved. The same reasoning shows that $\sigma_{ij}(x^{(j)}-x^{(i)})$ is in $N_2(E^n)$ with

$$N_{2,\delta}(\sigma_{ij}(x^{(j)}-x^{(i)}), E^n) \to 0 \ \text{as} \ \delta \to 0 \tag{9.17}$$

$$N_{2,x}(\sigma_{ij}(x^{(j)}-x^{(i)}), E^n) \to 0 \ \text{as} \ |x^{(j)}-x^{(i)}| \to \infty \tag{9.18}$$

for each i, j. Set

$$V(x) = \sum_{i,j} V_{ij}(x^{(j)}-x^{(i)})$$

and for each C

$$U_C(x) = V(x) - V_C(x).$$

If $Q_C(x)$ and $S_C(x)$ are defined by (9.1), then (9.2) follows from (9.17). Moreover property (d) of Theorem 9.1 follows from (9.18), since $U_C(x)$ contains only those $V_{ij}(x^{(j)}-x^{(i)})$ for which one of the indices is in C and the other is not. Thus all of the hypotheses of Theorem 9.1 are satisfied. This completes the proof.

Remark. If we interpret each operator $L_i(x^{(i)}, D^{(i)})$ pertaining to a "particle" and each $V_{ij}(x^{(j)}-x^{(i)})$ as the interaction between the ith and jth particle, then each C divides the particles into two "clusters" and L_C ignores the interaction between particles in different clusters. See §4 of ch. 10.

The following theorem is useful when one desires the reverse inclusion in (9.7).

Theorem 9.3. *Assume that the $b_j^{(i)}$ are constants and that there is a sequence $\{Z_{(m)}\}$ of vectors in E^{n_0} such that for each $y \in E^n$*

$$\int_{|y-z|<1} q_i(z+Z_{(m)})^2\,dz \to 0 \quad \textit{as} \quad m\to\infty\,, \qquad 1\leqslant i\leqslant N \tag{9.19}$$

$$\int_{|y-z|<1} V_{ij}(z+Z_{(m)})^2\,dz \to 0 \quad \textit{as} \quad m\to\infty\,, \qquad 1\leqslant i<j\leqslant N\,. \tag{9.20}$$

Then

$$\sigma_e(L) \supset [\varkappa_0, \infty)\,. \tag{9.21}$$

Proof. By Theorem 7.11 of ch. 1 it suffices to prove

$$\sigma(L) \supset [\varkappa_0, \infty)\,. \tag{9.22}$$

Let C be a proper subset of $\{1, \ldots, N\}$ such that $\varkappa_0 \in \sigma(L_C)$. Define $\xi_{(m)} \in E^n$ by

$$\begin{aligned} \xi^{(k)}_{(m)} &= 0\,, && k\in C \\ &= Z_{(m)}\,, && k\notin C\,, \end{aligned}$$

and put

$$T_m u(x) = u(x-\xi_{(m)})\,, \qquad m=1, 2, \ldots\,. \tag{9.23}$$

Note that

$$T_m L_C = L_C T_m\,, \qquad m=1, 2, \ldots\,, \tag{9.24}$$

since L_C contains no q_k for $k\notin C$ and the $b^{(i)}_j$ are constants. Since

$$V_C(x+t\xi_{(m)}) = V_C(x)\,, \qquad t>0, \quad m=1, 2, \ldots\,, \tag{9.25}$$

we can conclude from Theorem 7.3 that

$$\sigma(L_C) = [\varkappa_0, \infty)\,. \tag{9.26}$$

Let λ be any number $\geqslant \varkappa_0$. Then by (9.26) and Theorem 7.13 of ch. 1 there is a sequence $\{\varphi_k\}$ of functions in $C_0^\infty(E^n)$ such that

$$\|\varphi_k\| = 1, \quad \|(L_C-\lambda)\varphi_k\| \to 0 \text{ as } k\to\infty\,. \tag{9.27}$$

Then we have $\|T_m\varphi_k\| = 1$ and

$$\|(L-\lambda)T_m\varphi_k\| \leqslant \|(L-L_C)T_m\varphi_k\| + \|(L_C-\lambda)\varphi_k\|\,. \tag{9.28}$$

But

$$L-L_C = \sum_{j\notin C} q_j(x^{(i)}) + \sum_{\substack{i\in C\\ j\notin C}} V_{ij}(x^{(j)}-x^{(i)}) + \sum_{\substack{i\in C\\ j\notin C}} V_{ij}(x^{(j)}-x^{(i)})\,.$$

Furthermore for $\varphi\in C_0^\infty$

$$\|q_j T_m\varphi\|^2 \leqslant \|\varphi\|^2_\infty \int_\Omega q_j(z+Z_{(m)})^2\,dz\,, \qquad j\notin C$$

$$\|V_{ij}(x^{(j)}-x^{(i)})\,T_m\varphi\|^2 \leqslant \|\varphi\|_\infty^2 \int_\Omega V_{ij}(z+Z_{(m)})^2\,\mathrm{d}z\,, \qquad i\in C\,,\ \ j\notin C$$

$$\|V_{ij}(x^{(j)}-x^{(i)})\,T_m\varphi\|^2 \leqslant \|\varphi\|_\infty^2 \int_\Omega V_{ij}(z-Z_{(m)})^2\,\mathrm{d}z\,, \qquad i\notin C\,,\ \ j\in C\,,$$

where Ω is a bounded region of E^{n_0} determined by the support of φ. Since any bounded region can be covered by a finite number of spheres of radius one, we have by (9.19) and (9.20)

$$\|(L-L_C)\,T_m\,\varphi\| \to 0 \ \text{ as } \ m\to\infty \tag{9.29}$$

for each fixed $\varphi\in C_0^\infty(E^n)$. Now let $\varepsilon>0$ be given. By (9.27) we can take k so large that

$$\|(L_C-\lambda)\,\varphi_k\| < \tfrac{1}{2}\varepsilon\,.$$

Once φ_k is chosen, we can take m so large that

$$\|(L-L_C)\,T_m\,\varphi_k\| < \tfrac{1}{2}\varepsilon\,.$$

Combining the last two inequalities, we have by (9.28)

$$\|(L-\lambda)\,T_m\,\varphi_k\| < \varepsilon\,.$$

Since this is true for any $\varepsilon>0$, we see that $\lambda\in\sigma(L)$, and the proof is complete.

CHAPTER 10

APPLICATIONS

In this chapter we shall show how the mathematical theory developed in chs. 4–9 can be applied to questions in quantum mechanics. We shall begin with simpler situations and work up to more general systems.

§1. The Schrödinger operator for a particle

The energy operator corresponding to a single particle subjected to a static potential is of the form

$$H = -\frac{h^2}{8\pi^2 m}\Delta + q(x) = \frac{h^2}{8\pi^2 m}(\mathrm{D}_1^2 + \mathrm{D}_2^2 + \mathrm{D}_3^2) + q(x) \tag{1.1}$$

in E^3, where h is Planck's constant, m is the mass of the particle and $q(x)$ is a real valued function describing the potential energy. Here Δ represents the Laplacian in E^3 (recall that $\mathrm{D}_k = \partial/\mathrm{i}\,\partial x_k$).

Theorem 1.1. *Suppose that $q(x)$ is locally integrable, and define*

$$\rho(x)^2 = \max\,[q(x), 0]\,, \quad \sigma(x)^2 = \rho(x)^2 - q(x)\,. \tag{1.2}$$

Assume that $\sigma \in N_2(E^3)$ and that

$$N_{2,\delta}(\sigma) \to 0 \ \ as \ \ \delta \to 0\,. \tag{1.3}$$

Then H on $C_0^\infty \cap D(q)$ has a regularly accretive self-adjoint extension in L^2. If

$$N_{2,x}(\sigma) \to 0 \ \ as \ \ |x| \to \infty\,, \tag{1.4}$$

then it has such an extension B satisfying

$$\sigma_{\mathrm{e}}(B) \subset [0, \infty)\,. \tag{1.5}$$

If, in addition, $\rho \in N_4(E^3)$ and

$$N_{4,x}(\rho) \to 0 \ \ as \ \ |x| \to \infty\,, \tag{1.6}$$

then

$$\sigma_e(B) = [0, \infty). \tag{1.7}$$

If $q \in N_\alpha^{\text{loc}}$ for some $\alpha < 4$ then $D(q) \supset C_0^\infty$ and H on C_0^∞ is essentially self-adjoint. Thus B is the closure in L^2 of H on C_0^∞.

Proof. The first statement follows from Theorem 1.6 of ch. 8. The second and third follow from Theorems 2.7 and 2.3 of that chapter, respectively. The last statement follows from Theorem 2.1 of ch. 9.

We consider several examples.

(a) *Free particle.* In this case $q(x) = 0$. H is a constant coefficient operator with real coefficients. Its closure H_0 in L^2 is self-adjoint (Corollary 2.3 of ch. 4) and

$$\sigma_e(H_0) = \sigma(H_0) = [0, \infty) \tag{1.8}$$

(Corollary 3.3 of ch. 4 and Lemma 4.5 of ch. 9).

(b) *Coulomb potential.* In this case $q(x) = -e^2/|x|$, where e is a constant (charge). Here $\rho = 0$ and all of the hypotheses of Theorem 1.1 are satisfied. Thus H on C_0^∞ is essentially self-adjoint and its closure in L^2 satisfies

$$\sigma_e(H_0) = [0, \infty). \tag{1.9}$$

This of course is classical. The spectrum of H_0 represents the energy states of the hydrogen atom. We know that the spectrum of this operator consists of the non-negative real axis and the negative eigenvalues

$$\lambda_n = -2m\pi^2 e^4/h^2 n^2, \qquad n = 1, 2 \ldots, \tag{1.10}$$

with multiplicities n^2. Note that the lowest eigenvalue is

$$\lambda_1 = -2m\pi^2 e^4/h^2$$

(ground state) and the only limit point is 0 (escape energy).

(c) Consider the potential

$$q(x) = \sum_1^N c_k \mathrm{e}^{-a_k|x|}/|x|^{b_k}, \tag{1.11}$$

where the a_k, b_k, c_k are constants and $a_k \geqslant 0$, $b_k < 3$ for each k. Thus $q(x)$ is locally integrable. If $b_k < 2$ for each k such that $c_k < 0$ and $a_k = 0$, then (1.3) and (1.4) hold. In this case H on $C_0^\infty \cap D(q)$ has a self-adjoint extension B satisfying (1.5). If, in addition, $b_k > 0$ whenever $a_k = 0$, then (1.6) is also satisfied. In this case B satisfies (1.7). If, in addition, $b_k < 2$ for each k, then B is the closure in L^2 of H on C_0^∞.

Examples of such potentials arising in practice are e^{-ar}/r and $d(e^{-ar}/r)dr$ where $r=|x|$.

(d) *Hydrogen molecular ion.* Suppose we have two stationary positive charges $+e$ a distance R apart and one negative charge $-e$ in motion. We may assume that the first positive charge is at the origin and that the second is at a point r with $|r|=R$. The Hamiltonian (energy) operator for this system is

$$H = -\frac{h^2}{8\pi^2 m}\,\Delta + \frac{e^2}{R} - \frac{e^2}{|x|} - \frac{e^2}{|x-r|}\,. \tag{1.12}$$

We write this in the form

$$H = P(\mathrm{D}) + q(x)\,, \tag{1.13}$$

where

$$P(\mathrm{D}) = -\frac{h^2}{8\pi^2}\,\Delta + \frac{e^2}{R} \tag{1.14}$$

and

$$q(x) = -\frac{e^2}{|x|} - \frac{e^2}{|x-r|}\,. \tag{1.15}$$

Since $P(\mathrm{D})$ is a constant coefficient operator, we know that

$$\sigma(P_0) = \left[\frac{e^2}{R}, \infty\right) \tag{1.16}$$

(Corollary 3.4 of ch. 4). Moreover q is P_0-compact by Corollary 4.2 of ch. 6. Hence H on C_0^∞ is essentially self-adjoint and its closure B satisfies

$$\sigma_{\mathrm{e}}(B) = \left[\frac{e^2}{R}, \infty\right). \tag{1.17}$$

§ 2. Two particle systems

Consider a nucleus fixed at the origin in E^3 with a positive charge $2e$ and two moving particles each with negative charge $-e$ subject to Coulomb potentials. The corresponding energy operator is

$$H = \frac{h^2}{8\pi^2 m_1}(\mathrm{D}_1^2+\mathrm{D}_2^2+\mathrm{D}_3^2) + \frac{h^2}{8\pi^2 m_2}(\mathrm{D}_4^2+\mathrm{D}_5^2+\mathrm{D}_6^2) - \frac{2e^2}{|x^{(1)}|} - \frac{2e^2}{|x^{(2)}|} + \frac{e^2}{|x^{(1)}-x^{(2)}|} \tag{2.1}$$

where $x^{(1)}=(x_1, x_2, x_3)$ represents the coordinates of the first particle and $x^{(2)}=(x_4, x_5, x_6)$ those of the second. The masses of the particles are m_1 and m_2, respectively. Here H is an operator on $C_0^\infty(E^6)$ with E^6 represented by $x=(x^{(1)}, x^{(2)})=(x_1, \dots, x_6)$.

Theorem 2.1. *The operator H given by* (2.1) *is essentially self-adjoint on* C_0^∞ *and its closure* H_0 *satisfies*

$$\sigma_e(H_0) = \left[-\frac{8m\pi^2 e^4}{h^2}, \infty\right), \tag{2.2}$$

where $m=\max(m_1, m_2)$.

Proof. Set

$$H_1 = \frac{h^2}{8\pi^2 m_1}(D_1^2+D_2^2+D_3^2) - \frac{2e^2}{|x^{(1)}|} \tag{2.3}$$

$$H_2 = \frac{h^2}{8\pi^2 m_2}(D_4^2+D_5^2+D_6^2) - \frac{2e^2}{|x^{(2)}|}. \tag{2.4}$$

According to (1.10) the lowest eigenvalue of H_k on $C_0^\infty(E^3)$ is

$$-8m_k\pi^2 e^4/h^2, \qquad k=1, 2. \tag{2.5}$$

We now apply Lemma 8.1 and Theorem 8.2 of ch. 9. We take $T_i=H_i$, $i=1, 2$ and

$$U_i = V = \frac{e^2}{|x^{(1)}-x^{(2)}|}, \qquad i=1, 2. \tag{2.6}$$

Note that $V_i=Q_i=S_i=0$ and

$$q_i(x^{(i)}) = -\frac{2e^2}{|x^{(i)}|}, \qquad i=1, 2. \tag{2.7}$$

In order to apply the theorem we need

$$\int_{|y^{(i)}|<1} |q_i(x^{(i)}-y^{(i)})|\ |y^{(i)}|^{-1}\,dy^{(i)} \to 0 \text{ as } |x^{(i)}|\to\infty, \tag{2.8}$$

which is clearly satisfied. By (2.5)

$$\varkappa = \min_{i=1,2}\ \min_{\lambda\in\sigma(H_i)} \lambda = -8m\pi^2 e^4/h^2. \tag{2.9}$$

Hence we may conclude that

$$\sigma_e(H_0) \subset [\varkappa, \infty). \tag{2.10}$$

In order to obtain the other conclusion, we apply Theorem 8.3 of ch. 9. Let $\{\xi_{(m)}\}$ be any sequence of vectors in E^3 such that $|\xi_{(m)}|\to\infty$. Then

$$\int_{|x-y|<1} [q_i(x^{(i)}+\xi_{(m)})]^2\,\mathrm{d}x \to 0 \text{ as } m\to\infty \tag{2.11}$$

for each $y\in E^6$. This is all that is needed to conclude that

$$\sigma_e(H_0) \supset [\varkappa, \infty)\,. \tag{2.12}$$

The proof is complete.

Let us now consider a slightly different situation. Suppose that the second particle is free, i.e., has no interaction with either the nucleus or the first particle. The Hamiltonian in this case is

$$H = \frac{h^2}{8\pi^2 m_1}(\mathrm{D}_1^2+\mathrm{D}_2^2+\mathrm{D}_3^2) + \frac{h^2}{8\pi^2 m_2}(\mathrm{D}_4^2+\mathrm{D}_5^2+\mathrm{D}_6^2) - \frac{2e^2}{|x^{(1)}|}\,. \tag{2.13}$$

Theorem 2.2. *The operator H given by* (2.13) *is essentially self-adjoint on C_0^∞ and its closure H_0 satisfies*

$$\sigma(H_0) = [-8m_1\pi^2 e^4/h^2, \infty)\,. \tag{2.14}$$

Proof. Set

$$P(\mathrm{D}) = \frac{h^2}{8\pi^2 m_2}(\mathrm{D}_4^2+\mathrm{D}_5^2+\mathrm{D}_6^2)\,.$$

The spectrum of its closure P_0 satisfies (1.8). Since the lowest eigenvalue of H_1 given by (2.3) is $\varkappa_1 = -8m_1\pi^2 e^4/h^2$, we know that

$$\sigma(H_0) \supset [\varkappa_1, \infty)$$

by Lemma 7.1 of ch. 9. On the other hand for $\varphi\in C_0^\infty(E^6)$

$$\begin{aligned}(H\varphi,\varphi) &= (H_1\varphi, \varphi) \geqslant (P(\mathrm{D})\varphi, \varphi)\\ &= (H_1\varphi, \varphi) + \frac{h^2}{8\pi^2 m_2}(\|\mathrm{D}_4\varphi\|^2+\|\mathrm{D}_5\varphi\|^2+\|\mathrm{D}_6\varphi\|^2)\\ &\geqslant \varkappa_1\|\varphi\|^2\end{aligned}$$

(see Lemma 7.17 of ch. 1). This gives

$$\sigma(H_0) \subset [\varkappa_1, \infty)\,,$$

and the proof is complete.

As a contrast, consider the case of two moving particles, the first with positive charge e and the second with negative charge $-e$, attracting each other with Coulomb forces. The Hamiltonian is

$$H = \frac{h^2}{8\pi^2 m_1}(\mathrm{D}_1^2+\mathrm{D}_2^2+\mathrm{D}_3^2) + \frac{h^2}{8\pi^2 m_2}(\mathrm{D}_4^2+\mathrm{D}_5^2+\mathrm{D}_6^2) - \frac{e^2}{|x^{(1)}-x^{(2)}|}. \tag{2.15}$$

Theorem 2.3. *The operator H given by (2.15) is essentially self-adjoint on C_0^∞ and its closure H_0 satisfies*

$$\sigma(H_0) = \left[-\frac{2m_1 m_2 \pi^2 e^4}{(m_1+m_2)h^2}, \infty\right). \tag{2.16}$$

Proof. Let z be a non-vanishing vector in E^3 and set $\xi=(z, z)$. Since $x^{(1)}+tz-(x^{(2)}+tz)=x^{(1)}-x^{(2)}$, we see that

$$q(x+t\xi) = q(x), \qquad x \in E^6,\ t \text{ real},$$

where

$$q(x) = \frac{-e^2}{|x^{(1)}-x^{(2)}|}.$$

Thus we may apply Theorem 7.3 of ch. 9 to conclude that $\sigma(H_0)$ consists of an interval of the form $[\varkappa, \infty)$. It remains to compute $\varkappa$. To this end make the following coordinate transformation:

$$\tilde{x}^{(1)} = x^{(1)}-x^{(2)}, \quad \tilde{x}^{(2)} = m_1 x^{(1)}+m_2 x^{(2)}. \tag{2.17}$$

A simple calculation shows that H transforms into

$$H = \frac{h^2}{8\pi^2 m}(\tilde{\mathrm{D}}_1^2+\tilde{\mathrm{D}}_2^2+\tilde{\mathrm{D}}_3^2)+(m_1+m_2)\frac{h^2}{8\pi^2}(\mathrm{D}_4^2+\mathrm{D}_5^2+\mathrm{D}_6^2) - \frac{e^2}{|\tilde{x}^{(1)}|} \tag{2.18}$$

in the new coordinates, where $m=m_1 m_2/(m_1+m_2)$. This is of the form (2.13) with m_1 replaced by m, m_2 replaced by $1/(m_1+m_2)$ and e replaced by $e/\sqrt{2}$. Thus our result follows from Theorem 2.2.

§ 3. The existence of bound states

If H is the Hamiltonian corresponding to a system of particles and its closure H_0 satisfies

$$\sigma_e(H_0) = [v, \infty), \tag{3.1}$$

this interval is said to represent *free states* because they correspond to the situation when at least one of the particles is free, i.e., it has the energy to move far from the others. Any $\lambda \in \sigma(H_0)$ below v must be an eigenvalue of finite multiplicity. The corresponding eigenfunctions are called *bound* or *stationary states.* The eigenfunction corresponding to the lowest eigenvalue is called the *ground state.* Equation (3.1) does not tell us whether or not there exist bound states. Since the existence of bound states (and consequently of a ground state) is important in quantum theory (see §6), we shall give a simple method of demonstrating their existence and estimating the corresponding eigenvalues.

The theory is very simple. Suppose we can find a function u in $D(H_0)$ such that

$$(H_0 u, u) = \mu \|u\|^2, \tag{3.2}$$

where $\mu < v$. Then we know that the spectrum of H_0 extends below v at least down to μ (Theorem 7.17 of ch. 1). Since every point of $\sigma(H_0)$ below v must be an eigenvalue of finite multiplicity, the existence of bound states with energies $\leqslant \mu$ is established. In particular, there is a ground state with energy $\leqslant \mu$.

Let us illustrate the method for the two particles system described at the beginning of §2. We write

$$H = H_1 + H_2 + W, \tag{3.3}$$

where

$$H_1 = \frac{h^2}{8\pi^2 m_1}(D_1^2 + D_2^2 + D_3^2) - \frac{2e^2}{|x^{(1)}|} \tag{3.4}$$

$$H_2 = \frac{h^2}{8\pi^2 m_2}(D_4^2 + D_5^2 + D_6^2) - \frac{e^2}{|x^{(2)}|} \tag{3.5}$$

and

$$W = \frac{e^2}{|x^{(1)} - x^{(2)}|} - \frac{e^2}{|x^{(2)}|}. \tag{3.6}$$

From the discussion of §1 we know that the operators H_1 and H_2 in E^3 have ground states with energies

$$\lambda_1 = -8m_1 \pi^2 e^4/h^2, \quad \lambda_2 = -2m_2 \pi^2 e^4/h_2, \tag{3.7}$$

respectively. We note also that the corresponding normalized eigenfunctions (states) are

$$\varphi_1(x^{(1)}) = \frac{2^{\frac{9}{2}}\pi^{\frac{5}{2}}m_1^{\frac{3}{2}}e^3}{h^3}\exp\left\{-\frac{8\pi^2 m_1 e^2}{h^2}|x^{(1)}|\right\} \tag{3.8}$$

and

$$\varphi_2(x^{(2)}) = \frac{8\pi^{\frac{5}{2}}m_2^{\frac{3}{2}}e^3}{h^3}\exp\left\{-\frac{4\pi^2 m_2 e^2}{h^2}|x^{(2)}|\right\}, \tag{3.9}$$

respectively. Actually, all we need to know is that $\varphi_1(x^{(1)})$ depends only on $|x^{(1)}|$. Set

$$\varphi(x) = \varphi_1(x^{(1)})\varphi_2(x^{(2)}). \tag{3.10}$$

Then $\varphi(x)$ is in S and consequently in the domain of H_0. Furthermore

$$(H_1\varphi, \varphi) = (H_1\varphi_1, \varphi_1)_{E^3}\|\varphi_2\|_{E^3}^2 = \lambda_1 \tag{3.11}$$

and

$$(H_2\varphi, \varphi) = (H_2\varphi_2, \varphi_2)_{E^3} = \lambda_2\,. \tag{3.12}$$

We also have

$$(W\varphi, \varphi) = e^2\int|\varphi_2(x^{(2)})|^2\,\mathrm{d}x^{(2)}\int|\varphi_1(x^{(1)})|^2\left\{\frac{1}{|x^{(1)}-x^{(2)}|} - \frac{1}{|x^{(1)}|}\right\}\mathrm{d}x^{(1)}. \tag{3.13}$$

Let R be any positive number, and set

$$f(x^{(2)}) = \int_{|x^{(1)}|=R}\frac{\mathrm{d}S}{|x^{(1)}-x^{(2)}|}. \tag{3.14}$$

If $|x^{(2)}| > R$, then $1/|x^{(1)}-x^{(2)}|$ is harmonic in $|x^{(1)}| \leqslant R$ and consequently

$$f(x^{(2)}) = \frac{1}{|x^{(2)}|}\int_{|x^{(1)}|=R}\mathrm{d}S = \frac{4\pi R^2}{|x^{(2)}|}, \qquad |x^{(2)}| > R, \tag{3.15}$$

by the mean value theorem. If $|x^{(2)}| < R$, set

$$h(x^{(1)}) = \frac{1}{R} - \frac{1}{|x^{(1)}|}.$$

Since h is harmonic in $\varepsilon < |x^{(1)}| < R$ for $0 < \varepsilon < |x^{(2)}|$, we have

$$\begin{aligned} 4\pi h(x^{(2)}) = &\int_{|x^{(1)}|=\varepsilon}\frac{\partial h}{\partial|x^{(1)}|}\frac{\mathrm{d}S}{|x^{(1)}-x^{(2)}|} \\ &- \int_{|x^{(1)}|=\varepsilon}\frac{\partial h}{\partial|x^{(1)}|}\frac{\mathrm{d}S}{|x^{(1)}-x^{(2)}|} \\ &+ \left(\frac{1}{R} - \frac{1}{\varepsilon}\right)\int_{|x^{(1)}|=\varepsilon}\frac{\partial}{\partial|x^{(1)}|}\left(\frac{1}{|x^{(1)}-x^{(2)}|}\right)\mathrm{d}S, \end{aligned} \tag{3.16}$$

where we used the fact that $h(x^{(1)})=0$ for $|x^{(1)}|=R$. Since $1/|x^{(1)}-x^{(2)}|$ is harmonic in $|x^{(1)}|<\varepsilon$, we have

$$\int_{|x^{(1)}|=\varepsilon} \frac{\partial}{\partial |x^{(1)}|}\left(\frac{1}{|x^{(1)}-x^{(2)}|}\right) \mathrm{d}S = 0 .$$

Thus (3.16) yields

$$4\pi\left(\frac{1}{R} - \frac{1}{|x^{(2)}|}\right) = \frac{1}{R^2}\int_{|x^{(1)}|=R} \frac{\mathrm{d}S}{|x^{(1)}-x^{(2)}|} - \frac{4\pi}{|x^{(2)}|}$$

by (3.15). This gives

$$f(x^{(2)}) = 4\pi R , \qquad |x^{(2)}| < R . \tag{3.17}$$

On the other hand, we always have

$$\int_{|x^{(1)}|=R} \left\{\frac{1}{|x^{(1)}-x^{(2)}|} - \frac{1}{|x^{(1)}|}\right\} \mathrm{d}S \leqslant 0 . \tag{3.18}$$

We now use the fact that $\varphi_1(x^{(1)})$ depends only on $|x^{(1)}|$. Thus if we evaluate the integral in (3.13) by first integrating over $|x^{(1)}|=R$ and apply (3.18), we see that $(W\varphi, \varphi)\leqslant 0$. This in conjunction with (3.11) and (3.12) shows that

$$(H\varphi, \varphi) \leqslant \lambda_1 + \lambda_2 .$$

By Theorem 2.1, the smallest value of $\sigma_e(H_0)$ is λ_1 (note that $m_1 = \max(m_1, m_2)$). Since $\lambda_2 < 0$, we see that there must be bound states. In particular we have shown the existence of bound states for the helium atom.

§ 4. Systems of N particles

Consider N particles interacting with each other through static potentials and subjected to external electrostatic and magnetic fields. The energy operator for such a system is of the form

$$H = \sum_{k=1}^{N} \left\{\frac{h^2}{8\pi^2 m_k} \sum_{j=1}^{3} [\mathrm{D}_{3k-3+j} + b_j(x^{(k)})]^2 + q_k(x^{(k)})\right\} + \sum_{k=1}^{N}\sum_{j=1}^{k-1} V_{jk}(x^{(j)} - x^{(k)}) , \tag{4.1}$$

where $x^{(k)} = (x_{3k-2}, x_{3k-1}, x_{3k})$, $x = (x_1, \ldots, x_{3N}) = (x^{(1)}, \ldots, x^{(N)})$ and m_k is the mass of the kth particle. Here (b_1, b_2, b_3) is the external magnetic vector potential describing the magnetic field, the q_k are the external electrostatic potentials and the V_{jk} represent the interactions between the particles. Note that (4.1) is a special case of (1.1) of ch. 9. For $z \in E^3$ set

$$\rho_k(z)^2 = \max[q_k(z), 0], \ \sigma_k(z)^2 = \rho_k(z)^2 - q_k(z), \qquad 1 \leqslant k \leqslant N. \tag{4.2}$$

and

$$b(z) = b_1(z)^2 + b_2(z)^2 + b_3(z)^2, \quad e(z) = \mathrm{D}_1 b_1(z) + \mathrm{D}_2 b_2(z) + \mathrm{D}_3 b_3(z). \tag{4.3}$$

Theorem 4.1. *Assume that each* $\sigma_k \in N_2(E^3)$ *with*

$$N_{2,\delta}(\sigma_k, E^3) \to 0 \ \text{as} \ \delta \to 0, \qquad 1 \leqslant k \leqslant N, \tag{4.4}$$

$$N_{2,z}(\sigma_k, E^3) \to 0 \ \text{as} \ |z| \to \infty, \qquad 1 \leqslant k \leqslant N, \tag{4.5}$$

and that there is an $\alpha < 4$ *such that* b, e *and each* q_k, V_{jk} *are in* $N_\alpha^{\text{loc}}(E^3)$. *Assume further that*

$$V_{jk}(z) \geqslant 0, \quad z \in E^3, \qquad 1 \leqslant j < k \leqslant N. \tag{4.6}$$

Set

$$H_n = \sum_{k \neq n} \frac{h^2}{8\pi^2 m_k} \sum_{j=1}^{3} [\mathrm{D}_{3k-3+j} + b_j(x^{(k)})]^2 + q_k(x^{(k)}) + \sum_{j,k \neq n} V_{jk}(x^{(j)} - x^{(k)}), \qquad 1 \leqslant n \leqslant N, \tag{4.7}$$

and

$$\varkappa = \min_{1 \leqslant n \leqslant N} \ \min_{\lambda \in \sigma(H_{n0})} \lambda, \tag{4.8}$$

where H_{n0} *denotes the closure in* $L^2(E^{3N-3})$ *of* H_n *on* $C_0^\infty(E^{3N-3})$. *Then* H *on* $C_0^\infty(E^{3N})$ *is essentially self-adjoint in* $L^2(E^{3N})$ *and its closure* H_0 *satisfies*

$$\sigma_e(H_0) \subset [\varkappa, \infty). \tag{4.9}$$

If there is a vector $\eta \in E^3$ *and a sequence* $\{\xi_{(m)}\}$ *of vectors in* E^3 *such that*

$$\int_{|y-z|<1} [b_j(z + \xi_{(m)}) - \eta_j]^4 \mathrm{d}z \to 0 \ \text{as} \ m \to \infty, \qquad 1 \leqslant j \leqslant 3 \tag{4.10}$$

$$\int_{|y-z|<1} e(z + \xi_{(m)})^2 \mathrm{d}z \to 0 \ \text{as} \ m \to \infty \tag{4.11}$$

$$\int_{|y-z|<1} q_k(z + \xi_{(m)})^2 \mathrm{d}z \to 0 \ \text{as} \ m \to \infty, \qquad 1 \leqslant k \leqslant N \tag{4.12}$$

$$\int_{|y-z|<1} V_{jk}(z + \xi_{(m)})^2 \mathrm{d}z \to 0 \ \text{as} \ m \to \infty, \qquad 1 \leqslant j < k \leqslant N \tag{4.13}$$

for each $y \in E^3$, *then*

$$\sigma_e(H_0)=[\varkappa,\infty)\,. \tag{4.14}$$

Proof. We apply Theorems 8.2 and 8.3 of ch. 9. We take

$$V(x)=\sum_{j,k} V_{jk}(x^{(j)}-x^{(k)})$$

$$V_n(x)=\sum_{j,k\neq n} V_{jk}(x^{(j)}-x^{(k)})$$

$$U_n(x)=V(x)-V_n(x)$$

and note that $Q_n=S_n=0$ for each n. It is easily checked that the hypotheses of the theorems are satisfied.

In the special case when the electrostatic field is due to a charge e at the origin and all the interactions are governed by Coulomb potentials, the Hamiltonian is

$$H=\sum_{k=1}^{N}\left\{\frac{h^2}{8\pi^2 m_k}\sum_{j=1}^{3}[D_{3k-3+j}+b_j(x^{(k)})]^2+\frac{e\,e_k}{|x^{(k)}|}\right\}+\sum_{k=1}^{N}\sum_{j=1}^{k-1}\frac{e_j e_k}{|x^{(j)}-x^{(k)}|}, \tag{4.15}$$

where e_k is the (signed) charge of the kth particle.

Corollary 4.2. *Suppose that there are vectors η, $\{\xi_{(m)}\}$ in E^3 such that* (4.10) *and* (4.11) *hold for each $y\in E^3$. If all of the charges e_k are of the same sign, then the Hamiltonian operator H given by* (4.15) *is essentially self-adjoint on C_0^∞ and its closure H_0 satisfies* (4.14), *where $\varkappa$ is given by* (4.8).

Proof. All of the hypotheses of Theorem 4.1 are satisfied.

We now want to consider the situation when (4.6) does not hold. For this put

$$\sigma_{jk}(z)^2=-\min\,[V_{jk}(z),0]\,,\qquad 1\leqslant j<k\leqslant N\,. \tag{4.16}$$

We have

Theorem 4.3. *Suppose that $b(z)$, $e(z)$ and each $q_k(z)$, $V_{jk}(z)$ are in $N_\alpha^{\text{loc}}(E^3)$ for some $\alpha<4$, and that each $\sigma_k(z)$, $\sigma_{jk}(z)$ are in $N_2(E^3)$ with*

$$N_{2,\delta}(\sigma_{jk},E^3)\to 0 \text{ as } \delta\to 0\,,\qquad 1\leqslant j<k\leqslant N \tag{4.17}$$

$$N_{2,z}(\sigma_{jk},E^3)\to 0 \text{ as } |z|\to\infty\,,\qquad 1\leqslant j<k\leqslant N \tag{4.18}$$

holding in addition to (4.4) *and* (4.5). *For each proper subset C of $\{1,\ldots,N\}$ put*

$$V_C(x) = \sum_{j,k \in C} V_{jk}(x^{(j)} - x^{(k)}) + \sum_{j,k \notin C} V_{jk}(x^{(j)} - x^{(k)}), \tag{4.19}$$

$$H_C = \sum_{k=1}^{N} \frac{h^2}{8\pi^2 m_k} \sum_{j=1}^{3} [\mathrm{D}_{3k-3+j} + b_j(x^{(k)})]^2 + \sum_{k \in C} q_k(x^{(k)}) + V_C(x) \tag{4.20}$$

$$\varkappa_0 = \min_{C} \min_{\lambda \in \sigma(H_{C_0})} \lambda, \tag{4.21}$$

where the minimum is taken over all proper subsets C *of* $\{1, \ldots, N\}$ *and* H_{C0} *is the closure of* H_0 *in* $L^2(E^{3N})$*. Then the closure* H_0 *of* H *satisfies*

$$\sigma_e(H_0) \subset [\varkappa_0, \infty). \tag{4.22}$$

If the b_j *are constants and there is a sequence* $\{\xi_{(m)}\}$ *of vectors in* E^3 *such that* (4.12) *and* (4.13) *hold for each* $y \in E^3$*, then*

$$\sigma_e(H_0) = [\varkappa_0, \infty). \tag{4.23}$$

Proof. We apply directly Theorems 9.2 and 9.3 of ch. 9.

§ 5. The Zeeman effect

Suppose a particle of charge $-e$ is subject to a Coulomb potential $-e^2/|x|$ and a uniform magnetic field of strength b along the x_3-axis. The corresponding energy operator is

$$H = \frac{h^2}{8\pi^2 m} [(\mathrm{D}_1 + \beta x_2)^2 + (\mathrm{D}_2 - \beta x_1)^2 + \mathrm{D}_3^2] - e^2/|x|, \tag{5.1}$$

where $\beta = \pm 2\pi eb/hc$ (c is the speed of light and the sign of β depends on the direction of the magnetic field). If we set

$$H_1 = \frac{h^2}{8\pi^2 m} [(\mathrm{D}_1 + \beta x_2)^2 + (\mathrm{D}_2 - \beta x_1)^2 + \mathrm{D}_3^2], \tag{5.2}$$

then H and H_1 are both essentially self-adjoint on C_0^∞ (Theorem 2.1 of ch. 9). Moreover by Theorem 4.2 of ch. 9

$$\sigma_e(H_0) = \sigma_e(H_{10}), \tag{5.3}$$

where, as usual, H_0 and H_{10} denote the closures in L^2 of H and H_1 on C_0^∞, respectively. We write

$$S_1 = \frac{h^2}{8\pi^2 m} [(D_1 + \beta x_2)^2 + (D_2 - \beta x_1)^2] \tag{5.4}$$

$$S_2 = \frac{h^2}{8\pi^2 m} D_3^2 . \tag{5.5}$$

If we consider S_1 defined on $C_0^\infty(E^2)$ and S_2 defined on $C_0^\infty(E^1)$, then they are essentially self adjoint operators on $L^2(E^2)$ and $L^2(E^1)$, respectively. If S_{10} and S_{20} denote their closures, then $\sigma(S_{10})$ is clearly bounded from below and

$$\sigma(S_{20}) = [0, \infty) \tag{5.6}$$

(Lemma 4.5 of ch. 9). If

$$\lambda_0 = \min_{\lambda \in \sigma(S_{10})} \lambda , \tag{5.7}$$

then every $\lambda \geqslant \lambda_0$ is in $\sigma(H_{10})$ (Lemma 7.1 of ch. 9). Since

$$(H_1 \varphi, \varphi) \geqslant (S_1 \varphi, \varphi) \geqslant \lambda_0 \|\varphi\|^2 , \qquad \varphi \in C_0^\infty(E^3) ,$$

$\sigma(H_{10})$ cannot contain any $\lambda < \lambda_0$. Thus

$$\sigma(H_{10}) = [\lambda_0, \infty) , \tag{5.8}$$

and consequently

$$\sigma_e(H_0) = [\lambda_0, \infty) . \tag{5.9}$$

Thus in order to determine the essential spectrum of H_0 it suffices to compute λ_0 given by (5.7). It is to this that we now turn our attention.

If we introduce polar coordinates

$$x_1 = r \cos\theta , \qquad x_2 = r \sin\theta , \tag{5.10}$$

then (5.4) gives

$$S_1 u = - \frac{h^2}{8\pi^2 m} \left[\frac{1}{r} (r u_r)_r + \frac{1}{r^2} u_{\theta\theta} - 2\beta i u_\theta - \beta^2 r^2 u \right] . \tag{5.11}$$

Suppose $u(r, \theta)$ is of the form

$$u(r, \theta) = \sum_{k=-\infty}^{\infty} e^{ik\theta} \varphi_k(r) , \tag{5.12}$$

where the $\varphi_k(r)$ are infinitely differentiable in $[0, \infty)$ and vanish for r large. Then

$$S_1 u = -\frac{h^2}{8\pi^2 m} \sum_{k=-\infty}^{\infty} e^{ik\theta} \varphi_k(r) . \tag{5.13}$$

Consequently

$$(S_1 u, u) = \frac{h^2}{4\pi m} \sum_{k=-\infty}^{\infty} \int_0^\infty \left\{ -\frac{1}{r}(r\varphi_k')' + \left[\frac{k^2}{r^2} + \beta^2 r^2 - 2\beta k\right] \varphi_k \right\} \bar{\varphi}_k r \, dr$$

$$= \frac{h^2}{4\pi m} \sum_{k=-\infty}^{\infty} \int_0^\infty \left\{ |\varphi_k'|^2 + \left[\frac{k^2}{r^2} + \beta^2 r^2 - 2\beta k\right] |\varphi_k|^2 \right\} r \, dr . \tag{5.14}$$

Suppose λ_1 is such that

$$\int_0^\infty \left\{ |\varphi'|^2 + \left[\frac{k^2}{r^2} + \beta^2 r^2 - 2\beta k\right] |\varphi|^2 \right\} r \, dr \geqslant \lambda_1 \int_0^\infty |\varphi|^2 r \, dr \tag{5.15}$$

for all $\varphi(r)$ infinitely differentiable in $[0, \infty)$ and vanishing for large r and all integers k. Since

$$\|u\|^2 = 2\pi \sum_{k=-\infty}^{\infty} \int_0^\infty |\varphi_k|^2 \, r \, dr$$

for u given by (5.12), equation (5.14) implies

$$(S_1 u, u) \geqslant \frac{h^2}{8\pi^2 m} \lambda_1 \|u\|^2 \tag{5.16}$$

for all such u. Since every function in C_0^∞ can be put in this form and S_1 is essentially self-adjoint, we see that $\lambda_0 \geqslant \lambda_1 h^2/8\pi^2 m$.

In order to get an estimate for λ_1 make the transformation $\tau = |\beta| r^2$, and consider $w = r\varphi$ as a function of τ. Then

$$w_r = \varphi + r\varphi_r = 2|\beta| r w_\tau .$$

Thus

$$|\varphi_r|^2 = 4\beta |w_\tau|^2 - \frac{4|\beta|}{r} \operatorname{Re} w_\tau \bar{\varphi} + |\varphi|^2/r^2$$

$$= 4\beta^2 |w_\tau|^2 - \frac{4\beta^2}{\tau} \operatorname{Re} w_\tau \bar{w} + \beta^2 |w|^2/\tau^2 .$$

Now

$$\int_0^\infty (w_\tau \bar{w}/\tau) d\tau = -\int_0^\infty w(\bar{w}_\tau/\tau) dr + \int_0^\infty (|w|^2/\tau^2) d\tau .$$

Hence

$$2 \operatorname{Re} \int_0^\infty (w_\tau \bar{w}/\tau) d\tau = \int_0^\infty (|w|^2/\tau^2) d\tau .$$

Consequently

$$\int_0^\infty |\varphi_r|^2 r\,\mathrm{d}r = \frac{1}{2|\beta|}\int_0^\infty |\varphi_r|^2\,\mathrm{d}\tau$$

$$= 2|\beta|\int_0^\infty |w_\tau|^2\,\mathrm{d}\tau - \tfrac{1}{2}|\beta|\int (|w|^2/\tau^2)\,\mathrm{d}\tau\,.$$

Thus

$$\int_0^\infty \left\{|\varphi_r|^2 + \left[\frac{k^2}{r^2} + \beta^2 r^2 - 2\beta k\right]|\varphi|^2\right\} r\mathrm{d}r$$

$$= 2|\beta|\int_0^\infty |w_\tau|^2\,\mathrm{d}\tau + \tfrac{1}{2}|\beta|\int_0^\infty \left[\frac{k^2-1}{\tau^2} - \frac{2k}{\tau} + 1\right]|w|^2\,\mathrm{d}\tau$$

and

$$\int_0^\infty |\varphi|^2 r\,\mathrm{d}r = \tfrac{1}{2}\int_0^\infty (|w|^2/\tau)\,\mathrm{d}\tau\,.$$

Hence (5.15) is equivalent to

$$\int_0^\infty \left\{|w_\tau|^2 + \tfrac{1}{4}\left[\frac{k^2-1}{\tau^2} - \frac{2k}{\tau} + 1 - \frac{\lambda_1}{\beta\tau}\right]|w|^2\right\}\mathrm{d}\tau \geqslant 0\,. \tag{5.17}$$

Consider the ordinary differential operator

$$Tu = -\tau u_{\tau\tau} + \tfrac{1}{4}\tau\left[\frac{k^2-1}{\tau^2} - \frac{2k}{\tau} + 1\right]u \tag{5.18}$$

defined on $C_0^\infty(E^1)$. If we consider T as an operator on the space $L^2(0, \infty; 1/\tau)$ with norm given by

$$\|u\|^2 = \int_0^\infty |u(\tau)|^2\,\mathrm{d}\tau/\tau\,,$$

then it is known that T has a self-adjoint extension with discrete spectrum. If there is a smallest eigenvalue λ_2 of T, then we have

$$\int_0^\infty (Tu)\bar{u}\,\mathrm{d}\tau/\tau \geqslant \lambda_2\int_0^\infty |u|^2\,\mathrm{d}\tau/\tau$$

(Lemma 7.17 of ch. 1). This will give

$$\int_0^\infty \left\{|u_\tau|^2 + \tfrac{1}{4}\left[\frac{k^2-1}{\tau^2} - \frac{2k}{\tau} + 1\right]|u|^2\right\}\mathrm{d}\tau \geqslant \lambda_2\int_0^\infty |u|^2\,\mathrm{d}\tau/\tau\,. \tag{5.19}$$

This shows that we may take $\lambda_1 = |\beta|\lambda_2$. Now the solutions of $(T-\lambda)u=0$ are well known:

$$u(\tau) = AW_{\frac{1}{2}k+\frac{1}{4}\lambda,\frac{1}{2}k}(\tau) + BW_{-\frac{1}{4}\lambda-\frac{1}{2}k,\frac{1}{2}k}(-\tau)\,, \tag{5.20}$$

where $W_{h,m}(\tau)$ is the Whittaker function (confluent hypergeometric function) and A, B are arbitrary constants. However not all functions of the form (5.20) are in $L^2(0, \infty; 1/\tau)$. In fact, in order for $W_{h,m}(\tau)$ to be bounded near $\tau=0$ when $2m$ is an integer, it is necessary that $\Gamma(\pm m-h+\frac{1}{2})=\infty$. Hence in order for A not to vanish we need

$$\tfrac{1}{2}k-\tfrac{1}{4}\lambda-\tfrac{1}{2}k+\tfrac{1}{2} = -n_1$$

and

$$-\tfrac{1}{2}k-\tfrac{1}{2}k-\tfrac{1}{2}\lambda+\tfrac{1}{2} = n_2\,,$$

where n_1 and n_2 are non-negative integer. The first of these statements implies that $\lambda \geqslant 2$. Furthermore

$$W_{h,m}(-\tau) = e^{\frac{1}{2}\tau}(-\tau)^h[1+O(1/\tau)] \quad \text{as} \quad \tau \to \infty\,. \tag{5.21}$$

Hence we must always have $B=0$ in order that the function given by (5.20) be in $L^2(0, \infty; 1/\tau)$. Thus the smallest possible eigenvalue of T is 2. This shows that $\lambda_1 \geqslant 2|\beta|$, and consequently $\lambda_0 \geqslant |\beta|h^2/4\pi^2 m$.

We now show that $|\beta|h^2/4\pi^2 m$ is an eigenvalue of S_{10}. In fact set

$$g(r) = \exp\{-\tfrac{1}{2}|\beta|r^2\}\,. \tag{5.22}$$

Then

$$g' = -|\beta|\, rg$$

and

$$(rg')'/r = \beta^2 r^2 g - 2|\beta|\, g\,.$$

Thus

$$S_{10}\, g = -\frac{h^2}{8\pi^2 m}(-2|\beta|\, g) = \frac{h^2|\beta|}{4\pi^2 m}\, g\,.$$

This shows that

$$\lambda_0 = \frac{h^2|\beta|}{4\pi^2 m}\,. \tag{5.24}$$

Next we show the existence of a ground state for H_0. As explained in §3 it suffices to find a function $u \in D(H_0)$ such that

$$(H_0\, u, u) < \lambda_0 \|u\|^2\,. \tag{5.25}$$

To this end let u be given by

$$u(x) = g(r)\exp\{-\tfrac{1}{4}ax_3^2\}\,, \tag{5.26}$$

where a is a constant satisfying $0<a\leqslant 1$. Note that

$$\|u\|^2 = 2^{\frac{3}{2}}\pi C_1 C_2/|\beta|\, a^{\frac{1}{2}} ,$$

where

$$C_1 = \int_0^\infty t \exp\{-t^2\}\,\mathrm{d}t$$

and

$$C_2 = \int_{-\infty}^\infty \exp\{-u^2\}\,\mathrm{d}u = \pi^{\frac{1}{2}} .$$

Furthermore

$$S_{10}u = \lambda_0 u$$

and

$$S_{20}u = \frac{h^2}{8\pi^2 m}\left(\tfrac{1}{2}a - \tfrac{1}{4}a^2 x_3^2\right)u .$$

Hence

$$(H_{10}u, u) \leqslant \left(\lambda_0 + \frac{h^2 a}{16\pi^2 m}\right)\|u\|^2 . \tag{5.27}$$

On the other hand

$$\begin{aligned}
\|u/|x|\,\|^2 &= 2\pi \int_0^\infty r \exp\{-|\beta| r^2\} \int_{-\infty}^\infty \exp\{-\tfrac{1}{2}a x_3^2\}/(r^2+x_3^2)^{\frac{1}{2}}\,\mathrm{d}x_3\,\mathrm{d}r \\
&= 2\pi \int_0^\infty r \exp\{-|\beta| r^2\} \int_{-\infty}^\infty \exp\{-\tfrac{1}{2}a r^2 t^2\}/(1+t^2)^{\frac{1}{2}}\,\mathrm{d}t\,\mathrm{d}r \\
&= 2\pi \int_{-\infty}^\infty (1+t^2)^{-\frac{1}{2}} \int_0^\infty r \exp\{-(|\beta| + \tfrac{1}{2}a t^2) r^2\}\,\mathrm{d}r\,\mathrm{d}t \\
&= 4\pi C_1 \int_0^\infty \frac{\mathrm{d}t}{(1+t^2)^{\frac{1}{2}}(|\beta| + \tfrac{1}{2}a t^2)} \\
&\geqslant 4\pi C_1 \int_0^\infty \frac{\mathrm{d}t}{(1+t^2)^{\frac{1}{2}}(|\beta| + t^2)} .
\end{aligned}$$

Thus

$$(H_0 u, u) \leqslant \left(\lambda_0 + \frac{h^2 a}{16\pi^2 m} - e^2 C_3 a^{\frac{1}{2}}\right)\|u\|^2 , \tag{5.28}$$

where the constant C_3 depends only on $|\beta|$. If we now take

$$a^{\frac{1}{2}} < 16\pi^2 e^2 m C_3/h^2 ,$$

we see that (5.25) is satisfied. Thus the existence of a ground state is established. We summarize our results as follows.

Theorem 5.1. *If H is given by* (5.1), *then*

$$\sigma_e(H_0) = \left[\frac{ebh}{2\pi cm}, \infty\right), \tag{5.29}$$

and there exist bound states below this interval.

§ 6. Stability

Suppose we have a system of N particles governed by the operator (4.1). The spectrum of the operator H_0 consists of the possible values that the mean energy of the system can attain. If the hypotheses of Theorem 4.1 are satisfied, we know that the lowest point of $\sigma_e(H_0)$ is the smallest value that the mean energy can attain if we remove a particle from the system. If H_0 has bound states, it follows that the least mean value of the energy of the the system is less than the least mean value of the energy of any subsystem of $N-1$ particles. Thus if we remove a particle we are increasing the least mean energy value. Clearly, the only way this can happen is when energy is added to the system. Such a system is called *stable.* Thus we have

Theorem 6.1. *Under the hypotheses of Theorem* 4.1, *the system is stable if it has bound states.*

From these considerations we can see why it is important to know that bound states exist. In §§3, 5 we showed their existence in special cases. We now discuss some general criteria.

Consider the Hamiltonian operator H given by (4.1) with $b_1 = b_2 = b_3 = 0$. It is convenient to define $V_{ji}(z)$ for $i<j$ by

$$V_{ji}(z) = V_{ij}(-z)\,. \tag{6.1}$$

We assume that for each i, j and for each $w \in H^{1,2}(E^{3N-3})$ there are functions $h_i(z)$, $h_{ij}(z)$, $\psi_i(z)$ on E^3, a sequence $\{M_k\}$ of positive numbers and a constant $\lambda < 2$ such that $\psi_i \in C_0^\infty(E^3)$ and

$$\lim_{k\to\infty} M_k^\lambda \int V_{ij}(M_k x^{(i)} - x^{(j)})\,|w(\tilde{x}^{(i)})|^2\,|\psi_i(x^{(i)})|^2\,dx \leq \|w\|^2 \int h_{ij}(x^{(i)})\,|\psi_i(x^{(i)})|^2\,dx^{(i)}, \qquad 1 \leq i,\ j \leq N \tag{6.2}$$

(here $\tilde{x}^{(i)}$ represents the coordinate vector in E^{3N-3}),

$$\lim_{k\to\infty} M_k^{\lambda} \int q_i(M_k z)|\psi_i(z)|^2 \mathrm{d}z \leqslant \int h_i(z)|\psi_i(z)|^2 \mathrm{d}z\,, \qquad 1\leqslant i\leqslant N\,, \tag{6.3}$$

and

$$\int [h_i(z) + \sum_{j\neq i} h_{ij}(z)]\,|\psi_i(z)|^2 \mathrm{d}z < 0\,, \qquad 1\leqslant i\leqslant N\,. \tag{6.4}$$

We have

Theorem 6.2. *If the hypotheses of Theorem 4.1 are satisfied as well as the conditions stated above, then the system has bound states and consequently is stable.*

Proof. Set

$$L_i = \frac{h^2}{8\pi^2 m_i} \sum_{j=1}^{3} \mathrm{D}_{3i-3+j}^2 + q_i(x^{(i)})\,, \qquad 1\leqslant i\leqslant N\,. \tag{6.5}$$

Then the closure L_{i0} in $L^2(E^3)$ of L_i on $C_0^\infty(E^3)$ is self-adjoint (Theorem 2.1 of ch. 9). From (4.6) and (6.2) we see that

$$\int h_i(z)|\psi_i(z)|^2 \mathrm{d}z < 0\,, \qquad 1\leqslant i\leqslant N\,. \tag{6.6}$$

For each i set

$$\psi_{ik}(x^{(i)}) = \psi_i(x^{(i)}/M_k)/M_k^{\frac{3}{2}}\,, \qquad k=1,2,\ldots\,. \tag{6.7}$$

Then

$$(L_i\psi_{ik}, \psi_{ik}) = \frac{h^2}{8\pi^2 m_i M_k^2} \sum_{j=1}^{3} \|\mathrm{D}_{3i-3+j}\psi_i\|^2$$
$$+ \int q_i(M_k z)|\psi_i(z)|^2 \mathrm{d}z\,, \qquad 1\leqslant i\leqslant N,\ k=1,2,\ldots\,. \tag{6.8}$$

Let $\varepsilon > 0$ be given. By (6.3) we can take k so large that

$$M_k^{\lambda} \int q_i(M_k z)|\psi_i(z)|^2 \mathrm{d}z < \int h_i(z)\,|\psi_i(z)|^2 \mathrm{d}z + \varepsilon\,, \tag{6.9}$$

and if necessary we can take k even larger so that

$$\frac{h^2}{8\pi^2 m_i M_k^{2-\lambda}} \sum_{j=1}^{3} \|\mathrm{D}_{3i-3+j}\psi_i\|^2 < \varepsilon\,. \tag{6.10}$$

Combining (6.8)–(6.10) we get

$$(L_i\psi_{ik}, \psi_{ik}) < \frac{1}{M_k^{\lambda}} \left[\int h_i(z)|\psi_i(z)|^2 \mathrm{d}z + 2\varepsilon\right] < 0 \tag{6.11}$$

if ε is taken sufficiently small and k sufficiently large. Since

$$\sigma_e(L_{i0}) \subset [0, \infty), \qquad 1 \leqslant i \leqslant N \tag{6.12}$$

(Corollary 4.7 of ch. 9), we see that each L_{i0} has bound states (see §3).

We prove the theorem by induction. We have just shown that it is true for $N=1$. We now assume that it is true for $N-1$ and prove it for N. Let i be such that $\varkappa = \varkappa_i$ (we employ the notation of Theorem 4.1). By induction hypothesis, H_{i0} has bound states. Thus $\varkappa$ is an eigenvalue of H_{i0}. Let w be a corresponding normalized eigenfunction. Now

$$H = H_i + L_i + U_i \tag{6.13}$$

where

$$U_i = \sum_{j \neq i} V_{ij}(x^{(i)} - x^{(j)}). \tag{6.14}$$

Thus for ψ_{ik} given by (6.7) we have

$$(H[w\psi_{ik}], w\psi_{ik}) = (H_i w, w) + (L_1 \psi_{ik}, \psi_{ik}) + (U_i w\psi_{ik}, w\psi_{ik}). \tag{6.15}$$

Now $(H_i w, w) = \varkappa$. Moreover

$$(U_i w\psi_{ik}, w\psi_{ik}) = \sum_{j \neq i} \int V_{ij}(M_k x^{(i)} - x^{(j)}) |w(\tilde{x}^{(i)})|^2 |\psi_i(x^{(i)})|^2 \,dx. \tag{6.16}$$

By (6.4) there is an $\varepsilon > 0$ such that

$$\int [h_i(z) + \sum_{j \neq i} h_{ij}(z)] |\psi_i(z)|^2 \,dz + 2\varepsilon < 0. \tag{6.17}$$

By (6.2) and (6.16) we can take k so large that

$$M_k^{\lambda}(U_i w\psi_{ik}, w\psi_{ik}) < \sum_{j \neq i} \int h_{ij}(z) |\psi_i(z)|^2 \,dz + \varepsilon.$$

Furthermore, by (6.11) we can take k so large that

$$M_k^{\lambda}(L_i \psi_{ik}, \psi_{ik}) < \int h_i(z) |\psi_i(z)|^2 \,dz + \varepsilon.$$

Combining these inequalities with (6.15), we get

$$(H[w\psi_{ik}], w\psi_{ik}) < \varkappa + \frac{1}{M_k^{\lambda}} \left\{ \int [h_i(z) + \sum_{j \neq i} h_{ij}(z)] |\psi_i(z)|^2 \,dz + 2\varepsilon \right\} < \varkappa \tag{6.18}$$

by (6.17). Since $\sigma_e(H_0)$ is contained in the interval $[\varkappa, \infty)$, we see that H_0 must have bound states. Thus the theorem is true for N, and the proof is complete.

Let us now see how our criteria (6.2)–(6.4) are satisfied for the operator (4.15) with $b_1 = b_2 = b_3 = 0$. Here we have

$$q_i(x^{(i)}) = e\,e_i/|x^{(i)}|, \qquad 1 \leqslant i \leqslant N. \tag{6.19}$$

$$V_{ij}(x^{(i)} - x^{(j)}) = e_i e_j/|x^{(i)} - x^{(j)}|, \qquad 1 \leqslant i,\ j \leqslant N. \tag{6.20}$$

Theorem 6.3. *If all of the e_k have the same sign, e the opposite sign and $|e| \geqslant |\Sigma e_k|$, then the hypotheses of Theorem 6.2 are satisfied.*

In proving this theorem we shall make use of the following simple lemma.

Lemma 6.4. *Let i, j be given. Then for each $w \in L^2(E^{3N-3})$ and $\psi \in C_0^\infty(E^3)$ we have*

$$\int \frac{|w(\tilde{x}^{(i)})\,\psi(x^{(i)})|^2\,dx}{|x^{(i)} - \delta x^{(j)}|} \to \|w\|^2 \int \frac{|\psi(z)|^2\,dz}{|z|} \tag{6.21}$$

as $\delta \to 0$.

Proof. Set $v(x) = w(\tilde{x}^{(i)})\,\psi(x^{(i)})$. Let us first show that there is a constant K such that

$$\int \frac{|v(x)|^2\,dx}{|x^{(i)} - \delta x^{(j)}|^2} \leqslant K^2, \quad \text{all } \delta. \tag{6.22}$$

To see this note that

$$\int \frac{|\psi(x^{(i)})|^2\,dx^{(j)}}{|x^{(i)} - \delta x^{(j)}|^2} \leqslant \left\{ \int_{|x^{(i)} - \delta x^{(j)}| < 1} + \int_{|x^{(i)} - \delta x^{(j)}| > 1} \right\} \frac{|\psi(x^{(i)})|^2\,dx^{(j)}}{|x^{(i)} - \delta x^{(j)}|^2}$$

$$\leqslant \|\psi\|_\infty \int_{|z|<1} \frac{dz}{|z|^2} + \|\psi\|_2^2\,.$$

If we multiply by $|w(\tilde{x}^{(i)})|^2$ and integrate with respect to $\tilde{x}^{(i)}$, we obtain (6.22).

Let $\varepsilon > 0$ be given, and take M so large that

$$\int_{|x| > M} |v(x)|^2\,dx < \varepsilon^2/16K^2\,. \tag{6.23}$$

Since

$$\big|\,|x^{(i)} - \delta x^{(j)}| - |x^{(i)}|\,\big| \leqslant \delta |x^{(j)}| \leqslant \delta |x|\,,$$

we can take δ so small that

$$\big|\,|x^{(i)} - \delta x^{(j)}| - |x^{(i)}|\,\big| < \varepsilon/2K^2, \qquad |x| \leqslant M. \tag{6.24}$$

Now

$$\int |v(x)|^2 \left| \frac{1}{|x^{(i)} - \delta x^{(j)}|} - \frac{1}{|x^{(i)}|} \right| \mathrm{d}x$$

$$\leqslant \int_{|x|<M} |v(x)|^2 \frac{\big|\, |x^{(i)} - \delta x^{(j)}| - |x^{(i)}| \,\big|}{|x^{(i)}|\, |x^{(i)} - \delta x^{(j)}|} \mathrm{d}x$$

$$+ \int_{|x|>M} |v(x)|^2 \left[\frac{1}{|x^{(i)} - \delta x^{(j)}|} + \frac{1}{|x^{(i)}|} \right] \mathrm{d}x$$

$$\leqslant \frac{\varepsilon}{2K} \int_{|x|>M} \frac{|v(x)|^2 \,\mathrm{d}x}{|x^{(i)}|\, |x^{(i)} - \delta x^{(j)}|}$$

$$+ \left(\int_{|x|>M} |v|^2 \,\mathrm{d}x \right)^{\frac{1}{2}} \left\{ \left(\int \frac{|v(x)|^2 \,\mathrm{d}x}{|x^{(i)} - \delta x^{(j)}|^2} \right)^{\frac{1}{2}} + \left(\int \frac{|v(x)|^2 \,\mathrm{d}x}{|x^{(i)}|^2} \right)^{\frac{1}{2}} \right\}$$

$$< \frac{\varepsilon}{2K^2} \left(\int \frac{|v(x)|^2 \,\mathrm{d}x}{|x^{(i)}|^2} \right)^{\frac{1}{2}} \left(\int \frac{|v(x)|^2 \,\mathrm{d}x}{|x^{(i)} - \delta x^{(j)}|^2} \right)^{\frac{1}{2}} + \frac{\varepsilon}{4K} (2K)$$

$$< \frac{\varepsilon}{2K^2} K^2 + \tfrac{1}{2}\varepsilon = \varepsilon \,.$$

Since ε was arbitrary, the proof is complete.

We now turn to the

Proof of Theorem 6.3. Let $w \in H^{1,2}(E^{3N-3})$ be given. We take

$$h_i(z) = q_i(z) = e\, e_i / |x^{(i)}|$$
$$h_{ij}(z) = e_i e_j / |x^{(i)}|$$

and let ψ_i be any function in $C_0^\infty(E^3)$ such that $\|\psi_i\| = 1$. We set $M_k = k$ and $\lambda = 1$.

Since $k q_i(kz) = q_i(z)$, (6.3) is automatically satisfied. The inequality (6.2) is an immediate consequence of Lemma 6.4. Since

$$h_i(z) + \sum_{j \neq i} h_{ij}(z) = e_i (e + \sum_{j \neq i} e_j) / |z| \,,$$

inequality (6.4) follows immediately from the hypotheses of the theorem. This completes the proof.

CHAPTER 11

NOTES, REMARKS AND REFERENCES

In this chapter we shall discuss the background of some of the material presented in the book. References will be given together with comments on related matters. No attempt at completeness is being made; only those matters which come to our attention at this writing will be covered. We realize that it is impossible to give credit to all sources. Each of the ten sections of this chapter is devoted to one of the chapters 1 through 10.

§ 1. Chapter 1

The material in §§1, 2 can be found in most books on functional analysis or operator theory (see the bibliograply at the end of the book). Possible exceptions are Lemma 2.8 and Theorems 2.9, 2.12 and 2.13. Proofs of these are given. Theorem 2.5 is a consequence of the Hahn-Banach theorem. Theorem 2.6 states that finite dimensional subspaces of X' are saturated.

Theorem 2.14 is a form of the Riesz representation theorem. It is equivalent to the statement made in §1 that for any closed subspace V of a Hilbert space X one has $X = V \oplus V^{\perp}$. This latter statement is known as the projection theorem.

§3. The theory of Fredholm operators is fairly recent having its foundations in the work of Fredholm on integral operators. For a history of the development see Gokhberg-Krein [1957]. Proofs of Theorems 3.1–3.3 can be found in Gokhberg-Krein [1957], Goldberg [1966] and Schechter [1967b] or [1971]. For proofs of Theorems 3.5, 3.7 and 3.8 see Schechter [1966] or [1971]. Theorem 3.6 is based on

Theorem 1.1. *Let A be a closed linear operator from one Banach space to another. Then the range of A' (or A^*) is closed if and only if the range of A is closed.*

For a proof of this theorem see Kato [1958], Goldberg [1966] or Schechter [1971].

There are larger classes of operators which have similar properties. The class $\Phi_+(X, Y)$ is the set of closed, densely defined linear operators A from X to Y such that $\alpha(A) < \infty$ and $R(A)$ is closed in Y. The class of those closed, densely defined linear operators A from X to Y such that $R(A)$ is closed in Y and $\beta(A) < \infty$ is denoted by $\Phi_-(X, Y)$. Operators in either $\Phi_+(X, Y)$ or $\Phi_-(X, Y)$ are called *semi-Fredholm* from X to Y. We can extend the definition of index $i(A) = \alpha(A) - \beta(A)$ by including $+\infty$ and $-\infty$. Thus an operator in $\Phi_+(X, Y)$ which is not in $\Phi(X, Y)$ has index equal to $-\infty$. An operator in $\Phi_-(X, Y)$ which is not in $\Phi(X, Y)$ has index $+\infty$. (It would probably be desirable to change the sign of the index, but convention has resisted this. See Gokhberg-Krein [1957].) The statement corresponding to Theorem 3.3 of ch. 1 is

Theorem 1.2. *If* $A \in \Phi_+(X, Y)$ [*resp.* $\Phi_-(X, Y)$] *and* B *is* A*-compact, then* $A+B$ *is in* $\Phi_+(X, Y)$ [*resp.* $\Phi_-(X, Y)$] *and* $i(A+B) = i(A)$.

For a proof see Gokhberg-Krein [1957], Goldberg [1966] or Schechter [1971]. The classical theory of Riesz for operators of the form $I-K$ with $K \in K(X)$ can be found in most of the references for functional analysis.

§4. There are many definitions of essential spectrum found in the literature. We use the tabulation given in Gustafson-Weidmann [1969]. For any linear operator A on a Banach space X let Φ_{+A} [resp. Φ_{-A}] denote the set of these scalars λ such that $A-\lambda$ is in $\Phi_+(X)$ [resp. $\Phi_-(X)$]. Let $\Delta_1(A)$ denote the set of those scalars λ such that $R(A-\lambda)$ is closed in X, $\Delta_2(A) = \Phi_{+A} \cup \Phi_{-A}$, $\Delta_3(A) = \Phi_A$, $\Delta_4(A)$ the set of those $\lambda \in \Phi_A$ such that $i(A-\lambda) = 0$ and $\Delta_5(A)$ the set of those $\lambda \in \Delta_4(A)$ such that all scalars near λ are in $\rho(A)$. We let $\sigma_{ek}(A)$ denote the complements in the complex plane of the sets $\Delta_k(A)$, $1 \leqslant k \leqslant 5$, respectively. These are the essential spectrum according to (1) Dunford-Schwartz [1963], (2) Kato [1966], (3) Wolf [1959a, b], (4) Schechter [1965], (5) Browder [1961]. Note that by Theorem 4.5 of ch. 1, the definition that we have used in this book coincides with $\sigma_{e4}(A)$. If A is a self-adjoint operator in a Hilbert space, then

$$\sigma_{e2}(A) = \sigma_{e3}(A) = \sigma_{e4}(A) = \sigma_{e5}(A). \tag{1.1}$$

By Theorem 1.2 we have

Theorem 1.3. *If* A *is a closed linear operator in* X *and* B *is* A*-compact, then*

$$\sigma_{ek}(A+B) = \sigma_{ek}(A), \qquad 2 \leqslant k \leqslant 4. \tag{1.2}$$

This theorem does not hold for $\sigma_{e1}(A)$ or $\sigma_{e5}(A)$.

Another subset of $\sigma(A)$ which is invariant under compact perturbations and coincides with $\sigma_{ek}(A)$ for $2 \leqslant k \leqslant 5$ when A is self-adjoint is the complement of Φ_{+A}. This set coincides with the set of those scalars λ for which $A-\lambda$ has a singular sequence (see below). It has been given various names in the literature, including *limit points of the spectrum* (Riesz-Nagy [1955]), *condensation spectrum* (Birman [1961]), *limit spectrum* (Glazman [1965], Berezanskii [1968]), *concentration spectrum* (Glazman [1965]) and *continuous spectrum* (Glazman [1965]). It has most of the properties enjoyed by the sets $\sigma_{ek}(A)$. We shall denote it by $\sigma_{e\alpha}(A)$. Note that common usage usually reserves the term continuous spectrum for another concept. Balslev [1965c, 1966] refers to $\sigma_{e2}(A)$ as the *singular spectrum* of A.

By Theorem 1.2 we have

Theorem 1.4. *If A is a closed linear operator in X and B is A-compact, then*

$$\sigma_{e\alpha}(A+B) = \sigma_{e\alpha}(A) . \tag{1.3}$$

In Hilbert space Theorem 4.4 of ch. 1 is due to Wolf [1959a], who showed that $A-\lambda$ has a singular sequence if and only if $\lambda \notin \Phi_{+A}$. This was generalized to Banach space by Balslev-Schubert [1964].

For self-adjoint operators Theorem 4.5 is equivalent to a celebrated theorem of Weyl [1909] (see also Riesz-Nagy [1955]).

For self-adjoint operators Theorem 4.7 is due to Wolf [1959b] and Birman [1961] (see also Rejto [1966]). The general case is due to Schechter [1965]. We can generalize Lemma 4.6 and Theorem 4.7 as follows.

Lemma 1.5. *If $0 \in \rho(A)$, then $\lambda \neq 0$ is in Φ_{+A} (resp. Φ_{-A}) if and only if $1/\lambda$ is in $\Phi_{+A^{-1}}$ (resp. $\Phi_{-A^{-1}}$) and*

$$i(A^{-1}-\lambda^{-1}) = i(A-\lambda) .$$

Proof. Note that

$$A-\lambda = \lambda(\lambda^{-1}-A^{-1})A . \tag{1.5}$$

Since A is one–to–one and onto, this shows that $\alpha(A-\lambda)=\alpha(A^{-1}-\lambda^{-1})$ and that $R(A-\lambda)=R(A^{-1}-\lambda^{-1})$. This proves the lemma.

Theorem 1.6. *Let A, B be closed, densely defined operators in X. If $\lambda \in \rho(A) \cap \rho(B)$ and $(A-\lambda)^{-1}-(B-\lambda)^{-1}$ is a compact operator in X, then*

$$\sigma_{ek}(A) = \sigma_{ek}(B) , \qquad k=2, 3, 4, \alpha . \tag{1.6}$$

Proof. We may take $\lambda=0$. By Theorems 1.2 and 1.4, $\Delta_k(A^{-1})=\Delta_k(B^{-1})$ for $k=2, 3, 4, \alpha$. Now apply Lemma 1.5.

Theorem 4.8 appears to be new. In an analogous way we can prove

Theorem 1.5. *Let A, B be closed operators in X, and suppose that there are operators A_0, B_0 in $B(X)$ and $K_1, K_2 \in K(X)$ such that*

$$AA_0 = I - K_1 \tag{1.7}$$

$$BB_0 = I - K_2 \,. \tag{1.8}$$

If $A_0 - B_0$ is in $K(X)$, then

$$\sigma_{e3}(A) = \sigma_{e3}(B) \,. \tag{1.9}$$

The proof of Theorem 4.8 in ch. 1 applies if one ignores the consideration of the index.

§5. The interpolation theory of this section is known as the complex method. It was introduced by Calderón [1963], Lions [1961] and S.G. Krein [1960a, b]. Since the theory needed here is simple, we have included most of the proofs. Background material on analytic vector valued functions can be found in Hille-Phillips [1957]. For further work on the complex method see Calderón [1964], Krein-Petunin [1966] and Schechter [1967d]. For descriptions of other methods (known as *real methods*) see Lions-Magenes [1968], Butzer-Berens [1967] and Magenes [1964].

§6. The theory of intermediate extensions in Banach space originated in Schechter [1969a] as a generalization of the concept of regularly accretive operators introduced by Kato [1961]. The theory is simple and all proofs are included. Our definition of regularly accretive differs slightly from that of Kato [1961] in the presence of the constant N in (6.11). This does not change the theory but is convenient in some of the applications we consider. Theorem 6.7 as well as Lemma 6.8 are due to Kato [1961]. Lemma 6.8 is due to Lax-Milgram [1954]. Its proof is simple and can be found in Bers-John-Schechter [1964] or Friedman [1969]. Theorem 6.11 is due to Birman [1961]. It can be generalized to

Theorem 1.7. *Under the hypotheses of Theorem* 6.11 *of ch.* 1 *we have*

$$\sigma_{ek}(B) = \sigma_{ek}(A) \,, \qquad k=2, 3, 4, \alpha \,. \tag{1.10}$$

To prove this theorem we merely follow the proof of Theorem 6.11 of ch. 1 and apply Theorem 1.6 of the present chapter in place of Theorem 4.7 of ch. 1.

Theorems 6.12 and 6.13 appear to be new. Again we have a counterpart for σ_{e3}.

Theorem 1.8. *Suppose that A and B satisfy the hypotheses of Theorem* 6.12 *or Theorem* 6.13 *of ch.* 1 *without any assumptions on the indices of* $\hat{A}$ *or* $\hat{B}$. *Then we can conclude that* $\sigma_{e3}(A)=\sigma_{e3}(B)$.

To prove this theorem one can follow the proofs of Theorems 6.12 and 6.13 of ch. 1 without regard to considerations of the index.

§7. The theory of self-adjoint operators can be found in most books on functional analysis. Note that we do not use the spectral theorem until Theorem 7.15. Lemmas 7.3 and 7.4 are due to Kato [1951]. Theorem 7.6 is due to Kato [1961]. Theorem 7.7 does not seem to have been noticed in the literature. Theorem 7.9 is due to Weyl [1909], while Theorems 7.15 and 7.16 are due to Glazman [1952]. For a proof of Theorem 7.18 see Riesz-Nagy [1955]. For that of Theorem 7.19 see Glazman [1965].

§ 2. Chapter 2

§1. Lemma 1.1 is a special case of Lemma 1.3 of ch. 1. The proof of Theorem 1.2 is very simple. In fact let the function $j(x)$ be defined by (1.11) of ch. 2, and let Ω be a bounded open set containing U and such that $\bar{\Omega}\cap V$ is empty. The function

$$\varphi(x) = \int_{\Omega} j\left(\frac{x-y}{\delta}\right)\mathrm{d}y\,, \qquad \delta>0\,,$$

has the required properties for δ less than the distance from $\bar{\Omega}$ to V.

Theorems 1.3–1.6 can be found in most books on real variables or functional analysis. This is also true of the fact that continuous functions with finite supports are dense in $L^p(E^n)$. For a proof of Theorem 1.8 using this fact see Bers-John-Schechter [1964]. Once Theorem 1.8 is known, the proof of Theorem 1.7 is simple. In fact let $u\in L^p(\Omega)$ and $\varepsilon>0$ be given. There is a closed bounded set W in Ω such that

$$\int_{\Omega\setminus W} |u(x)|^p\,\mathrm{d}x < (\tfrac{1}{2}\varepsilon)^p\,.$$

Set

$$\begin{aligned} u_W(x) &= u(x)\,, \qquad & x\in W\\ &=0\,, & x\notin W\,. \end{aligned}$$

Clearly $u_W \in L^p(E^n)$. By Theorem 1.8

$$\begin{aligned}\|u-J_k u_W\|_p &\leqslant \|u-J_k u\|_p+\|J_k(u-u_W)\|_p\\ &\leqslant \|u-J_k u\|_p+\|u-u_W\|_p<\varepsilon\end{aligned}$$

for k sufficiently large. One checks easily that $J_k u_W \in C_0^\infty(\Omega)$ for k sufficiently large.

The operator J_k is called a *mollifier* and was introduced by Friedrichs [1944].

Theorem 1.9 is a special case of Young's inequality.

Theorem 1.10 is easily verified by differentiating under the integral sign and noticing that $\varphi * u$ vanishes for $|x|$ large. Theorems 1.12–1.15 can be found in most books on real variables or functional analysis.

§2. For the general theory of the Fourier transform see Treves [1967]. In particular this reference can be consulted for Theorems 2.1, 2.2, 2.4–2.6. Theorem 2.3 is due to Titchmarsh [1937] while Theorem 2.7 is due to Bochner [1959]. For Theorems 2.8 and 2.10 see Watson [1922]. Theorem 2.10 is a special case of formula due to Hankel [1875] (see §13.3 of Watson [1922]).

§3. The basic theory of L^p multipliers for Fourier transforms is due to Mikhlin [1956, 1965]. Modifications are due to Hirschman [1959], Hörmander [1960], Lizorkin [1963], Littman [1965], Peetre [1966], Kree [1966], Shamir [1966] and others. Theorem 3.3 is a special case of a theorem due to Littman [1965] (see also Peetre [1966]).

§4. For s a positive integer and $p=2$, the spaces $H^{s,p}$ were introduced by Sobolev [1937] and denoted by W_s^p. For real s and $1<p<\infty$ they were introduced by Calderón [1961] (who used the notation L_s^p) and Lions-Magenes [1961]. Theorems 4.1 and 4.3 can be found in Calderón [1961]. Theorem 4.5 is a simple consequence of Theorems 2.1 and 2.2 together with the identities (4.7) and (4.8). These identities are immediate consequences of the definitions of the spaces $H^{s,p}$ and $T^{s,p}$. Lemma 4.8 follows from

$$\bar{F}[(1+|\xi|^2)^{\frac{1}{2}s}F(D^\mu v)]=\bar{F}\{\xi^\mu/(1+|\xi|^2)^{\frac{1}{2}}F\bar{F}[(1+|\xi|^2)^{\frac{1}{2}(s+|\mu|)}Fv]\}$$

and the fact that $\xi^\mu/(1+|\xi|^2)^{\frac{1}{2}|\mu|}$ is a multiplier in L^p (see Theorems 2.1 and 3.3). The same reasoning gives

Lemma 2.1. *For s a positive integer the norm of $H^{s,p}$ is equivalent to*

$$|||v|||_{s,p}=\sum_{|\mu|\leqslant s}\|D^\mu v\|_p\,. \tag{2.1}$$

Theorem 4.9 follows from Theorems 2.1, 4.5 and Lemma 4.8.

Proof of Theorem 4.11. First assume $s \leqslant 0$. By Theorem 4.3 there is a constant C such that

$$\|v\|_{s,p} \leqslant C\|v\|_p, \qquad v \in S. \tag{2.2}$$

Let $u \in H^{s,p}$ and $\varepsilon > 0$ be given. Then there is a $v \in S$ such that $\|u-v\|_{s,p} < \frac{1}{2}\varepsilon$. Since C_0^∞ is dense in L^p, there is a function $\varphi \in C_0^\infty$ such that $\|v-\varphi\|_p < \varepsilon/2C$. Hence

$$\|u-\varphi\|_{s,p} \leqslant \|u-v\|_{s,p} + C\|v-\varphi\|_p < \varepsilon.$$

Thus C_0^∞ is dense in $H^{s,p}$. Next assume $s > 0$. Let k be an integer greater than $\frac{1}{2}s$. By Theorem 4.3 of ch. 2 and Theorem 3.1 of ch. 3 there is a constant C such that

$$\|v\|_{s,p} \leqslant C(\|\Delta^k v\|_p + \|v\|_p), \qquad v \in S, \tag{2.3}$$

where Δ is the Laplacian. Let $u \in H^{s,p}$ and $\varepsilon > 0$ be given. Then there is a $v \in S$ such that $\|u-v\|_{s,p} < \frac{1}{2}\varepsilon$. Since $v \in D[(\Delta^k)_0]$, there is a $\varphi \in C_0^\infty$ such that

$$\|\Delta^k(v-\varphi)\|_p + \|v-\varphi\|_p < \varepsilon/2C$$

(Lemma 2.1 of ch. 4). Hence by (2.3), $\|v-\varphi\|_{s,p} < \frac{1}{2}\varepsilon$. This completes the proof.

Once Theorem 4.11 is known, Theorem 4.12 follows from Theorem 4.1. Theorem 4.14 can be generalized to

Theorem 2.2. *Suppose that* $1 < p < \infty$, $s_0 < s_1$, $0 \leqslant \theta \leqslant 1$ *and* $s = (1-\theta)s_0 + \theta s_1$. *Then*

$$[H^{s_0,p}, H^{s_1,p}]_\theta = H^{s,p} \tag{2.4}$$

with equivalent norms.

For a proof of this theorem see Lions-Magenes [1961] or Schechter [1967d]. Once this is known, the proofs of the remaining theorems of this section become much easier. For instance consider Theorem 4.6. It clearly suffices to prove inequality (4.10) for u in S. This is very simple when s is a non-negative integer if we make use of Lemma 2.1 of the present chapter. We can then consider multiplication by v a bounded operator from $H^{s,p}$ to $H^{s,p}$ for each non-negative integer s. The inequality for positive real s now follows from Theorem 5.6 of ch. 1 and Theorem 2.2 just stated. For s real and negative we have

$$\begin{aligned}|(uv, \varphi)| &= |(u, \bar{v}\varphi)| \leqslant \|u\|_{s,p}\|\bar{v}\varphi\|_{-s,p'}\\ &\leqslant C\|u\|_{s,p}\|\varphi\|_{-s,p'}, \qquad \varphi \in C_0^\infty,\end{aligned}$$

by (4.2) of ch. 2 and what was just proved. Thus by Theorem 4.12

$$\|uv\|_{s,p} = \sup_{\varphi\in C_0^\infty} \frac{|(uv, \varphi)|}{\|\varphi\|_{-s,p'}} \leqslant C\|u\|_{s,p} .$$

This proves Theorem 4.6.

In a similar way we prove Theorem 4.7. It suffices to show that

$$\|u * v\|_{s,p} \leqslant C\|u\|_{s,p}\|v\|_1 , \qquad u, v \in C_0^\infty . \tag{2.5}$$

By Theorems 1.9 and 1.10 of ch. 2

$$\|D^\mu(u * v)\|_p \leqslant \|v\|_1 \|D^\mu u\|_p , \qquad u, v \in C_0^\infty . \tag{2.6}$$

In view of Lemma 2.1 of the present chapter, this implies (2.5) for s a non-negative integer. Again the inequality (2.5) for positive real numbers can be obtained by interpolation via Theorem 5.6 of ch. 1 and Theorem 2.2 stated above. To obtain (2.5) for negative s, note that

$$(u * v, \varphi) = \int v(x-y)u(y)\varphi(x)\mathrm{d}x\,\mathrm{d}y = (u, \varphi * \bar{v}) .$$

Moreover

$$\begin{aligned} |(u, \varphi * \bar{v})| &\leqslant \|u\|_{s,p}\|\varphi * \bar{v}\|_{-s,p'} \\ &\leqslant C\|u\|_{s,p}\|\varphi\|_{-s,p'}\|v\|_1 \end{aligned}$$

by what was just proved. As before this implies (2.5) for $s < 0$.

Theorem 4.10 follows from Theorems 2.6, 4.5 and 4.7. To prove Theorem 4.13 note that it is an immediate consequence of Theorem 4.7 when $t \leqslant 0$. For $t > 0$, let k be an integer such that $t \leqslant k$. Then

$$\begin{aligned} \|v * u\|_{s+t,p} &\leqslant \sum_{|\mu|\leqslant k} \|D^\mu(v * u)\|_{s,p} \\ &= \sum_{|\mu|\leqslant k} \|D^\mu v * u\|_{s,p} \leqslant C\|u\|_{s,p} , \qquad u \in S , \end{aligned}$$

by Theorems 2.5 and 4.7.

Finally we turn to the proof of Theorem 4.4. We shall use

Theorem 2.3. *If $sp > n$, then there is a constant C such that*

$$\|v\|_\infty \leqslant C\|v\|_{s,p} , \qquad v \in S . \tag{2.7}$$

Proof. Let v be any function in S, and set

$$f = \bar{F}[(1+|\xi|^2)^{\frac{1}{2}s} Fv] .$$

Then

$$\|f\|_p = \|v\|_{s,p} \tag{2.8}$$

and

$$v = G_s * f$$

(see ch. 6, §2). By Hölder's inequality (Theorem 1.5 of ch. 2)

$$\|v\|_\infty \leqslant \|G_s\|_{p'} \|f\|_p . \tag{2.9}$$

Note that

$$\|G_s\|_{p'} < \infty$$

if $ps > n$ by the properties of the functions G_s (see ch. 6, §3). The inequality (2.7) follows from (2.8) and (2.9).

We shall also need the following

Theorem 2.4. *Let* X_0, X_1 *be Banach spaces satisfying the hypotheses of Lemma* 5.4 *of ch.* 1. *Let* A *be an operator in* $B(X_0)$ *which is also in* $K(X_1, X_0)$. *Then* $A \in K(X_\theta, X_0)$ *for any* θ *satisfying* $0 < \theta < 1$. *Similarly, if* B *is a linear operator in* $B(X_1)$ *which is also in* $K(X_1, X_0)$, *then* B *is in* $K(X_1, X_\theta)$ *for each* θ *satisfying* $0 < \theta < 1$.

For a proof of this theorem see Lions-Peetre [1964]. Once these are known we can give the

Proof of Theorem 4.4. Let s and t be given with $0 \leqslant t < s$. Let $r > s$ satisfy $(r-1)p > n$. Then by Lemma 4.8 of ch. 2 and Theorem 2.3 just proved

$$\|v\|_\infty + \sum_{j=1} \|D_j v\|_\infty \leqslant C \|v\|_{r,p}, \qquad v \in S . \tag{2.10}$$

Thus if $\{v_k\}$ is a sequence in $H^{r,p}$ with bounded norms, then by (2.10) $\{v_k\}$ is a bounded equicontinuous sequence. By the Arzela-Ascoli theorem, there is a subsequence of $\{v_k\}$ which converges uniformly on the support of φ. This shows that φ is a compact operator from $H^{r,p}$ to L^p. By Theorem 4.6 of ch. 2, φ is a bounded operator from $H^{r,p}$. Hence by Theorem 2.4 just stated, φ is a compact operator from $H^{r,p}$ to $H^{t,p} = [L^p, H^{r,p}]_\theta$ with $\theta = t/r$. Furthermore φ is a bounded operator from $H^{t,p}$ to $H^{t,p}$. Hence it is a compact operator from $H^{s,p} = [H^{t,p}, H^{r,p}]_\theta$ to $H^{t,p}$, where $\theta = (s-t)/(r-t)$. If $s > 0$ and $t < 0$, the result holds a fortiori. If $t < s < 0$, note that φ is the adjoint of the operator $\bar{\varphi}$ mapping $H^{-t,p'}$ to $H^{-s,p'}$, which is compact by what was just proved (here we used Theorem 4.1). We now apply Theorem 2.17 of ch. 1. This completes the proof.

§ 3. Chapter 3

For most of the statements of this chapter one can consult any of the general references in the bibliography. Theorem 1.3 generalizes a result of Hörmander [1955]. Lemma 2.1 is a special case of a theorem having many names attached to it, including Gauss, Green, Stokes, divergence, etc. By this theorem

$$\int_{|x|<R} \frac{\partial v(x)}{\partial x_k}\,\mathrm{d}x = \int_{|x|=R} v(x)\cos\frac{x_k}{|x|}\,\mathrm{d}S \tag{3.1}$$

for each $R>0$. The right-hand side tends to 0 as $R\to\infty$. This gives (2.3) of ch. 3. Corollary 2.2 follows immediately, and Corollary 2.3 follows by repeated applications of Corollary 2.2.

Inequality (2.11) in Theorem 2.4 follows from Theorem 4.6 and Lemma 4.8 of ch. 2. Inequality (2.12) follows from (2.11) and inequality (4.2) of ch. 2. In fact we have

$$\begin{aligned}|(P(x,\mathrm{D})v, w)| &\leqslant \|P(x,\mathrm{D})v\|_{s-m,p}\|w\|_{m-s,p'}\\ &\leqslant C\|v\|_{s,p}\|w\|_{m-s,p'}\,.\end{aligned}$$

Theorem 3.2 is a special case of a theorem in Schechter [1963]. Our only use of the theorem is in the proof of Lemma 2.2 of ch. 9. In that case $P(\xi)$ is of second degree and satisfies

$$P(\xi)\geqslant c_0|\xi|^2\,,\qquad \xi\in E^n \tag{3.2}$$

for some $c_0>0$. In particular there is a scalar λ such that

$$P(\xi)\neq\lambda\,,\qquad \xi\in E^n\,. \tag{3.3}$$

Thus for our purposes it suffices to prove the theorem under the additional assumption (3.3). Clearly we may assume that $\lambda=0$ and that ψ is real valued. By Leibnitz's rule (Theorem 1.1 of ch. 3) for $v\in S$

$$(\psi u, P(\mathrm{D})v) = (u, P(\mathrm{D})[\psi v]) - \sum_{|\mu|\neq 0}(u, P^{(\mu)}(\mathrm{D})\psi\mathrm{D}^\mu v)/\mu!\,.$$

Hence

$$\begin{aligned}|(\psi u, P(\mathrm{D})v)| &\leqslant C_0\|\psi v\|_{t,2}+C_1\|v\|_{m-1,2}\\ &\leqslant C_0C_2\|v\|_{s,2}\,,\qquad v\in S\,,\end{aligned}$$

where $s=\max[t, m-1]$ and C_2 depends only on $P(\mathrm{D})$ and ψ. By Theorem 3.1 of ch. 3 this implies

$$|(\psi u, P(\mathrm{D})v)|\leqslant C_0C_3\|P(\mathrm{D})v\|_{s-m,2}\,,\qquad v\in S\,. \tag{3.4}$$

Since $P(\mathrm{D})$ is elliptic and $P(\xi)$ does not vanish for $\xi \in E^n$, there is a $c_0 > 0$ such that

$$|P(\xi)| \geqslant c_0, \qquad \xi \in E^n. \tag{3.5}$$

Hence by Lemma 1.3 of ch. 4 for each $f \in S$ there is a $v \in S$ such that $P(\mathrm{D})v = f$. Hence

$$|(\psi u, f)| \leqslant C_0 C_3 \|f\|_{s-m,2}, \qquad f \in S. \tag{3.6}$$

Now $(f, u\psi)$ is a linear functional in S. By (3.6) it is bounded in the $H^{s-m,2}$ norm. By the Hahn-Banach theorem (see any of the references for functional analysis) there is a bounded linear functional G on $H^{s-m,2}$ such that

$$Gf = (f, \psi u), \qquad f \in S. \tag{3.7}$$

But by Theorem 4.1 of ch. 2 there is an element $g \in H^{m-s,2}$ such that

$$Gf = (f, g), \qquad f \in H^{s-m,2} \tag{3.8}$$

and

$$\|g\|_{m-s,2} \leqslant C_0 C_3. \tag{3.9}$$

By (3.7) and (3.8)

$$(\psi u - g, f) = 0, \qquad f \in S.$$

This implies that $\psi u = g$. Thus we know that $\psi u \in H^{m-s,2}$ and

$$\|\psi u\|_{m-s,2} \leqslant C_0 C_3.$$

If $s = t$ we are finished. If not, $s = m-1$. Let ζ be any function in $C_0^\infty(\Omega)$ which equals one in the support of ψ. Then

$$\begin{aligned}(\psi u, P(\mathrm{D})v) &= (\psi\zeta u, P(\mathrm{D})v) \\ &= (u, P(\mathrm{D})[\psi v]) - \sum_{|\mu| \neq 0} (\zeta u, P^{(\mu)}(\mathrm{D})\psi \mathrm{D}^\mu v)/\mu!.\end{aligned}$$

Hence by (4.2) of ch. 2

$$|(\psi u, P(\mathrm{D})v)| \leqslant C_0 \|\psi v\|_{t,2} + C_4 \|\zeta u\|_{1,2} \|v\|_{m-2,2}$$

(by the result just proved $\zeta u \in H^{1,2}$). Thus

$$|(\psi u, P(\mathrm{D})v)| \leqslant C_0 C_5 \|v\|_{r,2}, \qquad v \in S,$$

where $r = \max[t, m-2]$ and C_5 depends only on $P(\mathrm{D})$ and ψ. We now repeat the previous argument to conclude that $\psi u \in H^{m-r,2}$ and

$$\|\psi u\|_{m-r,2} \leqslant C_0 C_6.$$

Clearly we can repeat the procedure until we obtain the desired result. This completes the proof.

§ 4. Chapter 4

§1, 2. The concepts of minimal and maximal extensions of a partial differential operator are closely related to those of strong and weak solutions of partial differential equations (with or without boundary conditions). These latter ideas have been found in the mathematical literature for many years usually related to specific problems. For $p=2$ the definitions given here are found in Hörmander [1955]. There they are given for arbitrary domains without boundary conditions. For general p see Browder [1962].

If $u(x)$ is in $D(P_p)$ and $P_p u=f$, then one can consider u as a solution of $P(\mathrm{D})u=f$ in the sense of distributions. For an exposition of this theory see Schwartz [1957], Gelfand-Shilov [1964] or Treves [1967].

Note that in the proof of Lemma 2.1 all we use is

$$\int_{R<|x|<2R} |P^{(\mu)}(\mathrm{D})v|^p \mathrm{d}x/R^{|\mu|p} \to 0 \text{ as } R\to\infty, \text{ each } \mu.$$

This suggest generalizations.

Theorem 2.1 is due to Goldstein [1966].

§3, 4. All of the theorems of these sections are due to Schechter [1968b], [1969b] and [1970a]. A slightly weaker form of Corollary 4.3 is due to Iha [1969] and Iha-Schubert [1970]. For the special case of elliptic operators most results were proved by Balslev [1965c]. Actually Theorem 3.1 can be strengthened to read

Theorem 4.1. *If* $1\leqslant p<\infty$ *and* $P(\xi)$ *is not bounded away from* λ *for all* $\xi\in E^n$, *then* $R(P_0-\lambda)$ *is not closed in* L^p.

Proof. First assume $1\leqslant p\leqslant 2$. If $u\in N(P_0-\lambda)$, then $Fu\in L^{p'}$ (Theorem 2.3 of ch. 2) and $[P(\xi)-\lambda]Fu=0$ (Theorem 4.9 of ch. 2). This means that $Fu=0$ and consequently that $u=0$. This means that $N(P_0-\lambda)=\{0\}$. In the proof of Theorem 3.1 of ch. 4 we constructed a sequence of functions $\{\varphi_k\}$ in C_0^∞ such that

$$\|\varphi_k\|=1,\ [P(\mathrm{D})-\lambda]\varphi_k\to 0 \text{ in } L^p, \qquad k=1,2,\ldots\ . \tag{4.1}$$

This shows that the inequality

$$\|\varphi\|\leqslant C\|[P(\mathrm{D})-\lambda]\varphi\|, \qquad \varphi\in C_0^\infty, \tag{4.2}$$

cannot hold. Consequently $R(P_0-\lambda)$ cannot be closed in L^p (Theorem 2.3 of ch. 1).

Next assume that $2<p<\infty$. Since $P(\xi)$ is not bounded away from λ, $\bar{P}(\xi)$ is not bounded away from $\bar{\lambda}$. Hence by what we have just shown, $R(\bar{P}_{0p'}-\bar{\lambda})$ is not closed in $L^{p'}$. But by Corollary 2.2 of ch. 4

$$\bar{P}_{0p'}-\bar{\lambda}=(P_{0p}-\lambda)^* . \tag{4.3}$$

This means that the range of the adjoint of $P_{0p}-\lambda$ is not closed. Theorem 1.1 of the present chapter now tells us that the range of $P_{0p}-\lambda$ is not closed either. This completes the proof.

For $P(\mathrm{D})$ elliptic this theorem was proved by Balslev [1965c].

If we refer to the various definitions of essential spectrum given in §1, we have the following immediate consequence of Theorem 4.1.

Corollary 4.2. *If* $1\leqslant p<\infty$ *and* $P(\xi)$ *is not bounded away from* λ *for all* $\xi\in E^n$, *then* $\lambda\in\sigma_{\mathrm{e}k}(P_0)$ *for* $k=1, 2, 3, 4, 5, \alpha$.

If $P(\xi)$ is a polynomial of degree m and

$$1/P(\xi)=\mathrm{O}(1/|\xi|^b) \text{ as } |\xi|\to\infty, \qquad b>0 , \tag{4.4}$$

then

$$1/P(\xi)^k=\mathrm{O}(1/|\xi|^{bk}) \text{ as } |\xi|\to\infty, \qquad k=1, 2, \dots . \tag{4.5}$$

By Corollary 4.3 of ch. 4 if we take $k>mn/b(n+2)$ and set $Q(\xi)=P(\xi)^k$, then

$$\sigma(Q_{0p})=\{Q(\xi), \xi\in E^n\}=\{P(\xi)^k, \xi\in E^n\} \tag{4.6}$$

for any p satisfying $1<p<\infty$. We would like to conclude from (4.6) that

$$\sigma(P_{0p})=\{P(\xi), \xi\in E^n\} . \tag{4.7}$$

In general this cannot be done. However, we have

Theorem 4.3. *Let p be any number satisfying* $1<p<\infty$ *and assume that* $P(\xi)$ *is a polynomial satisfying*

$$|P(\xi)|\to\infty \text{ as } |\xi|\to\infty . \tag{4.8}$$

If $\rho(P_{0p})$ *is not empty, then* (4.7) *holds.*

Proof. We first note that (4.8) implies (4.4) for some $b>0$. This is a special case of Theorem 3.2 of Hörmander [1958]. Let λ be a scalar satisfying $P(\xi)\neq\lambda$ for $\xi\in E^n$ and set $Q(\xi)=[P(\xi)-\lambda]^k$ for $k>mn/b(n+2)$. Then $0\in\rho(Q_0)$ by Corollary 4.3 of ch. 4. I claim that

$$Q_0 = (P_0 - \lambda)^k \,. \tag{4.9}$$

To see this note first that for $v \in S$

$$(P_0 - \lambda)^k v = (P(\mathrm{D}) - \lambda)^k v = Q(\mathrm{D})v \,.$$

Moreover by Theorem 2.16.4 of Hille-Phillips [1957], $(P_0 - \lambda)^k$ is a closed operator (this is where the fact that $\rho(P_0)$ is not empty is used). Hence every function in $D(Q_0)$ is also in $D[(P_0 - \lambda)^k]$ and consequently $(P_0 - \lambda)^k$ is an extension of Q_0. Next note that $N[(P_0 - \lambda)^k] = \{0\}$. For if $(P_0 - \lambda)^k u = 0$, we have

$$0 = ([P_0 - \lambda]^k u, v) = (u, [\bar{P}(\mathrm{D}) - \bar{\lambda}]^k v) \,, \qquad v \in S \,.$$

But $|P(\xi) - \lambda| \geqslant c_0$ for some $c_0 > 0$ since $P(\xi)$ satisfies (4.8). Thus for each $f \in S$ there is a $v \in S$ such that $[\bar{P}(\mathrm{D}) - \bar{\lambda}]^k v = f$ (Lemma 1.3 of ch. 4). Consequently $(u, f) = 0$ for all $f \in S$, which means that $u = 0$. Since $0 \in \rho(Q_0)$ and $(P_0 - \lambda)^k$ is a one–to–one operator which is an extension of Q_0, (4.9) must hold. In particular, $(P_0 - \lambda)^k$ is one–to–one. Since in general

$$R(P_0 - \lambda) \supset R[(P_0 - \lambda)^k] \,, \; N(P_0 - \lambda) \subset N[(P_0 - \lambda)^k] \,,$$

we see that the same is true of $P_0 - \lambda$. Thus $\lambda \in \rho(P_0)$. This shows that

$$\sigma(P_0) \subset \{P(\xi), \, \xi \in E^n\} \,.$$

Since the opposite inclusion is always true (Theorem 3.1 of ch. 4), the proof is complete.

This theorem is due to Iha [1969] and Iha-Schubert [1970].

As a consequence of Corollary 4.3 of ch. 4 we have

Theorem 4.4. *If* $P(\xi)$ *satisfies* (4.8), *then there is an* $\eta > 0$ *such that* (4.7) *holds for*

$$|1/p - 1/2| < \eta \,. \tag{4.10}$$

Proof. Again we use the fact that (4.8) implies that there is a $b > 0$ such that (4.4) holds.

Consider the polynomial

$$P(\xi) = (\xi_1 - \xi_2^2 - \xi_3^2 - i)(\xi_1 + \xi_2^2 + \xi_3^2 + i)$$

in E^3. Note that $P(\xi)$ satisfies (4.4) with $b = 1$ and (4.6) of ch. 4 with $a = -\frac{1}{2}$. Consequently (4.7) holds for

$$|1/p-1/2| < \tfrac{2}{9} \tag{4.12}$$

(Theorem 4.2 of ch. 4). It was shown by Iha-Schubert [1970] that (4.7) does not hold for

$$|1/p-1/2| > \tfrac{3}{8}. \tag{4.13}$$

It would be of interest to know the exact point of division. The polynomial (4.11) contains the factor

$$Q(\xi) = \xi_1 + \xi_2^2 + \ldots + \xi_n^2 + i \tag{4.14}$$

given by Littman-McCarthy-Riviere [1968], who showed that $1/Q(\xi)$ is not a multiplier in L^p for

$$|1/p-1/2| > 1/2(n+1). \tag{4.15}$$

Thus $\sigma(Q_0)$ is not equal to

$$\{Q(\xi),\ \xi \in E^n\} = \{z,\ \text{Im } z = i\}$$

for such values of p. In Iha-Schubert [1970] it is shown that $\sigma[(Q^{2N})_0]$ is the whole complex plane for each positive integer N.

§6. Theorem 6.4 is a generalization of Theorem 10, ch. 4 of Glazman [1965]. Note that our proof shows that $\{P(\xi),\ \xi \in E^n\}$ is contained in the condensation spectrum of L (i.e., in the complement of Φ_{+L}). Thus

$$\{P(\xi),\ \xi \in E^n\} \subset \sigma_{ek}(L), \qquad k=3, 4, \alpha. \tag{4.16}$$

§ 5. Chapter 5

The results of §§2–5 are due to Schechter [1968b]. The expression $M_{\alpha,p}(q)$ with $p=2$ was introduced by Stummel [1956] in his study of essential self-adjointness of the Schrödinger operator. Birman [1961] uses a similar expression with $\alpha=1$. Balslev [1965c] uses the expression for general p with different notation. Lemmas 2.5 and 2.6 are essentially due to Rejto [1966]. Lemma 3.3 is due to Balslev [1965c].

In view of Theorems 1.3 and 1.4 we have

Theorem 5.1. *Under the hypotheses of Theorem* 3.1 *of ch.* 5

$$\sigma_{ek}(P_0+q) = \sigma(P_0) = \{P(\xi),\ \xi \in E^n\}, \qquad k=2, 3, 4, \alpha. \tag{5.1}$$

Theorem 5.2. *Under the hypotheses of Theorem* 5.4 *of ch.* 5

$$\sigma_{ek}(L_0) = \sigma(P_0) = \{P(\xi),\ \xi \in E^n\}, \qquad k=2, 3, 4, \alpha. \tag{5.2}$$

It follows from these two theorems that if $P_0+q-\lambda$ (resp. $L_0-\lambda$) is semi-Fredholm, then it is Fredholm with index 0.

§7. Theorems 7.1, 7.3, 7.4 and 7.6 seem to be new. They are generalizations of results of Niznik [1963a, b]. Again as a consequence of Theorems 1.3 and 1.4 of the present chapter we have

Theorem 5.3. *Under the hypotheses of Theorems* 7.4 *of ch.* 5

$$\sigma_{ek}(L)=\sigma(P_0)=\{P(\xi),\ \xi\in E^n\}\,,\qquad k=2,3,4,\alpha\,. \tag{5.3}$$

See the remark at the end of the preceding paragraph. Lemmas 7.8 and 7.9 are due to Niznik [1963b]. It can be shown that (7.15) of ch. 5 is equivalent to saying that $\eta=0$ is the only vector in E^n which satisfies

$$P(\xi+t\eta)=P(\xi)\,,\ \xi\in E^n,\qquad -\infty<t<\infty \tag{5.4}$$

(see Hörmander [1955], Berezanskii [1968]). Such polynomials are called *complete.*

In contrast to the theorems of §§2, 7 of ch. 5, compare the following.

Theorem 5.4. *If both* $q(x)$ *and* $1/P(\xi)$ *are in* L^p *for* $1\leqslant p\leqslant 2$, *then*

$$\|qv\|_2\leqslant\|q\|_p\|1/P\|_p\|P(\mathrm{D})v\|_2\,,\qquad v\in S\,. \tag{5.5}$$

Proof. Put $h=P(\mathrm{D})v$. We have by Theorems 2.3, 2.4 and 4.10 of ch. 2 and Young's inequality

$$\begin{aligned}\|qv\|_2&=\|F(qv)\|_2=\|Fq*Fv\|_2\\&\leqslant\|Fq\|_{p'}\|Fh/P\|_s\leqslant\|q\|_p\|1/P\|_{st}\|Fh\|_{st'}\\&\leqslant\|q\|_p\|1/P\|_{st}\|h\|_2\,,\end{aligned}$$

where s and t are given by

$$\frac{1}{2}=\frac{1}{s}-\frac{1}{p},\qquad st'=2\,.$$

Thus $st=p$, and the proof is complete.

§ 6. Chapter 6

§2. Theorem 2.1 is due to Schechter [1969a]. See also Balslev [1965c] and Browder [1961].

§3. For the theory of Bessel potentials see Aronszajn-Smith [1961]. Our proof of Theorem 3.1 is slightly different from theirs. They show that

$$G_s(x) = \frac{|x|^{\frac{1}{2}(s-n)}}{2^{\frac{1}{2}(n+s-2)}\pi^{\frac{1}{2}n}\Gamma\left(\frac{1}{2}s\right)} K_{\frac{1}{2}(n-s)}(|x|) \tag{6.1}$$

for $s>0$, where $K_\lambda(z)$ is the modified Bessel function of the third kind. They then deduce the desired properties from those of $K_\lambda(z)$. We have been able to avoid this procedure. The fact that $G_s(x)=O\,(\exp\{-a|x|\})$ as $|x|\to\infty$ for some $a>0$ follows immediately from (6.1). For a very concise treatment see Donoghue [1969].

§4. Theorem 4.1 is due to Schechter [1969a]. It is a generalization of theorems due to Balslev [1965c]. The remaining results of this section are simple consequences of Theorems 2.1 and 4.1. By Theorems 1.3 and 1.4 of this chapter we have

Theorem 6.1. *Under the hypotheses of Theorem* 4.5 *of ch.* 6

$$\sigma_{ek}(L_0) = \sigma(P_0) = \{P(\xi),\ \xi\in E^n\}\,, \qquad k=2,3,4,\alpha\,.$$

§5. The theory of this section is due to Schechter [1967c, 1969a]. As a consequence of Theorem 1.8 of this chapter we have

Theorem 6.2. *Under the hypotheses of Theorem* 5.1 *of ch.* 6 *if* $\rho(P_0)$ *is not empty, then*

$$\sigma_{e3}(B) = \sigma(P_0)\,. \tag{6.2}$$

Theorem 6.3. *Under the hypotheses of Theorem* 5.9 *of ch.* 6 *if* $\rho(P_0)$ *is not empty, then*

$$\sigma_{e3}(E) = \sigma(P_0)\,. \tag{6.3}$$

§ 7. Chapter 7

The material in §§2–6 is due to Schechter [1970b]. We can improve (3.1) as follows.

Theorem 7.1. *Under the hypotheses of Theorem* 3.1 *of ch.* 7,

$$\sigma_{ek}(B) = \sigma(P_0)\,, \qquad k=2,3,4,\alpha\,. \tag{7.1}$$

The proof is the same. We merely make use of Theorem 1.6 of the present chapter. Similarly, we have

Theorem 7.2. *Under the hypotheses of Theorem* 4.3. *of ch.* 7, $P(\mathrm{D})+Q(x, D)$ *on* V *has a regularly accretive extension* B *satisfying* (7.1).

Theorem 7.3. *Under the hypotheses of Theorem* 6.3 *of ch.* 7, $P(\mathrm{D})+Q(x, \mathrm{D})$ *on* V *has a regularly accretive extension* B *such that* $\sigma_{ek}(B)$ *is contained in the half-plane* $\mathrm{Re}\,\lambda+M\geqslant 0$ *for* $k=2, 3, 4, \alpha$.

The results of §§7, 9 are due to Schechter [1968a] with some slight improvements due to Beals [1969]. We also have

Theorem 7.4. *Under the hypotheses of Theorem* 9.1 *of ch.* 7, $P(\mathrm{D})+q$ *on* $C_0^\infty \cap D(q)$ *has a regularly accretive extension* B *such that* (7.1) *holds.*

Again the proof is the same with Theorem 1.6 of the present chapter replacing Theorem 4.7 of ch. 1.

The results of §8 come from Schechter [1967a] and Beals [1969]. The theorems in §10 are slight improvements of those of Schechter [1968a] making use of idea in Beals [1969]. As before we have

Theorem 7.5 *Under the hypotheses of Theorem* 10.2 *of ch.* 7, $P(\mathrm{D})+Q(x, \mathrm{D})$ *on* V *has a regularly accretive extension* B *satisfying* (7.1).

Theorem 7.6. *Under the hypotheses of Theorem* 10.7 *of ch.* 7, $P(\mathrm{D})+Q(x, \mathrm{D})$ *on* V *has a regularly accretive extension* B *such that* $\sigma_{ek}(B)$ *is contained in the half-plane* $\mathrm{Re}\,\lambda+M\geqslant 0$ *for* $k=2, 3, 4, \alpha$.

§ 8. Chapter 8

Although they are not deep, most of the results of this chapter are new. Corollary 3.4 generalizes results of Glazman [1952, 1965]. Corollary 5.2 is essentially due to Schwinger [1961]. Theorem 5.3 is a special case of a result of Konno-Kuroda [1966]. Criteria similar to that of Corollary 5.2 are given by Birman [196]] and Brownell [1961]. The operators described in Theorem 5.4 are called Hilbert-Schmidt operators. For the theory of such operators see Riesz-Nagy [1955] or Kato [1966].

§ 9. Chapter 9

§1. Lemma 1.2 is essentially contained in Ikebe-Kato [1962].

§2. Essential self-adjointness of (1.1) has been the object of intensive investigations by many authors. See for instance Browder [1961], Brownell [1951, 1959], B. Hellwig [1964, 1965], Ikebe-Kato [1962], Jörgens [1964], Kato [1951, 1966], Povzner [1953], Rohde [1964, 1967], Stummel [1956], Walter [1967], Wienholtz [1958] and the references quoted in them. The impetus probably stems from the fact that Von Neumann [1932] showed that self-adjointness is necessary and sufficient for statistical statements to make sense for the Hamiltonian. Although some of the results of the authors just quoted are deeper than the theorem proved here, all of them make stronger regularity assumptions on the coefficients. Our argument is very close to that of Ikebe-Kato [1962].

§3. Most of the material in this section is new.

§4. The statements in this section generalize the work of Agudo-Wolf [1958], Balslev [1965c, 1966], Birman [1961], Jörgens [1965] and Rejto [1966]. Theorem 4.8 is due to Friedrichs [1934]. Corollary 4.9 is a special case of a theorem due to Molčanov [1953] (see also Balslev [1965a, b]). Theorem 4.11 and Lemma 4.12 improve on results of Glazman [1958, 1965].

§§5, 6. The results of these sections generalize the work of Zhislin [1960] and the refinements due to Jörgens [1965]. In the generality stated they appear to be new.

§7. Lemma 7.1 and Theorem 7.2 are due to Berezanskii [1968]. Theorem 7.3 is simple, but appears to be new. Lemma 7.4 is due to Jörgens [1965]. The idea for Theorem 7.5 comes from Hunziker [1966].

§8. Here again the results of Zhislin [1960] as refined by Jörgens [1965] are generalized.

§9. This section represents generalizations of the work of Van Winter [1964, 1965], Weinberg [1964] and Hunziker [1966].

§ 10. Chapter 10

For the background of the physical theory used in this chapter see Kemble [1958]. For special cases of Theorem 1.1 see Agudo-Wolf [1958], Balslev [1966], Birman [1961], Browder [1961], Glazman [1965], Kato [1951], Povzner [1953] and Rejto [1966]. For the derivation of (1.10) see Kemble [1958].

§3. The argument of this section is due to Kato [1951]. For the derivation of (3.8) and (3.9) see Kemble [1958]. For background in potential theory see Kellogg [1953].

§4. The theory of this section is due mainly to Zhislin [1960], Jörgens [1965] and Hunziker [1966]. However, the hypotheses are much weaker than those of any of these authors.

§5. The theory of this section is due to Jörgens [1967]. For a proof that the operator T given by (5.18) has a self-adjoint extension with discrete spectrum see Friedrichs [1948]. For the theory of Whittaker functions, see Whittaker-Watson [1915].

§6. The presentation of this section is a generalization of the work of Zhislin [1960]. In particular, Theorem 6.3 and Lemma 6.4 are due to this author.

TVSLB″O

BIBLIOGRAPHY

AGUDO, F. R. D. and F. WOLF, 1958, Rend. Acad. Naz. Lincei **25,** 273–275, 643–645.

AKHIEZER, N. I. and I. M. GLAZMAN, 1962, Theory of Linear Operators in Hilbert Space, vols. 1, 2 (Ungar, New York).

ARONSZAJN, N. and K. T. SMITH, 1961, Ann. Inst. Fourier Grenoble **11,** 385–475.

ARONSZAJN, N., F. MULLA and P. SZEPTYCKI, 1963, Ann. Inst. Fourier Grenoble **12**, 211–306.

BACHMAN, G. and L. NARICI, 1966, Functional Analysis (Academic Press, New York).

BALSLEV, E. and C. F. SCHUBERT, 1964, Essential Spectrum and Singular Sequences, unpublished manuscript, Univ. Calif., Los Angeles.

BALSLEV, E., 1965a, Discreteness of the Spectrum of Second Order Elliptic Differential Operators in $L^2(R_n)$, Aarhus Universitet, Matematisk Institute, 1–14.

BALSLEV, E., 1965b, Molčanov's Condition for Discreteness of the Spectrum of the Schrödinger Operator, Aarhus Universitet, Matematisk Institute, 1–18.

BALSLEV, E., 1965c, Trans. Amer. Math. Soc. **116,** 193–217.

BALSLEV, E., 1966, Math. Scand. **19,** 193–210.

BANACH, S., 1955, Théorie des Opérations Linéaires (Chelsea, New York).

BEALS, R., 1969, On Spectral Theory and Scattering for Elliptic Operators with Singular Potentials, Department of Mathematics, Yale University.

BERBERIAN, S. K., 1961, Introduction to Hilbert Space (Oxford, Fair Lawn, N.J.).

BERENS, H. and P. L. BUTZER, 1967, Semi-Groups of Operators and Approximation (Springer-Verlag, New York).

BEREZANSKII, JU. M., 1968, Expansions in Eigenfunctions of Selfadjoint Operators (American Mathematical Society, Providence, R.I.).

BERS, L., F. JOHN and M. SCHECHTER, 1964, Partial Differential Equations (Interscience, New York).

BIRMAN, M. S., 1961, Mat. Sb. (2) **55,** 125–174.

BOCHNER, S., 1959, Lectures on Fourier Integrals (Annals of Mathematics Studies, Princeton University Press, Princeton, N.J.).

BROWDER, F. E., 1961, Math. Ann. **142,** 22–130.

BROWDER, F. E., 1962, Math. Ann. **145,** 81–226.

BROWNELL, F. H., 1951, Ann. of Math. **54,** 554–594.

BROWNELL, F. H., 1959, Pacific J. Math. **9,** 953–977.

BROWNELL, F. H., 1961, Arch. Rat. Mech. Anal. **8**, 59–67.

CALDERÓN, A. P., 1961, Lebesgue Spaces of Differentiable Functions and Distributions, in: Morrey Jr., C. B., ed, Partial Differential Equations, Proc. Fourth Symp. in Pure Mathematics of the American Mathematical Society, Berkeley, California, 1960 (American Mathematical Society, Providence, R.I.) pp. 33–49.

CALDERÓN, A. P., 1963, Studia Math., Special Series, vol. I, pp. 31–34.

CALDERÓN, A. P., 1964, Studia Math. **24,** 113–190.

CARLEMAN, T., 1934, Ark. för Mat. Astr. Fys. **24B** (11), 1–7.

CARROLL, R. W., 1969, Abstract Methods in Partial Differential Equations (Harper and Row, New York).
COURANT, R. and D. HILBERT, 1953, 1962, Methods of Mathematical Physics, vols. 1, 2 (Interscience, New York).

DAY, M. M., 1962, Normed Linear Spaces (Academic Press, New York).
DONOGHUE JR., W. F., 1969, Distributions and Fourier Transforms (Academic Press, New York).
D-UDONNÉ, J., 1960, Foundations of Modern Analysis (Academic Press, New York).
DUNFORD, N. and J. T. SCHWARTZ, 1958, 1963, Linear Operators, Parts 1 and 2 (Interscience, New York).

EDWARDS, R. E., 1965, Functional Analysis (Holt, Rinehart and Winston, New York).

FADEEV, L. V., 1965, Mathematical Aspects of the Three-Body Problem in the Quantum Theory of Scattering (Israel Program of Scientific Translations, Jerusalem).
FRIEDMAN, A., 1969, Partial Differential Equations (Holt, Rinehart and Winston, New York).
FRIEDRICHS, K. O., 1934, Math. Ann. **109,** 465–487.
FRIEDRICHS, K. O., 1944, Trans. Amer. Math. Soc. **55,** 132–151.
FRIEDRICHS, K. O., 1948, Criteria for the Discrete Character of the Spectra of Ordinary Differential Operators, Studies and Essays Presented to R. Courant (Interscience, New York) pp. 145–160.

GAGLIARDO, E., 1965, Proc. Amer. Math. Soc. **16,** 429–434.
GELFAND, I. M., 1952, Uspehi Mat. Nauk **7,** 183–184.
GELFAND, I. M. and G. E. SHILOV, 1964, Generalized Functions (Academic Press, New York).
GLAZMAN, I. M., 1952, Dokl. Akad. Nauk SSSR **87,** 5–8, 171–174.
GLAZMAN, I. M., 1958, Dokl. Akad. Nauk SSSR **119,** 421–424.
GLAZMAN, I. M., 1959, Mat. Sb. (2) **35,** 231–246.
GLAZMAN, I. M., 1965, Direct Methods of the Qualitative Spectral Analysis of Singular Differential Operators (Israel Program of Scientific Translations, Jerusalem).
GOFFMAN, C. and G. PEDRICK, 1965, First Course in Functional Analysis (Prentice Hall, Englewood Cliffs, N.J.).
GOKHBERG, I. C. and M. G. KREIN, 1957, Uspehi Mat. Nauk **12,** 43–118.
GOLDBERG, S., 1966, Unbounded Linear Operators (McGraw-Hill, New York).
GOLDSTEIN, R. A., 1966, Proc. Amer. Math. Soc. **17,** 1031–1033.
GUSTAFSON, K. and J. WEIDMANN, 1969, J. Math. Anal. Appl. **25,** 121–127.

HALMOS, P. R., 1951, Introduction to Hilbert Space (Chelsea, New York).
HANKEL, H., 1875, Math. Ann. **8,** 469.
HELLWIG, B., 1964, Math. Z. **86,** 255–262.
HELLWIG, B., 1965, Math. Z. **89,** 333–344.
HELLWIG, G., 1967, Differential Operators of Mathematical Physics (Addison-Wesley, Reading, Mass.).
HILLE, E. and R. S. PHILLIPS, 1957, Functional Analysis and Semi-Groups (Amer. Math. Soc., Providence, R.I.).
HIRSCHMAN JR., I. J., 1959, Duke Math. J. **26,** 221–242.
HÖRMANDER, L., 1955, Acta Math. **94,** 161–248.
HÖRMANDER, L., 1958, Comm. Pure Appl. Math. **11,** 197–218.
HÖRMANDER, L., 1960, Acta Math. **104,** 93–140.
HÖRMANDER, L., 1963, Linear Partial Differential Operators (Springer-Verlag, Berlin).
HORVATH, J., 1966, Topological Vector Spaces (Addison-Wesley, Reading, Mass.).

HUNZIKER, W., 1966, Helv. Phys. Acta **39,** 451–462.
HUNZIKER, W., 1964, Phys. Rev. **135,** B800.

IHA, F. T., 1969, Spectral Theory of Partial Differential Operators, Doctoral Dissertation, University of California, Los Angeles.
IHA, F. T. and C. F. SHUBERT, 1970, Trans. Amer. Math. Soc.
IKEBE, T. and T. KATO, 1962, Arch. Rational Mech. Anal. **9,** 77–92.

JÖRGENS, K., 1967, Math. Z. **96,** 355–372.
JÖRGENS, K., 1964, Math. Scand. **15,** 5–17.
JÖRGENS, K., 1965, Über das wesentliche Spektrum elliptischer Differentialoperatoren vom Schrödinger-Typ, Institute für Angewandte Mathematik, Universität Heidelberg, 1–27.

KANTOROVICH, L. V. and G. P. AKILOV, 1964, Functional Analysis in Normed Spaces (Pergamon, Oxford).
KATO, T., 1951, Trans. Amer. Math. Soc. **70,** 195–211, 212–218.
KATO, T., 1958, J. d'Analyse Math. **6,** 273–322.
KATO, T., 1958, J. Math. Soc. Japan **13,** 246–274.
KATO, T., 1966, Perturbation theory for Linear Operators (Springer Verlag, New York).
KATO, T., 1967, Progr. Theoret. Phys. Supplement No. 40, 3–19.
KELLEY, J. L. and I. NAMIOKA, 1963, Linear Topological Spaces (Van Nostrand, Princeton, N.J.).
KELLOGG, O. D., 1953, Foundations of Potential Theory (Dover, New York).
KEMBLE, E. C., 1958, The Fundamental Principles of Quantum Mechanics (Dover, N.Y.).
KOLMOGOROV, A. N. and S. V. FOMIN, 1957, Elements of the Theory of Functions and Functional Analysis, vols. 1, 2 (Graylock Press, Rochester, N.Y.).
KONNO, R. and S. T. KURODA, 1966, J. Fac. Sci. Univ. Tokyo. Sect. I, **13,** 55–63.
KOREVAAR, J., 1968, Mathematical Methods (Academic Press, New York).
KREE, P., 1966, Ann. Inst. Fourier Grenoble **16,** 31–121.
KREIN, S. G., 1960a, Dokl. Akad. Nauk SSSR **130,** 491–494.
KREIN, S. G., 1960b, Dokl. Akad. Nauk SSSR **132,** 510–513.
KREIN, S. G. and YU. I. PETUNIN, 1966, Russian Math. Surveys **21,** 85–159.

LAX, P. D. and A. N. MILGRAM, 1954, Parabolic Equations, in: Bers, L., S. Bochner and F. John, eds., Contributions to the Theory of Partial Differential Equations, Annals of Math. Studies, No. 33 (Princeton Univ. Press, Princeton, N.J.) pp. 167–190.
LIDSKII, V. B., 1957, Dokl. Akad. Nauk SSSR **112,** 994–997; **113,** 28–31.
LIONS, J. L., 1961, C. R. Acad. Sci. Paris **251,** 1853–1855.
LIONS, J. L. and E. MAGENES, 1961, Ann. Scuola Norm. Sup. Pisa (3) **15,** 39–101.
LIONS, J. L. and J. PEETRE, 1964, Publ. Math. Inst. Hautes Études Sci. **19,** 5–68.
LIONS, J. L. and E. MAGENES, 1968, Problèmes aux Limites Non Homogènes (Dunod, Paris).
LITTMAN, W., 1965, Bull. Amer. Math. Soc. **71,** 764–766.
LITTMAN, W., C. MCCARTHY and N. RIVIÈRE, 1968, Studia Math. **30,** 219–229.
LIUSTERNIK, L. and V. SOBOLEV, 1961, Elements of Functional Analysis (Ungar, New York).
LIZORKIN, P. I., 1963, Dokl. Akad Nauk SSSR **152,** 808–811.

MAGENES, E., 1964, Spazi di Interpolazlone ed Equazioni a Derivate Parziali, in: Atti del VII Congresso, dell' Unione Matematica Italiana, Genoa, 1963 (Edizioni Cremonese, Rome).
MARTIROSJAN, R. M., 1961, Izv. Akad. Nauk Armjan. SSR Ser. Fiz-Mat. **14,** 9–19.
MARTIROSJAN, R. M., 1963, Izv. Akad. Nauk SSSR Ser. Mat. **27,** 677–700.
MARTIROSJAN, R. M., 1964, Izv. Akad. Nauk SSSR Ser. Mat. **28,** 79–90.
MARTIROSJAN, R. M., 1966, Izv. Akad. Nauk Armjan. SSR Ser. Mat. **1,** 192–216.

MAURIN, K., 1967, Methods of Hilbert Spaces (Polish Scientific Publishers, Warsaw).
MIKHLIN, S. G., 1956, Dokl. Akad. Nauk SSSR **109,** 701–703.
MIKHLIN, S. G., 1965, Multidimensional Singular Integrals and Integral Equations (Pergamon Press, Oxford).
MIKHLIN, S. G., 1967, Linear Equations of Mathematical Physics (Holt, Rinehart and Winston, New York).
MOLČANOV, A. M., 1953, Trudy Moskov. Mat. Obšč. **2,** 169–200.
MORREY JR., C. B., 1966, Multiple Integrals in the Calculus of Variations (Springer-Verlag, New York).

NILSSON, N., 1959, Kungl. Fysiogr. Sällsk. i Lund Förh. **29,** 1–19.
NIZNIK, L. P., 1959, Dokl. Acad. Nauk SSSR **124,** 517–519.
NIZNIK, L. P., 1963a, The Spectrum Properties of Self-Adjoint Partial Differential Operators Which Are Close to the Operators with Constant Coefficients, in: Outlines of the Joint Soviet-American Symposium on Partial Differential Equation, Moscow, 1963, pp. 195–199.
NIZNIK, L. P., 1963b, Ukrain. Mat. Z. **15,** 385–399.

PEETRE, J., 1966, Ricerche Mat. **15,** 3–36.
POVZNER, A. YA., 1953, Mat. Sb. **32,** 109–156.
PUTNAM, C. R., 1956, Quart. Appl. Math. **14,** 101.
PUTNAM, C. R., 1967, Commutation Properties of Hilbert Space Operators (Springer-Verlag, Berlin).

RAMM, A. G., 1965, Mat. Sb. (2) **66,** 321.
REJTO, P. A., 1966, Pacific J. Math. **19,** 109–140.
RIESZ, F. and B. SZ.-NAGY, 1955, Functional Analysis (Ungar, New York).
ROHDE, H. W., 1964, Math. Z. **86,** 21–34.
ROHDE, H. W., 1966, Math. Z. **91,** 30–49.
ROHDE, H. W., 1967, Dissertion, Rheinisch-Westfälischen Technischen Hochschule.
ROHDE, H. W., 1969, Arch. Rational Mech. Anal. **34,** 188–217.

SAGAN, H., 1961, Boundary and Eigenvalue Problems in Mathamatical Physics (Wiley, New York).
SCHECHTER, M., 1963, Amer. J. Math. **85,** 1–13.
SCHECHTER, M., 1965, Bull. Amer. Math. Soc. **71,** 365–367.
SCHECHTER, M., 1966, J. Math. Anal. Appl. **13,** 205–215.
SCHECHTER, M., 1967a,, Ricerche Mat. **16,** 3–26.
SCHECHTER, M., 1967b, Ann. Scoula Norm. Sup. Pisa (3) **21,** 361–380.
SCHECHTER, M., 1967c, Bull. Amer. Math. Soc. **73,** 567–572.
SCHECHTER, M., 1967d, Compositio Math. **18,** 117–147.
SCHECHTER, M., 1967e, Bull. Amer. Math. Soc. **73,** 979–985.
SCHECHTER, M., 1968a, Proc. Cambridge Philosophical Soc. **64,** 975–984.
SCHECHTER, M., 1968b, Israel J. Math. **6,** 384–397.
SCHECHTER, M., 1969a, J. d'Analyse Math. **22,** 87–115.
SCHECHTER, M., 1969b, Bull. Amer. Math. Soc. **75,** 548–549.
SCHECHTER, M., 1969c, J. London Math. Soc. (2) **1,** 343–347.
SCHECHTER, M., 1970a, Ann. Scoula Norm. Sup. Pisa. **24,** 201–207.
SCHECHTER, M., 1970b, Scripta Math. **19**.
SCHECHTER, M., 1971, Principles of Functional Analysis (Academic Press, N.Y.).
SCHWARTZ, L., 1957, Théorie des Distributions (Hermann, Paris).
SCHWINGER, J., 1961, Proc. Nat. Acad. Sci. U.S.A. **47,** 122–129.
SHAMIR, E., 1966, J. Math. Anal. Appl. **16,** 104–107.

SHILOV, G. E., 1968, Generalized Functions and Partial Differential Equations (Gordon and Breach, New York).
SOBOLEV, S. L., 1937, Mat. Sb. (2) **44,** 467–500.
STONE, M. H., 1932, Linear Transformation in Hilbert Space and Their Applications to Analysis (Amer. Math. Soc., Providence, R.I.).
STUMMEL, F., 1956, Math. Ann. **132,** 150–176.
STUMMEL, F., 1969, Rand- und Eigenwertaufgaben in Sobolewschen Räumen (Springer-Verlag, Berlin).

TAYLOR, A. E., 1958, Introduction to Functional Analysis (Wiley, New York).
TITCHMARSH, E. C., 1937, Introduction to the Theory of Fourier Integrals (Clarendon Press, Oxford).
TITCHMARSH, E. C., 1958, Eigenfunction Expansions Associated with Second Order Differential Equations, vol. 2 (Clarendon Press, Oxford).
TRÈVES, F., 1966, Linear Partial Differential Equations with Constant Coefficients (Gordon and Breach, New York).
TRÈVES, F., 1967, Topological Vector Spaces, Distributions and Kernels (Academic Press, New York).

UCHIYAMA, J., 1966, Publ. Res. Inst. Math. Sci. Kyoto Univ. Ser. A **2**, 117–132.

VULIKH, B. Z., 1963, Introduction to Functional Analysis for Scientists and Technologists (Pergamon Press, Oxford, Addison-Wesley, Reading, Mass.).

WALTER, J., 1967, Math. Z. **98,** 401–406.
WALTER, J., 1968, Math. Z. **106,** 149–152.
WATSON, G. N., 1922, A Treatise on the Theory of Bessel Functions (Cambridge Univ. Press).
WHITTAKER, E. T. and G. N. WATSON, 1915, A Course in Modern Analysis (Cambridge Univ. Press, London).
WEINBERG, S., 1964, Phys. Rev. **33,** B232.
WEYL, H., 1909, Rend. Circ. Mat. Palermo **27,** 373–392.
WIENHOLTZ, E., 1958, Math. Ann. **135,** 50–80.
WIENHOLTZ, E., 1959, Arch. Math. **10,** 126–133.
WILANSKY, A., 1964, Functional Analysis (Blaisdell, New York).
VAN WINTER, C., 1964, Kgl. Danske Videnskab. Selskab, Mat.-Fys. Skr. **2,** No. 8.
VAN WINTER, C., 1965, Kgl. Danske Videnskab. Selbskab. Mat.-Fys. Skr. **2,** No. 10.
WOLF, F., 1959a, Comm. Pure Appl. Math. **12,** 211–228.
WOLF, F., 1959b, Indag. Math. **21,** 142–315.
WOLF, F., 1960, Bull. Acad. Belg. **46,** 441–447.

YOSHIDA, K., 1965, Functional Analysis (Academic Press, New York).

ZAANEN, A. C., 1953, Linear Analysis (North-Holland, Amsterdam).
ZHISLIN, G. M., 1958, Dokl. Akad. Nauk SSR **122,** 331–334.
ZHISLIN, G. M., 1959, Dokl. Akad. Nauk SSR **128,** 231–234.
ZHISLIN, G. M., 1960, Trudy Moskov. Mat. Obšč. **9,** 81–120.

LIST OF SYMBOLS

The number opposite each symbol is the page on which it is defined or explained.

SUBJECT INDEX